Graduate Texts in Mathematics 126

Graduate Texts in Mathematics

1 TAKEUTI/ZARING. Introduction to Axiomatic Set Theory. 2nd ed.
2 OXTOBY. Measure and Category. 2nd ed.
3 SCHAEFFER. Topological Vector Spaces.
4 HILTON/STAMMBACH. A Course in Homological Algebra.
5 MAC LANE. Categories for the Working Mathematician.
6 HUGHES/PIPER. Projective Planes.
7 SERRE. A Course in Arithmetic.
8 TAKEUTI/ZARING. Axiomatic Set Theory.
9 HUMPHREYS. Introduction to Lie Algebras and Representation Theory.
10 COHEN. A Course in Simple Homotopy Theory.
11 CONWAY. Functions of One Complex Variable. 2nd ed.
12 BEALS. Advanced Mathematical Analysis.
13 ANDERSON/FULLER. Rings and Categories of Modules.
14 GOLUBITSKY/GUILLEMIN. Stable Mappings and Their Singularities.
15 BERBERIAN. Lectures in Functional Analysis and Operator Theory.
16 WINTER. The Structure of Fields.
17 ROSENBLATT. Random Processes. 2nd ed.
18 HALMOS. Measure Theory.
19 HALMOS. A Hilbert Space Problem Book. 2nd ed., revised.
20 HUSEMOLLER. Fibre Bundles. 2nd ed.
21 HUMPHREYS. Linear Algebraic Groups.
22 BARNES/MACK. An Algebraic Introduction to Mathematical Logic.
23 GREUB. Linear Algebra. 4th ed.
24 HOLMES. Geometric Functional Analysis and its Applications.
25 HEWITT/STROMBERG. Real and Abstract Analysis.
26 MANES. Algebraic Theories.
27 KELLEY. General Topology.
28 ZARISKI/SAMUEL. Commutative Algebra. Vol. I.
29 ZARISKI/SAMUEL. Commutative Algebra. Vol. II.
30 JACOBSON. Lectures in Abstract Algebra I: Basic Concepts.
31 JACOBSON. Lectures in Abstract Algebra II: Linear Algebra.
32 JACOBSON. Lectures in Abstract Algebra III: Theory of Fields and Galois Theory.
33 HIRSCH. Differential Topology.
34 SPITZER. Principles of Random Walk. 2nd ed.
35 WERMER. Banach Algebras and Several Complex Variables. 2nd ed.
36 KELLEY/NAMIOKA et al. Linear Topological Spaces.
37 MONK. Mathematical Logic.
38 GRAUERT/FRITZSCHE. Several Complex Variables.
39 ARVESON. An Invitation to C^*-Algebras.
40 KEMENY/SNELL/KNAPP. Denumerable Markov Chains. 2nd ed.
41 APOSTOL. Modular Functions and Dirichlet Series in Number Theory. 2nd ed.
42 SERRE. Linear Representations of Finite Groups.
43 GILLMAN/JERISON. Rings of Continuous Functions.
44 KENDIG. Elementary Algebraic Geometry.
45 LOÈVE. Probability Theory I. 4th ed.
46 LOÈVE. Probability Theory II. 4th ed.
47 MOISE. Geometric Topology in Dimensions 2 and 3.

continued after Index

Armand Borel

Linear Algebraic Groups

Second Enlarged Edition

Springer-Verlag
New York Berlin Heidelberg London
Paris Tokyo Hong Kong Barcelona

Armand Borel
School of Mathematics
Institute for Advanced Study
Princeton, New Jersey 08450 USA

First edition published by W.A. Benjamin, Inc., 1969

Library of Congress Cataloging-in-Publication Data

Borel, Armand.
Linear algebraic groups / Armand Borel.—2nd enl. ed.
p. cm.—(Graduate texts in mathematics; 126)
Includes bibliographical references and indexes.
ISBN 0-387-97370-2 (alk. paper).—ISBN 3-540-97370-2 (alk. paper)
1. Linear algebraic groups. I. Title. II. Series.
QA564.B58 1991
512'.5—dc20 90-19774

Printed on acid-free paper.

Typeset by Thomson Press (India) Ltd., New Delhi.
Printed and bound by R.R. Donnelley and Sons, Harrisonburg, Virginia.
Printed in the United States of America.

9 8 7 6 5 4 3 2 1

ISBN 0-387-97370-2 Springer-Verlag New York Berlin Heidelberg
ISBN 3-540-97370-2 Springer-Verlag Berlin Heidelberg New York

Introduction to the First Edition

These Notes aim at providing an introduction to the theory of linear algebraic groups over fields. Their main objectives are to give some basic material over arbitrary fields (Chap. I, II), and to discuss the structure of solvable and of reductive groups over algebraically closed fields (Chap. III, IV). To complete the picture, they also include some rationality properties (§§15, 18) and some results on groups over finite fields (§16) and over fields of characteristic zero (§7).

Apart from some knowledge of Lie algebras, the main prerequisite for these Notes is some familiarity with algebraic geometry. In fact, comparatively little is actually needed. Most of the notions and results frequently used in the Notes are summarized, a few with proofs, in a preliminary Chapter AG. As a basic reference, we take Mumford's Notes [14], and have tried to be to some extent self-contained from there. A few further results from algebraic geometry needed on some specific occasions will be recalled (with references) where used. The point of view adopted here is essentially the set theoretic one: varieties are identified with their set of points over an algebraic closure of the groundfield (endowed with the Zariski-topology), however with some traces of the scheme point of view here and there.

These Notes are based on a course given at Columbia University in Spring, 1968,* at the suggestion of Hyman Bass. Except for Chap. V, added later, Notes were written up by H. Bass, with some help from Michael Stein, and are reproduced here with few changes or additions. He did this with marvelous efficiency, often expanding or improving the oral presentation. In particular, the emphasis on dual numbers in §3 in his, and he wrote up Chapter AG, of which only a very brief survey had been given in the course. It is a pleasure to thank him most warmly for his contributions, without which these Notes would hardly have come into being at this time. I would also like to thank Miss P. Murray for her careful and fast typing of the manuscript, and J.E. Humphreys, J.S. Joel for their help in checking and proofreading it.

A. Borel
Princeton, February, 1969

---

*Lectures from May 7th on qualified as liberated class, under the sponsorship of the Students Strike Committee. Space was generously made available on one occasion by the Union Theological Seminary.

Introduction to the Second Edition

This is a revised and enlarged edition of the set of Notes: "Linear algebraic groups" published by Benjamin in 1969. The added material pertains mainly to rationality questions over arbitrary fields with, as a main goal, properties of the rational points of isotropic reductive groups. Besides, a number of corrections, additions and changes to the original text have been made. In particular:

§3 on Lie algebras has been revised.

§6 on quotient spaces contains a brief discussion of categorical quotients. The existence of a quotient by finite groups has been added to §6, that of a categorical quotient under the action of a torus to §8.

In §11, the original proof of Chevalley's normalizer theorem has been replaced by an argument I found in 1973, (and is used in the books of Humphreys and Springer).

In §14, some material on parabolic subgroups has been added.

§15, on split solvable groups now contains a proof of the existence of a rational point on any homogeneous space of a split solvable group, a theorem of Rosenlicht's proved in the first edition only for $\mathbf{GL}_1$ and $\mathbf{G}_a$.

§§19 to 24 are new. The first one shows that in a connected solvable k-group, all Cartan k-subgroups are conjugate under $G(k)$, a result also due to M. Rosenlicht. §§20, 21 are devoted to the so-called relative theory for isotropic reductive groups over a field k: Conjugacy theorems for minimal parabolic k-subgroups, maximal k-split tori, existence of a Tits system on $G(k)$, rationality of the quotient of G by a parabolic k-subgroup and description of the closure of a Bruhat cell. As a necessary complement, §22 discusses central isogenies.

§23 is devoted to examples and describes the Tits systems of many classical groups. Finally, §24 surveys without proofs some main results on classifications and linear representations of semi-simple groups and, assuming Lie theory, relates the Tits system on the real points of a reductive group to the similar notions introduced much earlier by E. Cartan in a Lie theoretic framework.

Many corrections have been made to the text of the first edition and my thanks are due to J. Humphreys, F.D. Veldkamp, A.E. Zalesski and V. Platonov who pointed out most of them.

I am also grateful to Mutsumi Saito, T. Watanabe and especially G. Prasad, who read a draft of the changes and additions and found an embarrassing number of misprints and minor inaccuracies. I am also glad to acknowledge help received in the proofreading from H.P. Kraft, who read parts of the proofs with great care and came up with a depressing list of corrections, and from D. Jabon.

The first edition has been out of print for many years and the question of a reedition has been in the air for that much time. After Addison–Wesley had acquired the rights to the Benjamin publications they decided not to proceed with one and released the publication rights to me. I am grateful to Springer-Verlag to have offered over ten years ago to publish a reedition in whichever form I would want it and to several technical editors (starting with W. Kaufmann–Bühler) and scientific editors for having periodically prodded me into getting on with this project. I am solely to blame for the procrastination.

In preparing the typescript for the second edition, use was made to the extent possible of copies of the first one, whose typography was quite different from the one present techniques allow one to produce. The insertions of corrections, changes and additions, which came in successive ways, presented serious problems in harmonization, pasting and cutting. I am grateful to Irene Gaskill and Elly Gustafsson for having performed them with great skill.

I would also like to express my appreciation to Springer-Verlag for their handling of the publication and their patience in taking care of my desiderata.

A. Borel

Contents

Conventions and Notation

1. Throughout these Notes, k denotes a commutative field, K an algebraically closed extension of k, k_s (resp. $\bar{k}$) the separable (resp. algebraic) closure of k in K, and p is the characteristic of k. Sometimes, p also stands for the chracteristic exponent of k, i.e. for one if $\mathrm{char}(k) = 0$, and p if $\mathrm{char(k)} = \mathrm{p} > 0$.

All rings are commutative, unless the contrary is specifically allowed, with unit, and all ring homomorphisms and modules are unitary.

If A is a ring, A^* is the group of invertible elements of A.

$\mathbb{Z}$ denotes the ring of integers, $\mathbb{Q}$ (resp. $\mathbb{R}$, resp. $\mathbb{C}$) the field of rational (resp. real, resp. complex) numbers.

2. *References.* A reference to section $(x.y)$ of Chapter AG is denoted by $(\mathrm{AG}.x.y)$. In the subsequent chapters $(x.y)$ refers to section $(x.y)$ in one of them.

There are two bibliographies, one for Chapter AG, on p. 83, one for Chapters I to V, on p. 391.

References to original literature in Chapters I and V are usually collected in bibliographical notes at the end of certain paragraphs. However, they do not aim at completeness, and a result for which none is given need not be new.

3. Let G be a group. If (X_i) $(1 \leqq i \leqq m)$ are sets and $f_i\colon X_i \to G$ maps, then the map

$f\colon X_1 \times \ldots \times X_m \to G$ defined by

$$(x_1, \ldots, x_n) \to f_1(x_1) \ldots . f_m(x_m), \qquad (x_i \in X_i;\ 1 \leqq i \leqq m),$$

is often called the product map of the f_i's.

Let N_i $(1 \leqq i \leqq n)$ be normal subgroups of G. The group G *is an almost direct product* of the N_i's if the product map of the inclusions $N_i \to G$ is a homomorphism of the direct product $N_1 \times \ldots \times N_m$ onto G, with finite kernel.

If M, N are subgroups of G, then (M, N) denotes the subgroup of G generated by the commutators $(x, y) = x.y.x^{-1}.y^{-1}$ $(x \in M,\ y \in N)$.

4. If V is a k-variety, and k' an extension of k in K, then $V(k')$ denotes the set of points of V rational over k'. $k'[V]$ is the k'-algebra of regular functions defined over k' on V, and $k'(V)$ the k'-algebra of rational functions defined over k' on V. If W is a k-variety, and $f\colon V \to W$ a k-morphism, then the map $k[W] \to k[V]$ defined by $\varphi \to \varphi \circ f$ is the *comorphism* associated to f and is denoted f°.

Chapter AG

Background Material from Algebraic Geometry

This chapter should be used only as a reference for the remaining ones. Its purpose is to establish the language and conventions of algebraic geometry used in these notes. The intention is to take, in so far as is practicable, the point of view of Mumford's chapter I. Thus our varieties are identified with their points over a fixed algebraically closed field K (of any characteristic). It is technically important for us, however, not to require (as does Mumford) that varieties be irreducible.

For the most part definitions and theorems are simply stated with references and occasional indications of proofs. There are two notable exceptions. We have given essentially complete treatments of the material presented on rationality questions (i.e. field of definition), in sections 11–14, and of the material on tangent spaces, in sections 15–16. This seemed desirable because of the lack of convenient references for these results (in the form used here), and because of the important technical role both of these topics play in the notes.

§1. Some Topological Notions
(Cf. [Class., exp. 1, no. 1].)

1.1 *Irreducible components.* A topological space X is said to be *irreducible* if it is not empty and is not the union of two proper closed subsets. The latter condition is equivalent to the requirement that each non-empty open set be dense in X, or that each one be connected.

If Y is a subspace of a topological space X then is irreducible if and only if its closure $\bar{Y}$ is irreducible. By Zorn's lemma every irreducible subspace of X is contained in a maximal one, and the preceding remark shows that the maximal irreducible subspaces are closed. They are called the irreducible components of X. Since the closure of a point is irreducible it lies in an irreducible component; hence X is the union of its irreducible components.

If a subspace Y of X has only finitely many irreducible components, say $Y_1, \ldots, Y_n$, then $\bar{Y}_1, \ldots, \bar{Y}_n$ are the irreducible components (without repetition) of $\bar{Y}$.

1.2 *Noetherian spaces.* A topological space X is said to be *quasi-compact* ("quasi-" because X is not assumed to be Hausdorff) if every open cover has a finite subcover. If every open set in X is quasi-compact, or, equivalently, if the open sets satisfy the maximum condition, then X is said to be *noetherian.* It is easily seen that every subspace of a noetherian space is noetherian.

Proposition. *Let X be a noetherian space.*

(*a*) *X has only finitely many irreducible components, say $X_1, \ldots, X_n$.*
(*b*) *An open set U in X is dense if and only if $U \cap X_i \neq \phi\ (1 \leqq i \leqq n)$.*
(*c*) *For each i, $X'_i = X_i - \bigcup_{j \neq i} (X_j \cap X_i)$ is open in X, and $U_o = \bigcup_i X'_i$ is an open dense set in X whose irreducible and connected components are $X'_1, \ldots, X'_n$.*

Part (a) follows from a standard "noetherian induction" argument.

Since X_i is irreducible the set $X'_i = X - \left(\bigcup_{j \neq i} X_j \right)$ is open in X and dense in X_i. Hence every open dense set U in X must meet X'_i. Conversely if U is open and meets each X_i then $U \cap X_i$ is dense in X_i, so $\bar{U}$ contains each X_i and hence equals X. It follows, in particular, that $U_o = \bigcup_i X'_i$ is open, dense. Since the X'_i are open, irreducible, and pairwise disjoint, they are the irreducible and connected components of U_o.

1.3 *Constructible sets.* A subset Y of a topological space X is said to be *locally closed* in X if Y is open in $\bar{Y}$, or, equivalently, if Y is the intersection of an open set with a closed set. The latter description makes it clear that the intersection of two locally closed sets is locally closed. A *constructible set* is a finite union of locally closed sets. The complement of a locally closed set is the union of an open set with a closed set, hence a constructible set. It follows that the complement of a constructible set is constructible. Thus, the constructible sets are a Boolean algebra (i.e. they are stable under finite unions and intersections and under complementation) In fact they are the Boolean algebra generated by the open and (or) closed sets.

If $f: X \to X'$ is a continuous map then f^{-1} is a Boolean algebra homomorphism carrying open and closed sets, respectively, in X' to those in X. Hence f^{-1} carries locally closed and constructible sets, respectively in X' to those in X.

Proposition. *Let X be a noetherian space, and let Y be a constructible subset of X. Then Y contains an open dense subset of $\bar{Y}$.*

Remark. Conversely, by a noetherian induction argument one can show that if Y is a subset of X whose intersection with every irreducible closed subset of X has the above property, then Y is constructible.

Proof. Write $Y = \bigcup_i L_i$ with each L_i locally closed. Then $\bar{Y} = \bigcup_i \bar{L}_i$, so, if $\bar{Y}$ is irreducible, $\bar{Y} = \bar{L}_i$ for some i. Moreover $L_i (\subset Y)$ is open in $\bar{L}_i$.

In the general case write $Y = \bigcup_j Y_j$ where the Y_j are the irreducible components of Y. The latter are closed in Y and hence constructible in X. Moreover the first case shows that Y_j contains a dense open set in $\bar{Y}_j$. Since the $\bar{Y}_j$ are the irreducible components of $\bar{Y}$ (see (AG.1.1)) it follows from (AG.1.2) that $Y = \cup Y_j$ contains a dense open set in $\bar{Y}$.

1.4 *(Combinatorial) dimension.* For a topological space X it is the supremum of the lengths, n, of chains $F_0 \subset F_1 \subset \cdots \subset F_n$ of distinct irreducible closed sets in X; it is denoted

$$\dim X.$$

If $x \in X$ we write

$$\dim_x X$$

for the infimum of $\dim U$ where U varies over open neighborhoods of x.

It follows easily from the definitions and the properties of irreducible closed sets that $\dim \phi = -\infty$, that

$$\dim X = \sup_{x \in X} \dim_x X,$$

and that $x \mapsto \dim_x X$ is an upper semi-continuous function. Moreover, if X has a finite number of irreducible components (e.g. if X is noetherian), say $X_1, \ldots, X_m$, then $\dim X$ is the maximum of $\dim X_i (1 \leqq i \leqq m)$.

§2. Some Facts from Field Theory

2.1 *Base change for fields* (cf. [C.-C., exp. 13–14]). We fix a field extension F of k. If k' is any field extension of k we shall write

$$F_{k'} = k' \bigotimes_k F.$$

This is a k'-algebra, but it is no longer a field, or even an integral domain, in general. However, each of its prime ideals is minimal (i.e. there are no inclusion relations between them) and their intersection is the ideal of nilpotent elements in $F_{k'}$ (see (AG.3.3) below). We say a ring is *reduced* if its ideal of nilpotent elements is zero.

Here are the basic possibilities:

(a) *k' is separable algebraic over k*: Then $F_{k'}$ is reduced, but it may have more than one prime ideal.

(b) *k' is algebraic and purely inseparable over k*: Then $F_{k'}$ has a unique prime ideal (consisting of nilpotent elements) but $F_{k'}$ need not be reduced.

(c) *k' is a purely transcendental extension of k*: Then $F_{k'}$ is clearly an integral domain.

2.2 *Separable extensions.* F is said to be *separable* over k if it satisfies the following conditions, which are equivalent: We write p for the characteristic exponent of k ($=1$ if $\mathrm{char}(k)=0$).

(1) F^p and k are linearly disjoint over k^p.
(2) $F_{(k^{1/p})}$ is reduced.
(3) $F_{k'}$ is reduced for all field extensions k' of k.

Suppose, for some extension L of k, that F_L is an integral domain, with field of fractions (F_L). Then *F is separable over $k \Leftrightarrow (F_L)$ is separable over L.* The implication $\Rightarrow$ follows essentially from the associativity of tensor products, using criterion (3). To prove the converse we embed a given extension k' of k in a bigger one, k'', containing L also. Since $F_{k'} \subset F_{k''}$ it suffices to show that $F_{k''}$ is reduced. But $F_{k''} = F_L \bigotimes_L k'' \subset (F_L)_{k''}$ and the latter is reduced, by hypothesis.

2.3 *Differential criteria.* (See [N.B., (a), §9], [Z.-S., v. I, Ch. II, §17], or [C.-C., exp. 13].) A *k-derivation* $D: F \to F$ is a k-linear map such that

$$D(ab) = D(a)b + aD(b) \text{ for all } a, b \in F.$$

The set of them,

$$\mathrm{Der}_k(F, F)$$

is a vector space over F.

Theorem. *Suppose F is a finitely generated extension of k. Put*

$$n = \mathrm{tr\,deg}_k(F)$$

and

$$m = \dim_F \mathrm{Der}_k(F, F).$$

Then $m \geqq n$, with equality if and only if F is separable over k.

Let $D_1, \ldots, D_m$ be a basis of $\mathrm{Der}_k(F, F)$ *and let $a_1, \ldots, a_m \in F$. Then F is separable algebraic over $k(a_1, \ldots, a_m)$ if and only if* $\det(D_i(a_j)) \neq 0$.

If $m = n$ then a set $\{a_1, \ldots, a_m\}$ as above is called a *separating transcendence basis.*

2.4 Proposition. *Let G be a group of automorphisms of a field F. Then F is a separable extension of $k = F^G$, the fixed elements under G.*

We shall prove that F and $k^{1/p}$ are linearly disjoint over k, i.e. that if $a_1, \ldots, a_n \in k^{1/p}$ are linearly independent over k then they are linearly independent over F. The action of G extends uniquely to $F^{1/p}$ and G acts trivially on $k^{1/p}$. Suppose $a_1, \ldots, a_n$ are linearly dependent over F, but not over k; we can assume n is minimal. Let $a_1 + b_2 a_2 + \cdots + b_n a_n = 0$ be a dependence relation. If some b_i, say b_n, is not in k then it is moved by some

$g \in G$. Subtracting $a_1 + g(b_2)a_2 + \cdots + g(b_n)a_n$ from the relation above we obtain a shorter relation; contradiction.

2.5 On occasions, we shall need a generalization of 2.4. Let A be a reduced noetherian algebra over k, denote by $k(A)$ its ring of fractions (cf. 3.1, Ex. 1) and let G be a group of automorphisms of A. The action then extends to $k(A)$. By Prop. 10 in [N.B.(b):IV, §2, no. 5], $k(A)$ is uniquely a sum of fields K_i then necessarily permuted by G. Let e_i be the corresponding idempotents. Thus $1 = \sum e_i$ and the e_i's are permuted by G. If $\alpha \in A^G$ is non-divisor of zero in A^G, then it is one in A. In fact we can write $1 = \sum f_j$ where f_j is the sum of idempotents e_i forming an orbit of G; then we have $f_i \cdot \alpha \neq 0$ and therefore since $g(e_i \cdot \alpha) = g(e_i) \cdot \alpha$, $e_i \alpha \neq 0$ for all i's. Therefore $k(A^G)$ embeds in $k(A)^G$.

Proposition. *We keep the previous notation. Then $e_i \cdot k(A)^G = K_i^{G_i}$, where G_i is the isotropy group of e_i. If $k(A)^G = k(A^G)$, then K_i is a separable extension of $e_i k(A^G)$.*

If $a \in k(A)^G$ then $e_i \cdot \alpha$ is fixed under G_i. Conversely, if $b \in K_i$ is fixed under G_i, then the sum of the $g(b)$, where g runs through a set of representatives of G/G_i, is an element of $k(A)^G$ whose image under e_i is b. Then 2.4 shows that K_i is a separable extension of $e_i \cdot k(A)^G$. The second assertion is then obvious.

§3. Some Commutative Algebra

3.1 *Localization* [N.B., (b)]. Let S be a multiplicative set in a ring A, i.e. S is not empty and $s, t \in S \Rightarrow st \in S$. Then we have the "localization" $A[S^{-1}]$ consisting of fractions a/s ($a \in A$, $s \in S$), and the natural map $A \to A[S^{-1}]$ which is universal among homomorphisms from A rendering the elements of S invertible.

If M is an A-module we further have the localized $A[S^{-1}]$-module $M[S^{-1}]$, consisting of fractions $x/s (x \in M,\ s \in S)$, which is naturally isomorphic to $A[S^{-1}] \bigotimes_A M$.

If $x \in M$ and $s \in S$ then $x/s = 0$ in $M[S^{-1}]$ if and only if $tx = 0$ for some $t \in S$. It follows directly from this that, *if M is finitely generated $M[S^{-1}] = 0$ if and only if $tM = 0$ for some $t \in S$*, i.e. if and only if $S \cap \operatorname{ann} M \neq \phi$, where $\operatorname{ann} M$ is the annihilator of M in A.

The functor $M \mapsto M[S^{-1}]$ from A-modules to $A[S^{-1}]$-modules is exact, and it preserves tensors and Hom's in the following sense: If M and N are A-modules then the natural map $\left(M \bigotimes_A N\right)[S^{-1}] \to M[S^{-1}) \bigotimes_{A[S^{-1}]} N[S^{-1}]$ is an isomorphism, and the natural map $\operatorname{Hom}_A(M, N)[S^{-1}] \to \operatorname{Hom}_{A[S^{-1}]}(M[S^{-1}], N[S^{-1}])$ is an isomorphism if M is finitely presented.

Examples. (1) Let S be the set of all non-divisors of zero in A. Then $A \to A[S^{-1}]$ is injective, and the latter is called the *full ring of fractions* of A. When A is an integral domain it is the field of fractions.

(2) If $S = \{f^n | n \geqq 0\}$ for some $f \in A$ then we write A_f or $A[1/f]$, and M_f for the localizations.

(3) An ideal P in A is prime if $S_P = A - P$ is a multiplicative set. The corresponding localizations are denoted A_P and M_P. In this case A_P has a unique maximal ideal, PA_P, i.e. A_P is a *local ring.*

3.2 *Local rings.* Let A be a local ring with maximal ideal $\mathfrak{m}$ and residue class field $k = A/\mathfrak{m}$. Let M be a finitely generated A-module.

(a) If $\mathfrak{m}M = M$ then $M = 0$.

For let $x_1, \ldots, x_n$ be a minimal set of generators of M, and suppose $n > 0$. Write $x_1 = \sum a_i x_i (a_i \in \mathfrak{m})$. Then $(1 - a_1)x_1 = \sum_{i>1} a_i x_i$. But $1 - a_1$ is invertible, so $x_2, \ldots, x_n$ already generate M; contradiction.

(b) If $x_1, \ldots, x_n \in M$ then they generate M if and only if they do so modulo $\mathfrak{m}M$. Hence the minimal number of generators of M is $\dim_k(M/\mathfrak{m}M)$.

This follows by applying (a) to M/N, where N is the submodule generated by $x_1, \ldots, x_n$.

(c) If M is projective then M is free.

We can write $A^n = M \oplus N$, so that $k^n = (M/\mathfrak{m}M) \oplus (N/\mathfrak{m}N)$. Lift a basis of k^n to A^n so that it lies in $M \cup N$. The result is, by (b), a set of n generators of A^n. These must clearly be a basis of A^n, e.g. because the associated matrix has an invertible determinant. Hence M, being spanned by part of a basis of A^n, is free.

3.3 *Nil radical; reduced rings.* The set of nilpotent elements in a ring A is an ideal denoted nil A. We call A *reduced* if nil $A = (0)$.

If J is any ideal the ideal $\sqrt{J}$ is defined by $\sqrt{J}/J = \mathrm{nil}(A/J)$. Thus nil $A = \sqrt{(0)}$. Moreover, we have

$$\sqrt{J} = \text{the intersection of all primes containing } J.$$

If S is a multiplicative set then $\sqrt{J \cdot A[S^{-1}]} = \sqrt{J} \cdot A[S^{-1}]$. In particular this implies that *A is reduced if and only if the full ring of fractions of A is reduced.*

3.4 *spec*(A) [M, Ch. II, §1]. We let $X = \mathrm{spec}(A)$ be the set of all prime ideals in A, equipped with the *Zariski topology*, in which the closed sets are those of the following form for some $J \subset A$:

$$V(J) = \{P \in X | J \subset P\}.$$

If $Y \subset X$ we put $I(Y) = \bigcap_{P \in Y} P$, and then $V(I(Y))$ is just the closure of Y.

Moreover, if J is an ideal of A it follows from 3.3 that

$$I(V(J)) = \sqrt{J}.$$

Thus closed sets correspond bijectively (with inclusions reversed) to ideals J for which $J = \sqrt{J}$. It follows that if A is noetherian then spec(A) is a noetherian space.

The map $P \mapsto \overline{\{P\}}$ is a bijection from X to the set of irreducible closed sets in X. Thus the irreducible components of X correpond to the minimal primes in A. Moreover the (combinatorial) dimension of X (measured by chains of irreducible closed sets) is called the (Krull) *dimension of* A, and it is denoted $\dim A$. Thus

$$\dim A = \dim X.$$

If $f \in A$ and $P \in X$ one sometimes writes $f(P)$ for the image of f in the residue class field of A_P (which is the field of fractions of A/P). With this notation the complement of $V(fA)$ is

$$X_f = \{P \in X \mid f(P) \neq 0\}.$$

This is called a *principal open set*. For any J we have $V(J) = \bigcap_{f \in J} V(f)$ so the principal open sets are a base for the topology.

Suppose $\alpha_o: A \to B$ is a ring homomorphism. Then α_o induces a continuous map $\alpha: Y = \text{spec}(B) \to X$, $\alpha(P) = \alpha_o^{-1}(P)$. In fact $\alpha^{-1}(V(J)) = V(\alpha_o(J))$.

Examples. (1) If J is an ideal then $A \to A/J$ induces a homeomorphism of spec(A/J) onto $V(J) \subset X$.

(2) If S is a multiplicative set then $\text{spec}(A[S^{-1}]) \to \text{spec}(A)$ induces a homeomorphism onto the set of $P \in X$ such that $P \cap S = \phi$.

(i) If $f \in A$ then we obtain a homeomorphism $\text{spec}(A_f) \to X_f$.

(ii) If $P \in X$ it follows that $\dim_P X = \dim \text{spec}(A_P) = \text{(Krull)} \dim A_P$.

3.5 *Support of a module.* Let $X = \text{spec}(A)$ where A is a noetherian ring, and let M be a finitely generated A-module. Then it follows from 3.1 that

$$\text{supp}(M) = \{P \mid M_P \neq 0\}$$

is the closed set $V(\text{ann}\, M)$. In particular $M = 0$ if and only if $\text{supp}(M) = \phi$.

Let $f: L \to M$ be a homomorphism of A-modules. Since localization is exact it follows that the set of P where f_P is an epimorphism is the (open) complement of supp(coker f). Applying this to $\text{Hom}_A(M, L) \to \text{Hom}_A(M, M)$, and using the fact that the Hom's localize properly (see 3.1) we conclude that the set U of $P \in X$ such that f_P is a split epimorphism is open, and f is a split epimorphism if and only if $U = X$.

Suppose f is surjective and L is free. Then we deduce from the last remark and 3.2(c) that:

$$U = \{P \in X \mid M_P \text{ is a free } A_P\text{-module}\}$$

is open, and M is a projective A-module if and only if $U = X$.

3.6 *Integral extensions* ([N.B., (b), Ch. 5] or [Z.-S., v. I, Ch. V]). Let $A \subset B$ be rings. A $b \in B$ is said to be *integral* over A if $A[b]$ is a finitely generated A-module, or, equivalently if b is a root of a monic polynomial with coefficients in A. The set B' of all elements of B integral over A is a subring, called the *integral closure* of A in B. We say B is *integral over* A if $B' = B$. We say A is *integrally closed* in B if $B' = A$. We call A *normal* if A is reduced and integrally closed in its full ring of fractions.

Suppose $A \subset B \subset C$ are rings. Then C is integral over A if and only if C and B are integral over B and A, respectively.

Suppose B is integral over A. Then $\operatorname{spec}(B) \to \operatorname{spec}(A)$ *is surjective and closed.* If B is a finitely generated A-algebra then B is a finitely generated A-module. If B is an integral domain then every non-zero ideal of B has non-zero intersection with A.

To see the latter let $b^n + a_{n-1}b^{n-1} + \cdots + a_o = 0$ be an integral equation of minimal degree over A of some $b \neq 0$ in B. Then $a_o = -b(a_{n-1}b^{n-2} + \cdots + a_1) \in bB \cap A$. Moreover $a_o \neq 0$; otherwise we could reduce the degree of the equation.

3.7 *Noether normalization* [M, Ch. I, p. 4]. *A k-algebra A is said to be affine* if it is finitely generated as a k-algebra. Such an A is a noetherian ring.

Theorem. *Let $R = k[y_1, \ldots, y_m]$ be an affine integral domain over k whose field of fractions, $k(y_1, \ldots, y_m)$, has transcendence degree n over k. Then there exist elements $x_1, \ldots, x_n \in R$, which are algebraically independent over k, and such that R is integral over the polynomial ring $k[x_1, \ldots, x_n]$. If $k(y_1, \ldots, y_m)$ is separable over k then $x_1, \ldots, x_n$ can be chosen to be a separating transcendence basis of $k(y_1, \ldots, y_m)$ over k.*

Except for the last assertion this theorem is essentially identical in statement and notation with that in Mumford, page 4. With the following modification, the proof in Mumford gives also the last assertion as well.

First, choose $y_1, \ldots, y_m$ so that the last n of them are a separating transcendence basis. Next, choose the integers $r_1, \ldots, r_m$ (as well as their analogues at other stages of the induction) to be divisible by p, the characteristic exponent of k. The proof in Mumford requires only that the r_i's be large and increase rapidly, so our additional restriction is harmless.

This done, the $x_1, \ldots, x_n$ produced by the proof will be congruent, modulo p^{th} powers, to the last n of the y_i's. Thus each x_i has the same image under every k-derivation as the corresponding y (if $p > 1$; otherwise there is no problem). It therefore follows that the x's, like the y's, are a separating transcendence basis (see (AG.2.3)).

3.8 *The Nullstellensatz* [M, Ch. I]. Let A be an affine K-algebra, and let $X = \max(A)$ be the subspace of maximal ideals in $\operatorname{spec}(A)$.

If $e: A \to K$ is a K-algebra homomorphism then $\ker(e) \in X$ so we have a natural map

$$\varphi: \mathrm{Mor}_{K\text{-alg}}(A, K) \to X.$$

Theorem. (Nullstellensatz).

(1) *φ is bijective.*
(2) *X is dense in* $\operatorname{spec}(A)$. *Moreover $F \mapsto F \cap X$ is a bijection from the set of closed sets in* $\operatorname{spec}(A)$ *to the set of closed sets in X. Therefore the analogous statement is valid for open sets also.*

If $x \in X$ we shall write e_x for the homomorphism $A \to K$ such that $x = \ker(e_x)$. If $f \in A$ we shall also use the functional notation

$$f(x) = e_x(f).$$

Thus each $f \in A$ determines a function $X \to K$. If f represents the zero function then $f \in I(X) = \bigcap_{x \in X} x$. It follows from part (2) that $I(x) = I(\operatorname{spec}(A)) = \operatorname{nil} A$. Thus, in general, the function on X associated with f determines f modulo $\operatorname{nil} A$. If A is reduced we can therefore view A as a ring of K-valued function on X.

We shall use for X the same notational conventions introduced for $\operatorname{spec}(A)$. For example, if $f \in A$ then $X_f = \{x \in X \mid f(x) \neq 0\}$. These principal open sets are a base for the topology on X.

If M is an A-module we also write $\operatorname{supp}_X(M) = \{x \in X \mid M_x \neq 0\}$, or simply $\operatorname{supp}(M)$ when the meaning is clear. In view of part (2) of the Nullstellensatz all the remarks of 3.5 remain valid with X in place of $\operatorname{spec}(A)$.

The correspondence in (2) also matches irreducible closed sets, clearly, and hence irreducible components. If $x \in X$, then $\dim_x X = \dim_x \operatorname{spec}(A) = \dim A_x$. Moreover $\dim X = \dim \operatorname{spec}(A)$.

3.9 *Regular local rings* [Z.-S., v. II, Ch. VIII, §11]. Let A be a noetherian local ring with maximal ideal $\mathfrak{m}$ and residue class field $k = A/\mathfrak{m}$. Then the minimal number of generators of $\mathfrak{m}$ is (see 3.2) the dimension over k of $\mathfrak{m}/\mathfrak{m}^2$. It is a basic fact that

$$\dim_k(\mathfrak{m}/\mathfrak{m}^2) \geqq \dim A,$$

where $\dim A$ is defined as in 3.4. When this inequality is an equality the local ring A is said to be *regular*.

Regularity has rather strong consequences for A, for example the fact that *A is then a unique factorization domain.*

We shall see in AG.17 that, when A is the local ring of a point x on a variety V, then regularity of A means that x is a simple point; hence the importance of the notion. A minimal set of generators of $\mathfrak{m}$ then gives the right number of local parameters at x on V, and $\mathfrak{m}/\mathfrak{m}^2$ is the cotangent space (see AG.16) of V at x.

§4. Sheaves
[*M*, Ch. I, §4]

4.1 *Presheaves.* Let X be a topological space. The open sets in X are the objects of a category, top(X), whose morphisms are inclusions. If C is a category then a *C-valued presheaf* on X is a contravariant functor $U \mapsto F(U)$ from top(X) to C. Thus, whenever $V \subset U$ are open sets in X we have a C-morphism

$$\operatorname{res}_V^U : F(U) \to F(V),$$

sometimes called "restriction." A morphism $\varphi: F \to F'$ of presheaves is just a morphism of functors. Thus it consists of morphisms $\varphi_U: F(U) \to F'(U)$ rendering the diagrams

$$\begin{array}{ccc} F(U) & \xrightarrow{\varphi_U} & F'(U) \\ {\scriptstyle \operatorname{res}_V^U}\downarrow & & \downarrow{\scriptstyle \operatorname{res}_V^U} \\ F(V) & \xrightarrow[\varphi_V]{} & F'(V) \end{array}$$

commutative.

Suppose C is a category of "sets with structure," like groups, rings, modules,.... Then we say F is a presheaf of groups, rings, modules,..., respectively, on X. If $x \in X$ then

$$F_x = \operatorname{ind\,lim}_U \quad F(U) \qquad (U \text{ nbhd. of } x)$$

is called the *stalk* of F over x.

If U is open in X then top(U) is a subcategory of top(X), to which we can restrict a presheaf F on X. The resulting presheaf on U is denoted $(U, F|U)$.

4.2 *Sheaves.* Let F be a C-valued presheaf, on X, where C is some category of "sets with structure." Then F is called a *sheaf* if it satisfies the following "sheaf axiom": Given an open cover $(U_i)_{i \in I}$ of an open set U in X, the sequence

$$F(U) \xrightarrow[\gamma]{\alpha} \prod_i F(U_i) \underset{\gamma}{\overset{\beta}{\rightrightarrows}} \prod_{i,j} F(U_i \cap U_j)$$

of sets is exact.

Explanation: "Exact" means that α induces a bijection from $F(U)$ to the set of elements on which β and γ agree. Thus, if F is a presheaf of abelian groups, for example, exactness means that α is the kernel of $(\beta - \gamma)$.

The map α is induced by the restrictions $F(U) \to F(U_i)(i \in I)$. Similarly, the restrictions $F(U_i) \to F(U_i \cap U_j)(j \in I)$ induce $F(U_i) \to \prod_j F(U_i \cap U_j)$. Taking the product of these over $i \in I$ we obtain β. The map γ is obtained similarly, starting from $F(U_j) \to F(U_i \cap U_j)$ to obtain $F(U_j) \to \prod_i F(U_i \cap U_j)$.

Explicitly, the sheaf axiom says that, given $s_i \in F(U_i)$ such that $s_i | U_i \cap U_j = s_j | U_i \cap U_j$ for all $i, j \in I$ (we write $s|V$ for $\text{res}_V^U(s)$) then there is a unique $s \in F(U)$ such that $s|U_i = s_i$ for all $i \in I$.

Example. Let $F(U)$ be the ring of continuous real valued functions on U. Then, with respect to restriction of functions, F is clearly a sheaf (of commutative rings).

4.3 *Sheafification.* Let F be a C-valued presheaf on X, where C is some category of "sets with structure." Then there is a sheaf, F', called the "sheafification" of F, or the *sheaf associated with* F, and a morphism $f: F \to F'$ through which all morphisms from F into sheaves factor uniquely. In other words the map

$$\text{Mor}(F', G) \to \text{Mor}(F, G)$$

induced by f is bijective whenever G is a sheaf.

Roughly speaking, F' can be constructed in two steps. First define $F_1(U)$ to be $F(U)$ modulo the equivalence relation which relates s and t if their restrictions agree on some open cover of U. Then form F' from F_1 by "adding" to $F_1(U)$ all elements obtainable from compatible local data on some covering of U. This process makes sense thanks to step 1.

If $x \in X$ the morphism of stalks $F_x \to F'_x$ is bijective.

Presheaves of abelian groups or modules form an abelian category, with the obvious notions of kernel, cokernel, exact sequence, etc. Thus, if $f: F \to G$ is a morphism of presheaves then $(\ker f)(U) = \ker(F(U) \to G(U))$, and similarly for $\text{coker}(f)$. If F and G are sheaves then $\ker(f)$ is also a sheaf. On the othe hand $\text{coker}(f)$ need not be a sheaf. The cokernel of f in the category of sheaves is the sheafification of the presheaf cokernel.

One can show that the category of sheaves of abelian groups is abelian. A sequence $F \to G \to H$ of sheaves is exact if and only if $F_x \to G_x \to H_x$ is exact for all $x \in X$.

§5. Affine K-Schemes; Prevarieties

5.1 A K-*space* is a topological space X together with a sheaf $\mathcal{O}_X$ of K-algebras on X whose stalks are local rings. If $x \in X$ we write $\mathcal{O}_{X,x}$ for the stalk over x, or simply $\mathcal{O}_x$ if X is clear from the context. Its maximal ideal is denoted $\mathfrak{m}_x$, and its residue class field by $K(x)$. One often writes X in place of $(X, \mathcal{O}_X)$ if this leads to no confusion.

A *morphism* $(Y, \mathcal{O}_Y) \to (X, \mathcal{O}_X)$ of K-spaces consists of a continuous function $\alpha: Y \to X$ together with K-algebra homomorphisms

$$\alpha_V^U: \mathcal{O}_X(U) \to \mathcal{O}_Y(V)$$

whenever $U \subset X$ and $V \subset Y$ are open sets such that $\alpha(V) \subset U$. These maps are required to be compatible with the respective restriction homomorphisms in $\mathcal{O}_X$ and $\mathcal{O}_Y$. For $y \in Y$ we can pass to the limit over neighborhoods V of y and U of $x = f(y)$ to deduce a homomorphism $\alpha_y: \mathcal{O}_x \to \mathcal{O}_y$. It is further required of a morphism that this always be a "local homomorphism," i.e. that $\alpha_y(\mathfrak{m}_x) \subset \mathfrak{m}_y$.

5.2 *The affine K-scheme* $\mathrm{spec}_K(A)$. An *affine K-algebra* A is one which if finitely generated as an algebra. For such an algebra the subspace $X = \max(A)$ of maximal ideals in $\mathrm{spec}(A)$ will be denoted

$$\mathrm{spec}_K(A).$$

Recall from the Nullstellensatz (AG. 3.8) that there is a canonical bijection $x \mapsto \ker(e_x)$

$$X = \mathrm{spec}_K(A) \text{ onto } \mathrm{Hom}_{K\text{-}alg}(A, K).$$

Moreover we adopt the functional notation

$$f(x) = e_x(f) \quad (x \in X, f \in A).$$

The resulting function $f: X \to K$ (for $f \in A$) determines f modulo the nil radical of A (see AG.3.8)) so, if A is reduced, we can thus identify A with a ring of K-valued functions on X.

We now introduce the K-space $(X, \tilde{A})$, where $\tilde{A}$ is the sheaf associated to the presheaf $U \mapsto A[S(U)^{-1}]$. Here, for U open in X, $S(U)$ is the set of $f \in A$ vanishing nowhere on U. It is easy to see that the stalk of $\tilde{A}$ at $x \in X$ is the local ring A_x, so that $(X, \tilde{A})$ is a K-space. The symbol $\mathrm{spec}_K(A)$ will be used both for X and for the K-space $(X, \tilde{A})$. A K-space isomorphic to one of this type will be called an *affine K-scheme*.

In case A is an integral domain with field of fractions L then the A_x's are subrings of L and we can describe $\tilde{A}$ directly by: $\tilde{A}(U) = \bigcap_{x \in U} A_x$.

A homomorphism $\alpha: A \to B$ of affine K-algebras induces a continuous function $\alpha': Y \to X$, where $Y = \mathrm{spec}_K(B)$. If $U \subset X$ and $V \subset Y$ are open and $\alpha'(V) \subset U$ then $\alpha(S(U)) \subset S(V)$ so there is a natural homomorphism $A[S(U)^{-1}] \to B[S(V)^{-1}]$. These induce a morphism on the associated K-spaces $(Y, \tilde{B}) \to (X, \tilde{A})$, thus making $A \mapsto \mathrm{spec}_K(A)$ a contravariant functor from affine K-algebras to K-spaces.

5.3 *K-schemes and prevarieties.* By a *K-scheme* we shall understand a K-space $(X, \mathcal{O}_X)$ such that X has a *finite* cover by open sets U such that $(U, \mathcal{O}_X|U)$ is an affine K-scheme. Note that X is thus a noetherian space. If

$(X, \mathcal{O}_X)$ is reduced, i.e. if, for each $x \in X$, the local ring $\mathcal{O}_{X,x}$ has no nilpotent elements $\neq 0$, then we call $(X, \mathcal{O}_X)$ a *prevariety*. In case $X = \mathrm{spec}_K(A)$ is affine then X is a prevariety if and only if A is reduced, in which case we call $\mathrm{spec}_K(A)$ an *affine variety*.

Caution. (1) A K-scheme is *not* a scheme in the usual sense. This would be the case if, in place of $\mathrm{spec}_K(A) = \max(A)$ we had used all of $\mathrm{spec}(A)$ (in the affine case). With this modification the definition of K-scheme above corresponds to the notion of a "scheme of finite type over K" (or over $\mathrm{spec}(K)$).

(2) Our notion of prevariety is essentially the same as that of Mumford (Chapter I) except that we have not required X to be irreducible.

Consider the affine K-scheme $\mathrm{spec}_K(K)$, consisting of one point with structure sheaf K. A morphism $\mathrm{spec}_K(K) \to X$ just picks a point $x \in X$ together with compatible K-algebra homomorphisms $\mathcal{O}_X(U) \to K$ for all neighborhoods U of x. The latter correspond to a K-algebra homomorphism $\mathcal{O}_x \to K$, and there is only one such: $f \mapsto f(x)$. Thus x determines the morphism, i.e. we can identify $\mathrm{Mor}_{K\text{-sch}}(\mathrm{spec}_K(K), X)$ with X (as sets).

5.4 Theorem. *Let $X = \mathrm{spec}_K(A)$ be an affine K-scheme and let Y be any K-scheme. The natural map $A \to \tilde{A}(X)$ is an isomorphism, and the map*

$$\mathrm{Mor}_{K\text{-sch}}(Y, X) \to \mathrm{Mor}_{K\text{-alg}}(A, \mathcal{O}_Y(Y))$$

is bijective. In particular $A \mapsto \mathrm{spec}_K(A)$ is a contravariant equivalence from the category of affine K-algebras to the category of affine K-schemes.

For this equivalence, see [M, Ch. II, §§1–2].

5.5 *Quasi-coherent modules* [M, Ch. III, §§1–2]. Let A be an affine K-algebra. If M is an A-module then the sheaf $\tilde{M}$ on $\mathrm{spec}_K(A)$ associated with the presheaf $u \mapsto A[S(U)^{-1}] \bigotimes_A M$ is a sheaf of $\tilde{A}$-modules, or, simply, an $\tilde{A}$-module. Moreover $M \mapsto \tilde{M}$ is an exact functor from A-modules to $\tilde{A}$-modules.

If Y is a K-scheme we say that an $\mathcal{O}_Y$-module (or sheaf of $\mathcal{O}_Y$-modules) F is *quasi-coherent* if Y can be covered by affine K-schemes $U = \mathrm{spec}_K(A)$ on which $F|U$ is isomorphic to some $\tilde{M}$ as above. If the U's can be chosen so that each M is a finitely generated (resp., free) A-module then we say F is *coherent* (resp., locally free).

If F is coherent then it follows easily from AG.3.5 that

$$\mathrm{supp}(F) = \{y \in Y \mid F_y \neq 0\}$$

is closed. Moreover AG.3.5 implies that, for F coherent, $\{y \in Y \mid F_y$ is a free $\mathcal{O}_y$-module$\}$ is open.

Theorem. *Let $X = \mathrm{spec}_K(A)$ be an affine K-scheme, and let $f \in A$. For any*

A-module M the natural map $M_f \to \tilde{M}(X_f)$ is an isomorphism. In particular

$$(\operatorname{spec}_K(A_f), \tilde{A}_f) \to (X_f, \tilde{A}|X_f)$$

is an isomorphism of K-schemes. Moreover $M \mapsto \tilde{M}$ is an equivalence from the category of A-modules to the category of quasi-coherent $\tilde{A}$-modules. $\tilde{M}$ is coherent if and only if M is finitely generated. In this case $\tilde{M}$ is locally free if and only if M is a projective A-module.

5.6 *Closed immersions* [M, Ch. II, §5]. A morphism $\alpha: Y \to X$ of K-schemes is called a *closed immersion* if α maps Y homeomorphically into a closed subspace of X and if the local homomorphisms $\mathcal{O}_{X,\alpha(y)} \to \mathcal{O}_{Y,y}$ are surjective for each $y \in Y$.

If $\mathcal{I}$ is a quasi-coherent sheaf of ideals in $\mathcal{O}_X$, and if $Y = \operatorname{supp}(\mathcal{O}_X/\mathcal{I})$ then Y is closed and $\mathcal{O}_X/\mathcal{I}$ is the "extension by zero" of a sheaf $\mathcal{O}_Y$ on Y for which there is a natural closed immersion $(Y, \mathcal{O}_Y) \to (X, \mathcal{O}_X)$. We then call Y the *closed subscheme* of X defined by $\mathcal{I}$.

In case $X = \operatorname{spec}_K(A)$ is affine every such $\mathcal{I}$ is of the form $\tilde{I}$ for some ideal I in A, and Y is just the affine subscheme

$$\operatorname{spec}_K(A/I) \hookrightarrow \operatorname{spec}_K(A).$$

Theorem. *The map $I \mapsto \operatorname{spec}_K(A/I)$ is a bijection from the ideals of A to the set of closed subschemes of* $\operatorname{spec}_K(A)$. *In particular every closed subscheme is affine.*

An *open immersion* is a morphism isomorphic to one of the form $(U, \mathcal{O}_X|U) \to (X, \mathcal{O}_X)$ where X is a K-scheme and U is an open subset. We call $(U, \mathcal{O}_X|U)$ an open subscheme of $(X, \mathcal{O}_X)$. A closed subscheme of an open subscheme is called a *locally closed subscheme.*

§6. Products; Varieties

6.1 *Products exist* [M, Ch. I, §6]. Let X and Y be K-schemes. The product $X \times Y$ is characterized by the property that morphisms from a K-scheme Z to $X \times Y$ are pairs of morphisms to the two factors. Applying this to $Z = \operatorname{spec}_K(K)$ we find that the underlying set of $X \times Y$ is the usual cartesian product. From AG.5.4 it follows immediately that the product of affine K-schemes $\operatorname{spec}_K(A)$ and $\operatorname{spec}_K(B)$ exists and equals

$$\operatorname{spec}_K\left(A \bigotimes_K B\right).$$

This is because $\bigotimes_K$ is the coproduct in the category of affine K-algebras.

More generally:

Theorem. *The product $X \times Y$ exists and the two projections are open maps.*

If $U \subset X$ and $V \subset Y$ are open subschemes then $U \times V \to X \times Y$ is an open immersion.

From this theorem and the description of the product in the affine case it is easy to show that the local ring of $X \times Y$ at (x, y) is the localization of $\mathcal{O}_x \bigotimes_K \mathcal{O}_y$ at $\mathfrak{m}_x \otimes \mathcal{O}_y + \mathcal{O}_x \otimes \mathfrak{m}_y$.

6.2 *Varieties.* Let X be a K-scheme. The pair $(1_X, 1_X)$ defines a diagonal morphism $d: X \to X \times X$, and one says X is *separated* if d is a closed immersion. A separated prevariety is called an (*algebraic*) *variety*.

For example:

(a) An affine variety is a variety.

(b) A locally closed subprevariety of a variety is a variety.

(c) A product of two varieties is a variety.

Let $\alpha, \beta: Y \to X$ be two morphisms of K-schemes, and let

$$\Gamma_{\alpha,\beta} = \{y \in Y \mid \alpha(y) = \beta(y)\}.$$

The pair (α, β) defines a morphism $\gamma: Y \to X \times X$ and $\Gamma_{\alpha,\beta} = \gamma^{-1}(d(X))$, clearly. Hence, if X is separated then $\Gamma_{\alpha,\beta}$ is closed. In particular, if α and β coincide on a dense set then they coincide at all points.

Applying the above remarks to $\alpha \circ pr_Y, pr_X: Y \times X \to X$ we see also that the graph of α is closed if X is separated.

6.3 *Regular functions and subvarieties.* Let $(X, \mathcal{O}_X)$ be an algebraic variety. If U is open in X we shall write

$$K[U] \text{ in place of } \mathcal{O}_X(U).$$

The elements f of $K[U]$ can be identified with K-valued functions on U, sometimes called *regular functions.* Moreover $\mathrm{res}^U_V: K[U] \to K[V]$ then corresponds to restriction of functions. For $x \in U$ the map $f \mapsto f(x) = e_x(f)$ is the composite of $K[U] \to \mathcal{O}_x$ with the map of $\mathcal{O}_x$ to its residue class field $K(x) = K$.

If U is open in X then $(U, \mathcal{O}_X | U)$ is a variety, called an *open subvariety* of X. In case U is affine we have $U = \mathrm{spec}_K(K[U])$.

If Y is a closed subspace of X then there is a unique *reduced* subscheme $(Y, \mathcal{O}_Y)$ of X. $\mathcal{O}_Y$ is the sheaf associated to the presheaf $(U \cap Y) \mapsto K[U]/I_U(Y)$, where $I_U(Y)$ is the ideal of all functions on U vanishing on $Y \cap U$. (Thus, in case U is affine, $Y \cap U$ is just $\mathrm{spec}_K(K[U]/I_U(Y))$.) In this way we can canonically regard a closed subspace Y of X as a *closed subvariety.*

A *locally closed subvariety* is then just a closed subvariety of an open subvariety.

Let $\alpha: Y \to X$ be a morphism of varieties. Then α is a continuous function and, whenever $U \subset X$ and $V \subset Y$ are open and $\alpha(V) \subset U$, there is a comorphism

$$\alpha^U_V: K[U] \to K[V]$$

such that

$$\alpha_V^U(f)(y) = f(\alpha(y)), \quad \text{or}$$
$$\alpha_V^U(f) = f \circ \alpha$$

for $f \in K[U]$ and $y \in V$. Since we are dealing here with rings of functions it follows that α (as a map of spaces) determines the sheaf homomorphisms α_V^U. We shall denote the latter simply by α^o (for all U and V) and call α^o the *comorphism*(s) of α.

Note that, for any set function $\alpha: Y \to X$, the comorphisms α^o can be defined as above on the rings of all K-valued functions. The condition that α be a morphism of varieties then can be reformulated as follows: (i) α is continuous, and (ii) if $U \subset X$ and $V \subset Y$ are open and if $\alpha(V) \subset U$ then $\alpha^o K[U] \subset K[V]$.

6.4 *The local rings on a variety.* Consider the local ring $\mathcal{O}_x$ of a point x on a variety V. It reflects the "local properties" of V near x. For example, by passing to a neighborhood of x we may assume $V = \mathrm{spec}_K(A)$, an affine variety. Then $\mathcal{O}_x$ is the local ring of A at the maximal ideal $\mathfrak{m} = \ker(e_x)$, and it follows from properties of localization that the prime ideal of $\mathcal{O}_x$ correspond bijectively to those of A contained in $\mathfrak{m}$, i.e. to the irreducible subvarieties of V passing through x. We see thus that $\dim_x V$ (in the sense of AG.1.4) is the Krull dimension of $\mathcal{O}_x$.

Note further that the irreducible components of V containing x correspond to the minimal primes of $\mathcal{O}_x$. Thus x lies on a unique irreducible component if and only if $\mathcal{O}_x$ is an integral domain.

6.5 Let $f: Y \to X$ be a morphism of varieties. It is said to be *finite* if X has an open cover by affine subvarieties X_i $(i \in I)$ such that $f^{-1}X_i$ is affine and $K[f^{-1}X_i]$ is a finitely generated $K[X_i]$-module. In that case, this condition is fulfilled by every open affine subset of X (cf [Ha: II, 3.2]). If f is finite, the fibre over each point of X is finite [EGA: II, 6.1.7] and f is closed [EGA: II, 6.1.10].

The morphism f is said to be *affine* if there exists an open affine finite cover $\{X_i\}$ [$i \in I$) of X such that $f^{-1}X_i$ is affine for all $i \in I$. Then $f^{-1}(U)$ is affine for every open affine subset U of X (see [Ha: II: 5.17] or EGA II, §1.2).

In particular, a finite morphism is affine by definition.

6.6 Let X and Y be two varieties. The Zariski topology on $X \times Y$ is finer than the product topology. We have already remarked that the two projections are open. Moreover, if $A \subset X$ and $B \subset Y$, then $(A \times B)^- = \bar{A} \times \bar{B}$: By using the continuity of the projections, we see that the right-hand side is closed and contains the left-hand side. On the other hand, for any $b \in B$, the closure of $A \times b$ is $\bar{A} \times b$, hence $\bar{A} \times B \subset (A \times B)^-$. Similarly, for any $a \in \bar{A}$, the product $a \times \bar{B}$ is contained in $(A \times B)^-$, whence our assertion. By induction, it follows that if X_i are varieties $(i = 1, \ldots, n)$ and $A_i \subset X_i$, then the closure of $A = A_1 \times \cdots \times A_n$ in $X_1 \times \cdots \times X_n = X$ is the product of the $\bar{A}_i$'s.

As a consequence, if $f: X \to Z$ is a morphism of varieties, then $f(\bar{A}_1 \times \cdots \times \bar{A}_n) \subset \overline{f(A)}$.

§7. Projective and Complete Varieties

7.1 *The affine spaces* V *and* K^n. Let V be a finite dimensional vector space (over K). Then the symmetric algebra $A = S_K(V^*)$ on the dual of V is the (graded) algebra of "polynomial functions" on V, generated by the linear functions V^* in degree one. The universal property of the symmetric algebra implies that

$$\mathrm{Hom}_{K\text{-alg}}(S_K(V^*), K) = \mathrm{Hom}_{K\text{-mod}}(V^*, K) = (V^*)^* = V.$$

In this way we can identify V with the points of the affine variety $\mathrm{spec}_K(A)$.

In case $V = K^n$ we have $A = K[T_1, \ldots, T_n]$, the polynomial ring in n variables, where $T_i(t) = t_i$ for $t = (t_1, \ldots, t_n) \in K^n$.

7.2 *The projective spaces* $\mathbf{P}(V)$ *and* $\mathbf{P}_n$ [M, Ch. I, §5]. The set of lines in V can be given the structure of a variety, denoted $\mathbf{P}(V)$, and called the projective space on V. We also write $\mathbf{P}_n = \mathbf{P}(K^{n+1})$.

It is convenient to describe the set $\mathbf{P}(V)$ as the set of equivalence classes, $[x]$, of non-zero vectors $x \in V$, where $[x] = [y]$ means $y = tx$ for some $t \in K^*$. Let $\pi: V - \{0\} \to \mathbf{P}(V)$ denote the projection, $\pi(x) = [x]$. We topologize $\mathbf{P}(V)$ so that π is continuous and open, where $V - \{0\}$ is viewed as an open subvariety of V. Thus $U \subset \mathbf{P}(V)$ is open if and only if $\pi^{-1}(U)$ is open.

Let $A = S_K(V^*)$ a above, and let S be the multiplicative set of all homogeneous elements $\neq 0$ in A. Then $A[S^{-1}]$ is still a graded ring whose degree zero term is

$$L = \{f/g \mid f \text{ and } g \text{ are homogeneous of the same degree in } A \text{ and } g \neq 0\}.$$

If $[x] \in \mathbf{P}(V)$ we shall write

$$\mathcal{O}_{[x]} = \{f/g \in L \mid g(x) \neq 0\}.$$

First note that the condition $g(x) \neq 0$ depends only on $[x]$, for if g is of degree d we have $g(tx) = t^d g(x)$ for $t \in K^*$. This shows further that $f(x)/g(x)$ depends only on $[x]$ because f also has degree d. Thus a given $f/g \in L$ can be viewed as a function on the set of $[x] \in \mathbf{P}(V)$ for which $g(x) \neq 0$. Moreover $\mathcal{O}_{[x]}$ is the *local* ring of all such functions defined at $[x]$.

If U is open in $\mathbf{P}(V)$ we put

$$\mathcal{O}_{\mathbf{P}(V)}(U) = \bigcap_{[x] \in U} \mathcal{O}_{[x]}$$

and define restriction maps to be inclusions whenever $U' \subset U$. This is a sheaf on $\mathbf{P}(V)$, and $(\mathbf{P}(V), \mathcal{O}_{\mathbf{P}(V)})$ is the algebraic variety promised above.

Suppose $V = K^{n+1}$ so that $A = K[T_o, T_1, \ldots, T_n]$. Here we have $T_i(t) = t_i$ for $t = (t_o, \ldots, t_n) \in K^{n+1}$. Even though T_i is not a function on $\mathbf{P}_n = \mathbf{P}(K^{n+1})$ the set $\mathbf{P}_{n,T_i} = \{[t] \in \mathbf{P}_n \mid T_i(t) \neq 0\}$ still makes sense. Moreover, there is a bijection $\mathbf{P}_{n,T_i} \to K^n$ sending $[t_o, \ldots, t_n]$ to $(t_o/t_i, \ldots, \hat{t}_i/t_i, \ldots, t_n/t_i) = (s_1, \ldots, s_n)$. It is easily shown that this is an isomorphism from the open subvariety $\mathbf{P}_{n,T_i}$ of $\mathbf{P}_n$ to the affine space K^n. Since the $\mathbf{P}_{n,T_i}$ $(0 \leqq i \leqq n)$ cover $\mathbf{P}_n$ this shows why $\mathbf{P}_n$ is at least a prevariety.

Consider the open set U^o in K^{n+1} of all $(t_o, \ldots, t_n)$ such that $t_o \neq 0$. Then we have an isomorphism of varieties

$$K^* \times K^n \to U$$

$$(s_o, s_1, \ldots, s_n) \mapsto s_o \cdot (1, s_1, \ldots, s_n).$$

The composition of this with $U \to \mathbf{P}_n$ is just projection on the factor K^n followed by the inverse of the isomorphism $\mathbf{P}_{n,T_0} \to K^n$ constructed above. In this way we see that $V - \{0\} \to \mathbf{P}(V)$ looks, like a projection from a cartesian product as above.

7.3 *Projective varieties.* A *projective variety* is one isomorphic to a closed subvariety of a projective space. A *quasi-projective variety* is an open subvariety of a projective variety. Since affine spaces are open subvarieties of projective spaces it follows that all affine varieties are quasi-projective.

Products of projective varieties are projective. To see this it suffices to show that each $\mathbf{P}_n \times \mathbf{P}_m$ is projective. For this, in turn, one has the explicit closed immersion

$$\mathbf{P}_n \times \mathbf{P}_m \to \mathbf{P}_{(n+1)(m+1)-1} = \mathbf{P}_{nm+n+m}$$

defined by:

$$([x_i], [y_j]) \mapsto ([x_i y_j]).$$

7.4 *Complete varieties* [M, Ch. I, §9]. A variety V is *complete* if, for any variety X, the projection $pr_X: X \times V \to X$ is a closed map. (In the category of Hausdorff topological spaces the analogous property characterizes compact spaces. Thus "complete" for varieties is the analogue of "compact" for topological spaces.)

It follows immediately from the definition that *a closed subvariety of a complete variety is complete*, and that *a product of complete varieties is complete.*

Let $\alpha: V \to X$ be a morphism of varieties with V complete. Then the graph $\Gamma_\alpha \subset V \times X$ is closed, so its projection into X, which is $\alpha(V)$, is closed in X. If α is surjective then it follows directly from the definition that X is also complete. Applying this to $\alpha(V)$ we conclude that *the image of a morphism from a complete variety is closed and complete.*

The affine line K is an open but not closed subset of the projective line $\mathbf{P}_1$, so K is not complete. The only other closed subsets of K are the finite ones, so a connected complete subvariety of K consists of a single point.

If V is a connected complete variety then $K[V]=K$, i.e. every regular function f on V is constant. This follows from the last paragraph because $f(V)$ is a connected complete subvariety of K.

Combining the observation above we conclude easily that *a morphism from a connected complete variety into an affine variety must be constant.* For the image, being closed, is affine as well as complete. But an affine variety with only constant regular functions is a point.

That complete varieties exist in abundance follows from the:

Theorem. *Projective varieties are complete.*

§8. Rational Functions; Dominant Morphisms

8.1 *Rational functions.* Let V be an algebraic variety. The open dense sets U in V form an inverse system, under inclusion, so their rings of functions, $K[U]$, form an inductive system. The inductive limit

$$K(V) = \text{ind. lim. } K[U], \qquad (U \text{ open dense in } V)$$

is called the ring of *rational functions* on V. The following properties are easily established.

(a) If U is open dense in V then $K[U] \to K(V)$ is injective; we shall regard it as an inclusion. Moreover $K(U)=K(V)$.

(b) If $f \in K(V)$ we say f is *regular at x* if $f \in K[U]$ for some neighborhood U of x (which is open dense). The set of all such x is then a dense open set U_o called the domain of definition of f. U_o is the largest dense open set for which $f \in K[U_o]$.

(c) Suppose V is irreducible. Then each dense open U is irreducible also. If $f \in K[U]$ is not zero then $U_f = \{x \in U \mid f(x) \neq 0\}$ is non-empty and open, hence dense (by irreducibility), and $1/f \in K[U_f]$. It follows that $K(V)$ is a field, called the *function field* of V.

(d) In general, let $V_1, \ldots, V_n$ be the irreducible components of V. It follows from (AG.1.2) that there is a dense open U such that the $U_i = U \cap V_i$ $(1 \leqq i \leqq n)$ are open in V and pairwise disjoint. It follows, using (a) and (c) above, that

$$K(V) = K(U) = \prod K(U_i) = \prod K(V_i),$$

the product of the function fields of the irreducible components of V.

(e) If $V = \text{spec}_K(A)$ is affine, where $A = K[V]$, then $K(V)$ is just the full ring of fractions of A.

8.2 *Dominant morphisms.* The ring $K(V)$ of rational functions on V is not functorial. For if $\alpha: V \to W$ is a morphism of varieties, and if U is open dense in W, then $\alpha^{-1}(U)$ need not be dense in V. But if this is always true, and if $\overline{\alpha(V)} = W$, we say α is *dominant*. Such an α induces an *injective* comorphism $\alpha^o: K(W) \to K(V)$.

If V, and therefore also W, are irreducible then this makes $K(V)$ a field extension of $K(W)$. We then say that α is *separable* if this extension is separable. Similarly we call α *purely inseparable* if $K(V)$ is a purely inseparable algebraic extension, and α is said to be *birational* if $K(V) = \alpha^0 K(W)$.

The local rings of V and W can be viewed as subrings of $K(V)$ and of $K(W)$, respectively, and α^0 induces an injection $\alpha^0: \mathcal{O}_x \to \mathcal{O}_{\alpha(x)}$ for $x \in V$. Identifying $K(W)$ with $\alpha^0 K(W)$ we see that the sheaf morphism corresponding to α is just induced by the inclusions of local rings in $K(V)$.

In general, if V is not irreducible but $\overline{\alpha(V)} = W$, then it is easy to see that $\alpha: V \to W$ is dominant if and only if, for each irreducible component V' of V, $\overline{\alpha(V')} = W'$ is an irreducible component of W. We then say that α is separable (purely inseparable, birational,...) if, for each such V', the induced morphism $V' \to W'$ (which is dominant) has the corresponding property.

If V' is an irreducible component of V then $\overline{\alpha(V')} = W'$ is an irreducible subvariety of W, and it will be an irreducible component of W provided it contains a non-empty open set in W. Since V' contains such an open set in V this remark shows that: *If α is surjective and open then α is dominant.*

As a converse, if W and V are irreducible (for convenience), given an injective homomorphism $\beta: K(W) \to K(V)$, there is a dominant morphism α of a Zariski open subset U of V into W, such that $\beta = \alpha^0$. We postpone the discussion of this point to 13.4, where we can add some complement pertaining to fields of definition.

§9. Dimension
[M, Ch. I, §7]

9.1 *The dimension of a variety V.* We have the combinatorial dimension of V, denoted by $\dim V$, introduced in (AG.1.4). It is the supremum of the dimensions of the irreducible components of V. In case V is irreducible we have the function field $K(V)$, and the basic fact is that, in this case,

$$\dim V = \text{tr.deg.}_K K(V).$$

9.2 *Hypersurfaces.* Let V be an irreducible variety and let $f \in K[V]$ be a non-constant function whose set $Z(f) = \{x \in V \mid f(x) = 0\}$ of zeros is not empty. Then the dimension of each irreducible component of $Z(f)$ is $\dim \mathrm{V} - 1$.

9.3 *Products.* The dimension of $V \times W$ is $\dim V + \dim W$.

§10. Image and Fibres of a Morphism
[M, Ch. I, §8]

10.1 *The basic theorem.* Let $\alpha: X \to Y$ be a morphism of varieties. The *fibre* of α over $y \in Y$ is the subvariety $\alpha^{-1}(\{y\})$ of X. To study the non-empty fibres there is no harm in shrinking Y to the closure of the *image*, $\alpha(X)$, i.e. we may as well assume $\alpha(X)$ is dense in Y. If X (and Y) are irreducible this means that α is dominant.

Theorem. *Let $\alpha: X \to Y$ be a dominant morphism of irreducible varieties, and put $r = \dim X - \dim Y$. Let W be an irreducible closed subvariety of Y and let Z be an irreducible component of $\alpha^{-1}(W)$.*

(1) *If Z dominates W then $\dim Z \geqq \dim W + r$. In particular, if $W = \{y\}$, then $\dim Z \geqq r$.*
(2) *There is an open dense $U \subset Y$ (depending only on α) such that*
 (i) *$U \subset \alpha(X)$, and*
 (ii) *If $Z \cap \alpha^{-1}(U) \neq \phi$ then*

$$\dim Z = \dim W + r.$$

In particular, if $W = \{y\} \subset U$ then $\dim Z = r$.

10.2 Corollary (Chevalley). *Let $\alpha: X \to Y$ be any morphism of varieties. Then the image of any constructible set is constructible. In particular $\alpha(X)$ contains a dense open subset of $\overline{\alpha(X)}$.*

The last assertion follows from the first using AG.1.3. The proof of the first assertion can be reduced easily to the case of a dominant morphism of irreducible varieties. Then it is deduced, by induction on $\dim Y$, from part (2)(i) of the theorem.

10.3 Corollary. *Let $\alpha: X \to Y$ be a morphism of varieties. If $x \in X$ let $e(x)$ be the maximum dimension of an irreducible component, containing x, of the fibre of α through x (i.e. of $\alpha^{-1}(\alpha(x))$). Then $x \mapsto e(x)$ is upper semi-continuous, i.e. the sets $\{x \in X \mid e(x) \geqq n\}$ are closed for each integer n.*

§11. k-Structures on K-Schemes

This and the following two sections contain the basic notions required here for the treatment of rationality questions. Recall that k denotes a subfield of the algebraically closed field K.

11.1 *k-structures on vector spaces.* A *k-structure* on a (not necessarily finite dimensional) vector space V (over K) is a k-module $V_k \subset V$ such that the homomorphism $K \bigotimes_k V_k \to V$, induced by the inclusion, is an isomorphism. The surjectivity means that V_k spans V (over K), and the injectivity means that elements of V_k linearly independent over k are also linearly independent over K. The elements of V_k are said to be *rational over k*.

If U is a subspace of V we put $U_k = U \cap V_k$, and we say U is *defined* (or *rational*) *over k* if U_k is a k-structure on U. This is equivalent to U_k spanning U.

If $W = V/U$ we write W_k for the projection of V_k into W, and we say W is *defined over* k if this is a k-structure on W. This happens if and only if U is defined over k, or if and only if elements of W_k linearly independent over k are linearly independent over K.

Let $f: V \to W$ be a K-linear map of vector spaces with k-structures. We say that f is *defined over* k, or that f is a *k-morphism* if $f(V_k) \subset W_k$. The k-morphisms from V to W form a k-submodule

$$\mathrm{Hom}_K(V, W)_k \subset \mathrm{Hom}_K(V, W),$$

and this is even a k-structure provided that W is finite dimensional. In particular, when $W = K$, we have a k-structure on the dual V^* of V.

Similarly $V_k \bigotimes_k W_k$ is a k-structure on $V \bigotimes_K W$, and there are natural k-structures on the exterior and symmetric algebras of V.

11.2 *k-structures on K-algebras.* A k-structure on a K-algebra A is a k-structure A_k which is a k-subalgebra.

If J is an ideal in A then J is defined over k if and only if $J_k(= J \cap A_k)$ generates J as an ideal. This is easily seen.

If S is a multiplicative set in A_k then $A_k[S^{-1}]$ is easily seen to be a k-structure on $A[S^{-1}]$.

If B is another K-algebra with k-structure then we write

$$\mathrm{Mor}_{K\text{-alg}}(A, B)_k$$

for the K-algebra homomorphisms defined over k. The map $f \mapsto l_K \otimes f$ is a bijection from $\mathrm{Mor}_{k\text{-alg}}(A_k, B_k)$ to this set.

11.3 *k-structures on K-schemes.* A *k-structure* on a K-scheme $(X, \mathcal{O}_X)$ consist of

(1) a k-topology k-top$(X) \subset$ top(X), and
(2) a k-structure on $\mathcal{O}_X(U)$ for each k-open U, such that the restriction homomorphisms are defined over k.

(Condition (2) just says that the restriction of $\mathcal{O}_X$ to k-top(X) is a sheaf of K-algebras-with-k-structures.) It is further required that, on k-open affine subschemes, the induced k-structure be of the following type:

A k-structure on an affine K-scheme $X = \mathrm{spec}_K(A)$ is one defined by a k-structure A_k on A as follows: A set is k-closed if it is of the form supp(A/J) for some ideal J defined over k. For example, if $f \in A_k$ then X_f is k-open, and any k-open set is covered by a finite number of these. Moreover $A_f = \tilde{A}(X_f)$ has the k-structure $(A_k)_f$ (see 11.2).

If U is k-open we can cover U by X_{f_i}'s for a family of $f_i \in A_k$. Moreover $X_{f_i} \cap X_{f_j} = X_{f_i f_j}$. By the sheaf axiom we have an exact sequence

$$\tilde{A}(U) \to \prod_i \tilde{A}(X_{f_i}) \overset{\alpha}{\underset{\beta}{\rightrightarrows}} \prod_{i,j} \tilde{A}(X_{f_i f_j}).$$

Therefore $\tilde{A}(U)$ acquires a natural k-structure as the kernel of

$$\prod_i A_{f_i} \xrightarrow{\alpha-\beta} \prod A_{f_i f_j},$$

which is a k-morphism of vector spaces with k-structures.

It is not difficult to check that this k-structure on $\tilde{A}(U)$ is well defined, and that the above construction satisfies the requirements of (1) and (2) above.

Note that we recover A_k as the k-structure on $\tilde{A}(X)$.

Let $\alpha: X \to Y$ be a morphism of K-schemes with k-structures. We say α is *defined over* k or that α is a *k-morphism* if (i) α is continuous relative to the k-topologies, and (ii) if $U \subset Y$ and $V \subset X$ are k-open such that $\alpha(V) \subset U$ then $\alpha_V^U: \mathcal{O}_Y(U) \to \mathcal{O}_X(V)$ is defined over k. The set of morphisms defined over k will be denoted

$$\mathrm{Mor}(X, Y)_k.$$

A homomorphism $\alpha^0: B \to A$ of K-algebras with k-structures induces a morphism $\alpha: \mathrm{spec}_K(A) \to \mathrm{spec}_K(B)$ and it is clear that α is defined over k if and only if α^0 is defined over k. Thus the category of affine K-schemes with k-structures, and k-morphisms, is contravariantly equivalent to the category of affine K-algebras with k-structures, and k-morphisms, and the latter is clearly equivalent to the category of affine k-algebras.

11.4 *Subschemes defined over* k. Let $(X, \mathcal{O}_X)$ be a K-scheme with k-structure. If $U \subset X$ is k-open then $(U, \mathcal{O}_X | U)$ has an induced k-structure.

Suppose $(Z, \mathcal{O}_Z)$ is a closed subscheme of X. We say it is defined over k if (i) Z is k-closed, and (ii) the sheaf $\mathcal{I}$ of ideals such that $\mathcal{O}_X/\mathcal{I}$ is the extension by zeros of $\mathcal{O}_Z$ is defined over k, i.e. $\mathcal{I}(U) \subset \mathcal{O}_X(U)$ is defined over k for all k-open U. Condition (ii) is equivalent to the condition that, for all k-open affine U, the kernel $\mathcal{I}(U)$ of the epimorphism of affine rings, $\mathcal{O}_X(U) \to \mathcal{O}_Z(U \cap Z)$, is defined over k. Thus we see that $(Z, \mathcal{O}_Z)$ acquires a unique k-structure such that the closed immersion $Z \to X$ is defined over k.

It further follows easily that $(Z, \mathcal{O}_Z)$ is defined over k if and only if, for some cover of X by k-open affine U's $(Z \cap U, \mathcal{O}_Z | Z \cap U)$ is defined over k in $(U, \mathcal{O}_X | U)$ for each U.

§12. k-Structures on Varieties

12.1 *Affine k-varieties.* A variety V with a k-structure will be called a *k-variety*. Let $V = \mathrm{spec}_K(A)$ be an affine k-variety with k-structure defined by $A_k = k[V]$ in $A = K[V]$.

Let $Z = \mathrm{spec}_K(A/J)$ be a closed subvariety of V, where J is the ideal of all functions vanishing on Z. Then we have an exact sequence

$$0 \to J_k \to k[V] \to k[Z] \to 0,$$

where $J_k = J \cap k[V]$ and where $k[Z]$ is the restriction to Z of $k[V]$. Thus $k[Z]$ is a reduced affine k-algebra, and we denote its full ring of fractions by $k(Z)$. We have $K \bigotimes_k k(Z) = K[V]/J_k \cdot K[V]$ so that the kernel of the epimorphism $K \bigotimes_k k[Z] \to K[Z]$ is $J/J_k \cdot K[V]$.

Now Z is k-closed when it is the set of zeros of some ideal defined over k. It follows that

$$Z \text{ is } k\text{-closed} \Leftrightarrow J = \sqrt{J_k \cdot K[V]}.$$

In this case then the kernel above is the nil radical of $K \bigotimes_k k[Z]$.

We conclude therefore that the following conditions on a k-closed Z are equivalent:

(a) Z is defined (as a subvariety) over k, i.e. $J = J_k \cdot K[V]$.
(b) $k[Z]$ and K are linearly disjoint over k in $K[Z]$.
(c) $K \bigotimes_k k[Z]$ is reduced.
(d) $K \bigotimes_k k(Z)$ is reduced.

The equivalence of (c) and (d) follows from AG.3.3 because $K \bigotimes_k k(Z)$ is a ring of fractions of $K \bigotimes_k k[Z]$ with respect to a multiplicative set of non-divisors of zero.

We can look at these conditions also from the following point of view. Suppose we are given a reduced affine k-algebra B_k. Then B_k is a k-structure on $B = K \otimes_k B_k$ and hence defines one on the affine K-scheme $Z = \mathrm{spec}_K(B)$. Z is a variety if and only if B is reduced. Thus we can think of k-closed subsets of V as the underlying spaces of closed sub*schemes* of V which are defined over k, but not necessarily as subvarieties defined over k.

Suppose $\mathrm{char}(k) = p > 0$. Then the zeros of $f \in A$ and of f^p coincide. If $f \in k^{1/p}[V]$ then $f^p \in k[V]$. Thus any $k^{1/p}$-closed set is also k-closed. It follows that *the k-topology coincides with the $k^{p^{-\infty}}$-topology.*

12.2 *Subvarieties defined over k.* Let V be any (not necessarily affine) k-variety, and let Z be a k-closed subvariety. If U is k-open in V we write $k[Z \cap U]$ for the restriction to $Z \cap U$ of $k[U]$. Passing to the inductive limit over k-open U for which $Z \cap U$ is dense in Z we obtain the ring $k(Z)$ of "rational functions on Z defined over k." In case V is affine this notation is consistent with that introduced in 12.1 above (cf. (AG.8.1). It follows; from AG.11.4 and 12.1 that *Z is defined over k if and only if $K \bigotimes_k k(Z)$ is reduced.*

Now $k(Z)$ is the product of a finite number of finitely generated field extensions of k. Using the results of AG.2.2 we therefore conclude that the following conditions are equivalent:

(a) Z is defined over k.
(b) $K \bigotimes_k k(Z)$ is reduced.
(c) $k^{p^{-\infty}} \bigotimes_k k(Z)$ is reduced.
(d) Each factor of $k(Z)$ is a separable field extension of k.

In particular we see that:

A k-closed subvariety is defined over $k^{p^{-\infty}}$, and hence over k if k is perfect.

12.3 *Irreducible components are defined over k_s.* Consider the irreducible components of a k-variety V. To show that each one is defined over k_s there is no loss in assuming that $k = k_s$. It suffices further to check this on a cover of V by k-open affine subvarieties, so we may assume V is affine. Then we must show that, if $P_1, \ldots, P_n$ are the minimal primes of $k[V]$, each $P_i \cdot K[V]$ is still a prime ideal. Since k is separably closed it follows from AG.2.1 that $K[V]/(P_i \cdot K[V])$, which equals $K \bigotimes_k (k[V]/P_i)$, has a unique minimal prime, so it remains to be shown that $K \bigotimes_k (k[V]/P_i)$ is reduced.

We have $k[V] \subset \Pi(k[V]/P_i)$, because $k[V]$ is reduced, and both of these rings have the same full ring of fractions, $k(V)$. Since $K \bigotimes_k k(V) = K(V)$ is reduced it follows, as claimed, that each $K \bigotimes_k (k[V]/P_i)$ is reduced.

12.4 Let X, Y be two k-varieties. Then

$$k[X \times Y] = k[X] \otimes k[Y].$$

More precisely, the obvious map of the right-hand side in the left-hand side is injective. There remains to check the other inclusion. If X and Y are affine, it holds by definition 6.1. Assume now X to be affine and let $Y = \bigcup_i Y_i$ $(i \in I)$ be a finite open affine cover of Y. Let f be a regular function on $X \times Y$. Its restriction to $X \times Y_i$ belongs to $k[X] \otimes k[Y_i]$. Since I is finite, there exists a finite dimensional subspace $V \subset k[X]$ such that $f|X \times Y_i$ belongs to $V \otimes k[Y_i]$ for each i. Let f_j $(j \in J)$ be a basis of V. Then we can write uniquely

$$f|X \times Y_i = \sum_j f_j \otimes g_{i,j}, \quad \text{with } g_{i,j} \in k[Y_i].$$

By the uniqueness, $g_{i,j}$ and $g_{k,j}$ have the same restriction to $Y_i \cap Y_k$ $(i, k \in I)$. Therefore, for given $j \in J$, the $g_{i,j}$ $(i \in I)$ match to define an element of $k[Y]$, hence $f \in k[X] \otimes k[Y]$. If now X is not affine, argue similarly using a finite open cover of X.

We shall use this when one factor is affine and the other *quasi-affine*, i.e., by definition, isomorphic to a k-open sub-set of an affine k-variety $\bar{X}$. Note that if X is quasi-affine and irreducible, then $k(X)$ is the quotient field of $k[\bar{X}]$, (since $k(X)=k(\bar{X})$).

§13. Separable Points

13.1 *The functor of points.* Let V be a k-variety. For any affine K-algebra B we shall write

$$V(B)=\mathrm{Mor}_{K\text{-sch.}}(\mathrm{spec}_K(B), V).$$

If B has a k-structure B_k we also write $V(B_k)=V(B)_k$ for the set of morphisms as above which are defined over k.

If $V=\mathrm{spec}_K(A)$ is affine then

$$V(B)=\mathrm{Mor}_{K\text{-alg.}}(A, B)=\mathrm{Mor}_{k\text{-alg}}(A_k, B),$$

and $V(B_k)=\mathrm{Mor}_{k\text{-alg}}(A_k, B_k)$. From these descriptions it is clear that one can extend the definitions to any K-algebra B, not necessarily affine. (For example B might be a large field extension of K.) In this way we obtain a functor $B_k\mapsto V(B_k)$ from k-algebras to sets. It is called the *functor of points* of the k-variety V.

$V(B_k)$ is also functorial in V. If $\alpha: V\to W$ is a k-morphism of k-varieties then α induces a map $V(B_k)\to W(B_k)$.

In the special case $B=K$ we have $V(K)=\mathrm{Mor}_{K\text{-sch.}}(\mathrm{spec}_K(K), V)$, which we can, and will canonically identify with the points of V. Moreover, for any subfield k' of K containing k we have $V(k')\subset V$. These are the *k'-rational points* of V. In particular we have $V(k)\subset V(k_s)\subset V(\bar{k})\subset V$. The points of $V(k_s)$ are called *separable points.*

If W is any locally closed subvariety of V, not necessarily defined over k, we shall permit ourselves to write $W(k')$ for the k'-rational points of V which lie in W.

Examples. If $V=K^n=\mathrm{spec}_K(K[t_1,\dots,t_n])$ with the standard k-structure, given by $k[t_1,\dots,t_n]$, then $V(k)=k^n$.

If V is a vector space with k-structure V_k then $\mathbf{P}(V)$ acquires a k-structure so that $\mathbf{P}(V)(k)$ is the image of $V_k-\{0\}$ under the canonical projection $V-\{0\}\to\mathbf{P}(V)$.

We remark, finally, that the definitions above apply without change to any K-scheme V (resp. K-scheme with k-structure).

13.2 Theorem. *Let $\alpha: V\to W$ be a k-morphism of k-varieties which is dominant and separable. Then there is an open dense set $W_o\subset W$ such that $W_o\subset\alpha(V)$ and such that, for each $w\in W_o(k_s)$, the fibre $\alpha^{-1}(w)$ has a dense set of separable points.*

We shall carry out the proof in several steps.

(a) There is clearly no loss in assuming that $k=k_s$.

(b) This done, it follows from AG.12.3 that the irreducible components of a k-variety are defined over k.

(c) There is no harm in replacing W by a dense k-open set W', and V by $\alpha^{-1}(W')$. Thus we can easily reduce to the case when W is irreducible and affine. Then cover V by irreducible k-open affines V_i. This is possible, using (b). If W_{oi} answers the requirements of the theorem for $\alpha_i: V_i \to W$ then $W_o = \bigcap W_{oi}$ will work for α. Hence we may assume that *V and W are irreducible and affine.* furthermore, with the aid of AG.10.1 we can, after shrinking W, assume that α is surjective and that *all irreducible components of all fibres have the same dimension.*

(d) α is induced by the comorphism $k[W] \to k[V]$ which we can regard as an inclusion. Since $K(V)$ is separable over $K(W)$, by hypothesis, and since K is linearly disjoint, over k, from $k(W)$ and from $k(V)$, it follows that $k(V)$ is separable over $k(W)$. Hence we can apply the (separable) normalization lemma (AG.3.7) to the affine $k(W)$-algebra $k(W) \bigotimes_{k[W]} k[V]$. This permits us to consider the latter as a finite integral extension of some polynomial ring $k(W)[t_1, \ldots, t_n]$ over whose field of fractions $k(V)$ is (finite and) separable. Since $k[V]$ has a finite number of generators we can find a "common denominator" $f \neq 0$ in $k[W]$ for each of the t_i as well as for the coefficients of the integral equations of the generators of $k[V]$ over the polynomial ring. Then if, using (c), we replace $k[W]$ by $k[W]_f = k[W_f]$, and V by $V_f = \alpha^{-1}(W_f)$, we can already write $k[V]$ as a finite integral extension of the polynomial ring $k[W][t_1, \ldots, t_n] = k[W \times K^n]$. Thus we have reduced our problem to the case where α admits a factorization

$$V \xrightarrow{\beta} W \times K^n \xrightarrow{\pi} W.$$

Here π is the coordinate projection, and β is a finite integral morphism such that $k(V)$ is separable over $k(W \times K^n)$.

(e) We claim that there is a dense open set $U_o \subset W \times K^n$ such that $\beta_o: V_o = \beta^{-1}(U_o) \to U_o$ has the following property: Each fibre of β_o over a separable point consists entirely of separable points.

Write $A = k[W \times K^n]$ and say $k[V] = A[b_1, \ldots, b_m]$. Let $P_i(b_i) = 0$ be the minimal polynomial equation of b_i over the field of fractions, $k(W \times K^n)$, of A. Since P_i is a separable polynomial its derivative, P_i', does not vanish at b_i.

The P_i all have coefficients in A_g for some $g \neq 0$ in A. Put $b = \prod_i P_i'(b_i)(\neq 0)$.

Since $k[V]_g$ is integral over A_g it follows from AG.3.6 that there is a non-zero multiple h of b in A_g. Then $k[V]_{gh}$ is integral over A_{gh}, and each residue class field of the former is generated by roots of polynomials which are separable over the corresponding residue class field of A_{gh}. Thus $U_o = (W \times K^n)_{gh}$ has the property described above.

(f) We conclude the proof now by showing that $W_o = \pi(U_o)$ satisfies the requirements of the theorem. Since π is an open map W_o is open in W. We must show, for $w \in W(k)$ (recall $k = k_s$), that $\alpha^{-1}(w)$ has a dense set of separable points.

Since the irreducible components of $\alpha^{-1}(w)$ are equidimensional, and since β is a closed surjective map, it follows that $\beta: \alpha^{-1}(w) \to \beta(\alpha^{-1}(w)) = \pi^{-1}(w)$ is dominant. Clearly $\pi^{-1}(w)$ is a subvariety defined over k and k-isomorphic to K^n. Therefore β maps each irreducible component, X, of $\alpha^{-1}(w)$ *onto* $\pi^{-1}(w)$. Let X' denote the closure of the set of separable points in X. It follows from (d) that $\beta(X')$ contains all separable points in $U_o \cap \pi^{-1}(w)$, which form a dense set in (the irreducible variety) $\pi^{-1}(w)$. Since β is closed it follows that $\beta(X') = \pi^{-1}(w)$. Therefore, since β is finite, $\dim X' = \dim \pi^{-1}(w) = \dim X$. But X is irreducible so $X' = X$. Q.E.D.

13.3 Corollary. *Let V be a k-variety. Then $V(k_s)$ is dense in V.*

We just apply the theorem to the projection of V onto a point.

13.4 *Dominant morphisms.* Assume V and W are irreducible k-varieties. We know (§3, 1.3) that if the k morphism $\alpha: V \to W$ is dominant, then the comorphism $\alpha^o: K(W) \to K(V)$ is an injective homomorphism defined over k. Let now $\beta: K(W) \to K(V)$ be such a homomorphism. Let Y and Z be non-empty Zariski k-open affine subvarieties of V and W respectively. Then $k(V)$ and $k(W)$ are the quotient fields of $k[Y]$ and $k[Z]$ respectively. Let $\{f_i\}$ $(i \in I)$ be a finite generating set for $k[Z]$. Then $\beta(f_i) = u_i/v_i$ with $u_i, v_i \in k[Y]$. Let U be the subset of Y on which all the v_i are nowhere zero. It is a non-empty Zariski k-open affine subset of V, with coordinate ring over k equal to $k[Y][S^{-1}]$, where S is the product of the v_i's $(i \in I)$. We have an injective homomorphism $k[Z] \to k[U]$, whence, canonically a surjective k-morphism of U into Z with dense image, hence a dominant morphism, whose associated comorphism is β. Thus, as a converse to 8.2, we see that an injective k-homomorphism $\beta: K(W) \to K(V)$ is associated to a dominant k-morphism of a non-empty Zariski k-open subvariety U of V into W.

13.5 Assume here that K is a "universal field" (over k), i.e. has infinite transcendence degree over k (besides being algebraically closed, as usual). Let V be an irreducible k-variety. A point $x \in V(K)$ is *generic over* k if $k(x) = k(V)$, i.e. if the evaluation at x yields an isomorphism of $k(V)$ into K. Generic points always exist: Let $r = \dim V$. By 3.7, we may write the coordinate ring $k[U]$ of an affine k-open subset U of V in the form $k[x_1, \ldots, x_t]/J$ where J is the ideal of U, $t \geqq r$ and the x_i $(1 \leqq i \leqq r)$ are algebraically independent over k. Choose $\xi_1, \ldots, \xi_r$ in K algebraically independent over k. Since $k(V)$ is a finite algebraic extension of $k(x_1, \ldots, x_r)$, the map $x_i \mapsto \xi_i (1 \leqq i \leqq r)$ extends to an isomorphism of $k(V)$ into K. The images of the x_i $(1 \leqq i \leqq t)$ are then the coordinates of a generic point over k. In fact, this construction shows easily that the generic points form a Zariski dense (but not open if $r \geqq 1$) subset.

Generic points were ubiquitous in earlier formulations of algebraic geometry, consequently rather prominent in [2], but are less talked about nowadays. In this book, we use them only in the following proposition. When we draw some consequences of it later, it is tacitly understood that K is universal.

13.6 Proposition. *Let $\alpha: V \to W$ be a k-morphism of (absolutely) irreducible k-varieties. Assume that $\alpha(V(E)) = W(E)$ for every extension of E of k contained in K. Then there exists a non-empty k-open subset U of W and a k-morphism $\beta: U \to V$ such that $\alpha \circ \beta = \mathrm{Id}$.*

By restricting V and W if necessary, we may assume that V and W are affine and α is surjective. Let now x be a generic point of W over k (13.5). By assumption, there exists $y \in \alpha^{-1}(x) \cap V(k(x))$. Let f_i $(i \in I)$ be a finite generating set for $k[V]$ over k. We can write $f_i = u_i/v_i$, with u_i, $v_i \in k[W]$ and $v_i(x) \neq 0$ $(i \in I)$. Let $U \subset W$ be the set of points on which all the v_i's are non-zero. We have now a homomorphism $\gamma: k[V] \to k[U]$ and obviously, $\gamma \circ \alpha^o(f) = f$ if $f \in k[U]$. Therefore the unique k-morphism $\beta: U \to V$ such that $\beta^o = \gamma$ satisfies our condition.

13.7 *Rational and unirational varieties.* Let W be an irreducible k-variety. It is said to be *rational over* k if $k(W)$ is a purely transcendental extension of k, *unirational over* k if there exists an injective homomorphism $\beta: k(W) \to L$, where L is a finitely generated purely transcendental extension of k. Let n be the transcendence degree of L. Then L can be viewed as the field of rational functions defined over k of the affine n-space $\mathbf{A}^n$ over k.

Therefore, W is a rational k-variety if and only if it contains a Zariski k-open subset which is k-isomorphic to a Zariski k-open subset of affine space. By 8.3, 13.4, W is unirational over k if and only if there exists a dominant k-morphism of a Zariski k-open subset of affine space into W.

Let k be infinite. Then $\mathbf{A}^n(k)$ is obviously Zariski dense in $\mathbf{A}^n$. Since the image of a Zariski dense subset under a dominant morphism is Zariski dense, we see that *if W is unirational over k and k is infinite, then $W(k)$ is Zariski dense in W.*

§14. Galois Criteria for Rationality

The Galois group $\mathrm{Gal}(k_s/k)$ of k_s over k will be denoted by Γ.

14.1 *Galois actions on vector spaces.* Let V be a vector space with k-structure V_k. Then Γ operates on $V_{k_s} = k_s \bigotimes_k V_k$ through the first factor, and it is clear that V_k is the set $V_{k_s}^\Gamma$ of fixed points under Γ. If W is another vector space with a k-structure, then Γ operates on

$$\mathrm{Hom}_K(V, W)_{k_s} = \mathrm{Hom}_{k_s}(V_{k_s}, W_{k_s})$$

by

$$({}^{\sigma}f)(v) = \sigma(f(\sigma^{-1}v)).$$

Here $\sigma \in \Gamma$, $f: V \to W$ is defined over k_s, and $v \in V_{k_s}$. It is easily seen that the following conditions on such an f are equivalent:

(i) f is defined over k.
(ii) $f: V_{k_s} \to W_{k_s}$ is Γ-equivariant.
(iii) $f \in \mathrm{Hom}(V, W)_{k_s}^{\Gamma}$.

14.2 *The k-structure defined by a Galois action.* Consider a vector space V with a k_s-structure V_{k_s} on which Γ operates semi-linearly, i.e.

$$\sigma(ax) = \sigma(a)\sigma(x) \quad (a \in k_s, x \in V_{k_s}).$$

Suppose further that the stability group of each $x \in V_{k_s}$ is an open subgroup (of finite index) in Γ. Then we claim that

$$V_k = V_{k_s}^{\Gamma}$$

is a k-structure on V.

Certainly V_k is a k-subspace, and the natural map $k_s \bigotimes_k V_k \to V_{k_s}$ is Γ-equivariant. Its kernel is therefore a Γ-invariant k_s-subspace having zero intersection with $1 \otimes V_k$. Therefore the proposition below implies that the map is a monomorphism.

It remains to show that V_k spans V_{k_s}. Let $x \in V_{k_s}$ and Γ_x be the stability group of x. It contains a normal open subgroup Γ' of Γ. The fixed point set k' of Γ' in k_s is a Galois extension of finite degree of k. Let

$$\Gamma'' = \Gamma/\Gamma' = \{\sigma_1, \ldots, \sigma_n\} \cong \mathrm{Gal}(k'/k)$$

and let $a_1, \ldots, a_n$ be a k-basis of k'. The elements $y_i = \Sigma_j \sigma_j(a_i x)$ clearly belong to V_k. Since the elements of Γ' are linearly independent over k', the matrix $(\sigma_j(a_i))$ is invertible, say with inverse (b_{rs}). Then

$$\sum_i b_{ih} y_i = \sum_i b_{ih} \sum_j \sigma_j(a_i)\sigma_j(x) = \sum_j \left(\sum_i \sigma_j(a_i) b_{ih} \right) \sigma_j(x) = \sum_j \delta_{jh} \sigma_j(x) = \sigma_h(x).$$

Some σ_h is the identity, so x is indeed a linear combination of fixed elements.

Proposition. *Let W be a subspace of a vector space V with k-structure. Then W is defined over k if and only if (i) W is defined over k_s, and (ii) W_{k_s} is Γ-stable.*

Proof. The "only if" is clear, and the "if" follows if we prove that the subspace W' spanned by W_k coincides with W. In any case we can pass to V/W' and the subspace W/W' and so reduce to the case $W_k = 0$. We claim $W = 0$. Choose a k-basis (e_i) for V and, if $W \neq 0$, choose a $w \neq 0$ in W_{k_s} so that w is a linear combination of the least possible number of e_i's. After renumbering the e_i's and multiplying w by an element of k_s^* we can write $w = e_1 + a_2 e_2 + \cdots$ with each coefficient in k_s, but $a_2 \notin k$. Then there is a $\sigma \in \Gamma$ such that $\sigma(a_2) \neq a_2$, so $w - \sigma w \in W_{k_s}$ is non-zero and is a linear combination of fewer of the e_i's; contradiction.

14.3 *Galois actions on k-varieties.* Let V be a k-variety. We know from AG.13.3 that $V(k_s)$ is dense in V. We shall introduce now an action of Γ on $V(k_s)$. It will leave $U(k_s)$ stable for all k-open U, so it suffices to describe the action when V is affine. Then we can match $V(k_s)$ with $\mathrm{Mor}_{k_s\text{-alg}}(k_s[V], k_s)$ so that $x \in V(k_s)$ corresponds to the algebra homomorphism e_x. If $\sigma \in \Gamma$ then $\sigma(x)$ is defined by

$$e_{\sigma(x)} = \sigma \circ e_x \circ \sigma^{-1}.$$

Here the left hand σ operates on k_s and the right hand one on $k_s[V] = k_s \bigotimes_k k[V]$. If we denote the latter action by $f \mapsto \sigma_f$ for $f \in k_s[V]$ then the equation above reads

$$f(\sigma(x)) = \sigma({}^{\sigma^{-1}}f)(x), \quad \text{or} \quad ({}^{\sigma}f)(x) = \sigma(f(\sigma^{-1}x)).$$

Writing $V(f)$ for the variety of zeros of f we see that σ maps the separable points of $V(f)$ to those of $V({}^{\sigma}f)$. The same applies to $V(J)$ for any ideal J in $k_s[V]$. In this way we can define the *conjugate variety* ${}^{\sigma}W$ of any closed subvariety W of V defined over k_s. Such a definition is allowable because of the density of separable points. In the affine case ${}^{\sigma}W$ is just the variety obtained by applying σ to the coefficients of equations defining W over k_s.

Let $\alpha: V \to W$ be a morphism of k-varieties, and assume α is defined over k_s. Then, for $\sigma \in \Gamma$, we define a k_s-morphism ${}^{\sigma}\alpha: V \to W$ as follows:

$${}^{\sigma}\alpha(x) = \sigma(\alpha(\sigma^{-1}x)) \quad (x \in V(k_s)).$$

By density of separable points there is at most one k_s-morphism with this property. To see that there is one it suffices to exhibit, for k-open $V' \subset V$ and $W' \subset W$ such that $\alpha V' \subset W'$, the comorphism $({}^{\sigma}\alpha)^o: k_s[W'] \to k_s[V']$. It is defined by the commutativity of

$$\begin{array}{ccc} k_s[V'] & \xleftarrow{({}^{\sigma}\alpha)^o} & k_s[W'] \\ \downarrow{\scriptstyle \sigma} & & \downarrow{\scriptstyle \sigma} \\ k_s[V'] & \xleftarrow[\alpha^o]{} & k_s[W'], \end{array}$$

i.e. $({}^{\sigma}\alpha)^o = \sigma^{-1} \circ \alpha^o \circ \sigma$. Thus, for $f \in k_s[W']$, $({}^{\sigma}\alpha)^o(f) = {}^{\sigma^{-1}}(\alpha^o({}^{\sigma}f))$. Thus Γ acts on $\mathrm{Mor}(V, W)_{k_s}$.

The following conditions on α are easily seen to be equivalent

(i) α is defined over k;
(ii) $\alpha: V(k_s) \to W(k_s)$ is Γ-equivariant;
(iii) $\alpha \in \mathrm{Mor}(V, W)_{k_s}^{\Gamma}$.

14.4 Theorem. *Let V be a k-variety and let Z be a closed subvariety. The following conditions are equivalent:*

(1) *Z is defined over k.*
(2) *Z is defined over k_s and $Z(k_s)$ is Γ-stable.*
(3) *There is a subset $E \subset Z \cap V(k_s)$ such that E is Γ-stable and dense in Z.*

Proof. (1)$\Rightarrow$(2) is clear, and (2)$\Rightarrow$(3) follows from the density of $Z(k_s)$. (AG. 13.3).

(3)$\Rightarrow$(1): By covering V with k-open affine varieties, we can reduce to the case when V is affine. Then $J = \bigcap_{x \in E} \mathfrak{m}_x = \mathrm{I}(\bar{E}) = \mathrm{I}(Z)$ is the ideal of functions vanishing on Z. Since $E \subset V(k_s)$ it follows that J is defined (as a subspace of $K[V]$) over k_s. If $\sigma \in \Gamma$, then

$$^{\sigma}J_{k_s} = \bigcap_{x \in E} {}^{\sigma}\mathfrak{m}_{x,k_s} = \bigcap_{x \in E} \mathfrak{m}_{\sigma(x),k_s} = \bigcap_{x \in E} \mathfrak{m}_{x,k_s} = J_{k_s},$$

the latter because E is Γ-stable. Hence, by (14.2), J is defined over k, as claimed.

14.5 Corollary. *Let $\alpha: V \to W$ be a k-morphism of k-varieties. Then $\overline{\alpha(V)}$ is defined over k.*

Proof. Since $V(k_s)$ is dense in V (AG.13.3), $\alpha(V(k_s))$ is dense in $\overline{\alpha(V)}$, so that we may apply criterion (3) to it.

14.6 Corollary. *Let (Z_i) be a family of subvarieties of V defined over k, and let Z be the closure of $\bigcup_i Z_i$. Then Z is defined over k.*

Proof. Apply criterion (3) to $E = \bigcup_i Z_i(k_s)$.

14.7 Corollary. *Let $\alpha: V \to W$ be a k-morphism of k-varieties which is dominant and separable. Then there is a dense open set W_0 in W such that every fibre of α over a k-rational point of W_0 is defined over k.*

Proof. Let W_0 be as in (AG.13.2). If $w \in W_0(k)$ then the set E of separable points in $\alpha^{-1}(w)$ is Γ-stable. Moreover AG.13.2 implies that E is dense in $\alpha^{-1}(w)$ so the corollary follows from 14.4, criterion (3).

§15. Derivations and Differentials (Cf. [EGA, Ch. 0, §20].)

This section contains the algebra which is preliminary to the discussion of tangent spaces, to follow in (AG.16).

15.1 $\Omega_{A/k}$. We shall work with k-algebras, even though most of the discussion applies when k is a commutative ring, not necessarily a field.

Since a k-algebra A is commutative we can regard an A-module M as a bimodule, so that $ax = xa$ for $x \in M$ and $a \in A$. With this convention a *k-derivation* from A to M is a k-linear map $X: A \to M$ such that

$$X(ab) = (Xa)b + a(Xb) \quad (a, b \in A).$$

Since $X(ab) = aX(b)$ for $a \in k$ we can take $b = 1$ to conclude that $Xa = 0$ for $a \in k$.

The set

$$\mathrm{Der}_k(A, M)$$

of all such k-derivations is an A-module which is functorial in M.

There is a *universal* k-derivation

$$d(= d_{A/k}): A \to \Omega(= \Omega_{A/k})$$

obtained by taking Ω to be the A-module defined by generators, da $(a \in A)$, and relations, $d(ab) = (da)b + a(db)$ $(a, b \in A)$ and $dc = 0 (c \in k)$. Its universality is expressed by the natural isomorphism

$$\mathrm{Hom}_{A\text{-mod}}(\Omega, M) \to \mathrm{Der}_k(A, M)$$

sending f to $f \circ d$.

(There is a well known construction of Ω which we shall not need: Let J be the kernel of $A \bigotimes_k A \to A$, $a \otimes b \mapsto ab$. Then $a \otimes 1 - 1 \otimes a \to da$ induces an isomorphism $J/J^2 \to \Omega$.)

If $f: A \to B$ is a k-algebra homomorphism, it induces a semi-linear map $df: \Omega_A \to \Omega_B$ sending $d_A a$ to $d_B f(a)$. (We drop k from the notation when k is fixed by the discussion.) This corresponds to the map

$$\mathrm{Der}_k(B, M) \to \mathrm{Der}_k(A, M)$$

defined by:

$$X \mapsto X \circ f,$$

for each B-module (and hence also A-module) M. In this way Ω_A is functorial in A.

15.2 *Polynomial rings.* If $A = k[T_1, \ldots, T_n]$ is a polynomial ring, then Ω is a free A-module with basis $dT_1, \ldots, dT_n$. Moreover $d: A \to \Omega$ is given by

$$df = \sum \frac{\partial f}{\partial T_i} dT_i$$

for $f \in A$. These assertions translate the fact that a derivation $X: A \to M$ is determined by the XT_i, which can be arbitrarily prescribed.

15.3 *Residue class rings.* Let $A' = A/J$ for some ideal J, and let M be an

A'-module (or A-module annihilated by J). Then, since $JM = 0$, we have

$$\mathrm{Der}_k(A, M) = \mathrm{Hom}_{A\text{-mod}}(\Omega_A, M) = \mathrm{Hom}_{A'\text{-mod}}(\Omega_A/J\Omega_A, M)$$

We can identify $\mathrm{Der}_k(A', M)$ with the k-derivations $A \to M$ which kill J, i.e.

$$\mathrm{Der}_k(A', M) = \mathrm{Hom}_{A'\text{-mod}}(\Omega_A/A\cdot d_A J, M).$$

Thus $\Omega_{A'}$ is Ω_A modulo the A-module generated by all $df\,(f \in J)$. It even suffices to vary the f's over a set of generators of J.

For example, suppose $A = k[T_1, \ldots, T_n]$ is a polynomial ring, so that $A' = k[t_1, \ldots, t_n]$ (t_i = image of T_i). Then if $f_1, \ldots, f_m$ generate J we conclude from above that $\Omega_{A'}$ is defined by generators dt_i ($1 \leqq i \leqq n$) and relations

$$\sum_i \left(\frac{\partial f_j}{\partial T_i}\right)(t)\, dt_i = 0 \quad (1 \leqq j \leqq m).$$

Here $g(t)$ denotes the image in A' of a polynomial $g(T) = g(T_1, \ldots, T_n)$ in A.

15.4 Proposition. *Suppose above that $A = k \oplus J$, i.e. that k is mapped onto $A' = A/J$. Then d_A induces an isomorphism of A'-modules*

$$J/J^2 \to \Omega_A/J\cdot\Omega_A.$$

Proof. It suffices to show that these modules have the same homomorphisms into any A'-module M, i.e. that $\mathrm{Der}_k(A, M) \cong \mathrm{Hom}_{A'\text{-mod}}(J/J^2, M)$. If $X: A \to M$ is a k-derivation then $X(k) = 0$ so, since $A = k \oplus J$, X is determined by $X|J$. Since $JM = 0$ we must have $X(J^2) = 0$, so X is determined by a homomorphism $h: J/J^2 \to M$. Conversely, given such an h, it induces $J \to J/J^2 \to M$, and hence an $X: A \to M$ so that $X(k) = 0$. A routine calculation shows that X is a k-derivation.

15.5 *Localization.* Let S be a multiplicative set in A. Then $\Omega_{A[S^{-1}]} = \Omega_A[S^{-1}]$, and we have

$$d\left(\frac{a}{s}\right) = \frac{(da)s - a(ds)}{s^2} \quad (a \in A, s \in S).$$

In particular, it follows that, if M is an $A[S^{-1}]$-module, i.e. an A-module on which the elements of S act invertibly, then

$$\mathrm{Der}_k(A, M) = \mathrm{Der}_k[A[S^{-1}], M).$$

For example, if M is a module over one of the local rings A_P of A then $\mathrm{Der}_k(A, M) = \mathrm{Der}_k(A_P, M)$.

Here is another important consequence of the localizability of Ω: Suppose V is a K-scheme. Then there is a coherent sheaf $\Omega_{V/K}$ of $\mathcal{O}_V$-modules such that, on any affine open subscheme $U = \mathrm{spec}_K(A)$, the sheaf $\Omega_{U/K} = \Omega_{V/K}|U$ is the sheaf $\tilde{\Omega}_{A/K}$ corresponding to $\Omega_{A/K}$. If $x \in U$ then the stalk Ω_x of $\Omega_{V/K}$ is therefore just the localization of $\Omega_{A/K}$ at the local ring $\mathcal{O}_x$ of A, or, alternatively, $\Omega_{\mathcal{O}_x/K}$.

15.6 *Separable field extensions* (see AG.2.3). Suppose A is a finitely generated field extension of k of transcendence degree n. Then

$$\dim_A \Omega_A \geqq n$$

with equality if and only if A is separable over k. In this case $a_1, \ldots, a_n \in A$ are a separating transcendence basis of A over k if and only if $da_1, \ldots, da_n$ are an A-basis of Ω_A ([N.B] (a): V, §16, no. 6, Cor. 1).

If B is a finitely generated field extension of A which is separable over k then it follows from the exact sequence ([N.B] (a): III, §10, no. 7, Prop. 7)

$$0 \to \mathrm{Der}_A(B, B) \to \mathrm{Der}_k(B, B) \to \mathrm{Der}_k(A, B),$$

by counting B-dimensions, that B is separable over $A \Leftrightarrow \mathrm{Der}_k(B, B) \to \mathrm{Der}_k(A, B)$ is surjective $\Leftrightarrow B \bigotimes_A \Omega_A \to \Omega_B$ is injective.

15.7 *Tensor products.* Suppose $A = A_1 \bigotimes_k A_2$, and write $\Omega_i = \Omega_{A_i}$. Then

$$\Omega_A \cong \left(\Omega_1 \bigotimes_k A_2\right) \oplus \left(A_1 \bigotimes_k \Omega_2\right).$$

Equivalently, if M is any A-module, we have

$$\mathrm{Der}_k(A, M) \cong \mathrm{Der}_k(A_1, M) \oplus \mathrm{Der}_k(A_2, M).$$

The map from left to right is induced by the homomorphisms $A_i \to A$. For the inverse we must produce a k-derivation $X: A \to M$ from a given pair of them $X_i: A_i \to M$. The formula is:

$$X(a_1 \otimes a_2) = (X_1 a_1 \otimes a_2) + (a_1 \otimes X_2 a_2).$$

15.8 *Base change.* For any base change $k \to k'$ we have a natural isomorphism

$$k' \bigotimes_k \Omega_{A/k} \to \Omega_{k' \bigotimes_k A/k'}.$$

15.9 *The tangent bundle lemma.* We consider k-algebras A and D where D is of the form $D = B \oplus M$ with B a subalgebra and M an ideal of square zero. If $f: A \to B$ is an algebra homomorphism we write M_f for the resulting A-module M with A operating via f.

The projection $D \to B = D/M$ induces a map

$$\mathrm{Hom}_{k\text{-alg}}(A, D) \xrightarrow{p} \mathrm{Hom}_{k\text{-alg}}(A, B).$$

We assert that, for f as above, *there is a canonical bijection*

$$\mathrm{Der}_k(A, M_f) \to p^{-1}(f).$$

In fact, any element of $p^{-1}(f)$ can be written uniquely in the form $f + X$, for some k-linear map $X: A \to M$, with the understanding that $(f + X)(a) =$

$f(a) + X(a) \in D = B \oplus M$. The assertion above can then be translated: $f + X$ is multiplicative if and only if X is a derivation. To see this take $a, b \in A$. Then

$$\begin{aligned}(f(a) + X(a))(f(b) + X(b)) &= f(a)f(b) + X(a)f(b) + f(a)X(b) + X(a)X(b) \\ &= f(ab) + (X(a)f(b) + f(a)X(b)),\end{aligned}$$

because f is multiplicative and $M^2 = 0$.

§16. Tangent Spaces

16.1 *The Zariski tangent space.* Let x be a point on a variety (or even a K-scheme) V. Recall that $K(x) = \mathcal{O}_x/\mathfrak{m}_x$ denotes the residue field of the local ring of x. It coincides with K, but the notation refers, more precisely, to its $\mathcal{O}_x$-module structure.

The *tangent space* of V at x is

$$T(V)_x = \operatorname{Der}_K(\mathcal{O}_x, K(x)).$$

It follows from (AG. 15.4) that this is canonically isomorphic to

$$\operatorname{Hom}_{K\text{-mod}}(\mathfrak{m}_x/\mathfrak{m}_x^2, K).$$

If $f \in \mathcal{O}_x$ write $(df)_x$ for the image modulo $\mathfrak{m}_x^2$ of $f - f(x)$. Then the "tangent vector" $X \in T(V)_x$ corresponding to $h: \mathfrak{m}_x/\mathfrak{m}_x^2 \to K(x)$ is defined by $Xf = h((df)_x)$.

Suppose V has a k-structure and $x \in V(k)$. Then $\mathcal{O}_x$ has a natural k-structure $\mathcal{O}_{x,k}$, whose residue class field $k(x)$ is a k-structure on $K(x)$. Thus we obtain a k-structure $\operatorname{Der}_k(\mathcal{O}_{x,k}, k(x))$ on $T(V)_x$. As above this k-structure is isomorphic to

$$\operatorname{Hom}_{k\text{-mod}}(\mathfrak{m}_{x,k}/\mathfrak{m}_{x,k}^2, k).$$

Let $\alpha: V \to W$ be a morphism of varieties (or of K-schemes). Then we have the comorphism

$$\alpha^o: \mathcal{O}_{\alpha(x)} \to \mathcal{O}_x.$$

$K(x)$, thus viewed as an $\mathcal{O}_{\alpha(x)}$-module coincides with $K(\alpha(x))$. Therefore we have a natural map

$$\operatorname{Der}_K(\mathcal{O}_x, K(x)) \to \operatorname{Der}_K(\mathcal{O}_{\alpha(x)}, K(\alpha(x)))$$

which we denote by

$$(d\alpha)_x: T(V)_x \to T(W)_{\alpha(x)}.$$

Explicitly, if $X \in T(V)_x$ and if $f \in \mathcal{O}_{\alpha(x)}$, then

$$(d\alpha)_x(X)(f) = X(\alpha^o(f)).$$

In case α is a k-morphism relative to k-structures on V and W and if $x \in V(k)$, then $\alpha(x) \in W(k)$ and it is easy to see that $(d\alpha)_x$ is defined over k, relative to the k-structure described above on the tangent spaces.

The *differential*, $(d\alpha)_x$, behaves functorially in the following sense:

$$(d1_V)_x = 1_{T(V)_x}.$$

If $\beta: W \to Z$ then

$$d(\beta \circ \alpha)_x = (d\beta)_{\alpha(x)} \circ (d\alpha)_x \quad \text{(chain rule).}$$

Suppose $V = V_1 \times V_2$ is a *product* and that $x = (x_1, x_2)$. Define $\alpha_i: V_i \to V$ $(i = 1, 2)$ by $\alpha_1(u) = (u, x_2)$ and $\alpha_2(y) = (x_1, y)$. We claim that

$$(d\alpha_1)_{x_1} + (d\alpha_2)_{x_2}: T(V_1)_{x_1} \oplus T(V_2)_{x_2} \to T(V)_x$$

is an isomorphism. Since this is a local matter we can assume the V_i to be affine, say $V_i = \operatorname{spec}_K(A_i)$. Then $V = \operatorname{spec}_K(A)$ where $A = A_1 \bigotimes_K A_2$. It follows from (AG.15.5) that we can compute the tangent spaces as $T(V)_x = \operatorname{Der}_K(A, K(x))$ and $T(V_i)_{x_i} = \operatorname{Der}_K(A_i, K(x_i))$. As an A-module we have $K(x) = K(x_1) \bigotimes_K K(x_2)$, (both sides being isomorphic to K). Hence it follows from AG. 15.7 that $T(V)_x = T(V_1)_{x_1} \oplus T(V_2)_{x_2}$, and it is easily checked that this identification admits the description given above.

16.2 *The tangent bundle.* At each point of a K-scheme V we have a tangent space. We shall now construct the *tangent bundle*, $T(V)$, which fits all of these vector spaces into a coherent family parametrized by V.

Write $K[\delta] = K \oplus K\delta$ for the *dual numbers*, the algebra with one generator, δ, and one relation, $\delta^2 = 0$. We have the inclusion i and projection p,

$$K[\delta] \underset{i}{\overset{p}{\rightleftarrows}} K,$$

defined by $p(\delta) = 0$. As a set we define $T(V)$ to be $V(K[\delta])$, the points of V in $K[\delta]$ (see AG.13.1). It therefore comes equipped with maps

$$\begin{array}{ccc} T(V) & = & V(K[\delta]) \\ {\scriptstyle p_V}\downarrow\uparrow{\scriptstyle i_V} & & \downarrow\uparrow \\ V & & V(K) \end{array}$$

induced by p and i above. Moreover $T(V)$ is functorial: If $\alpha: V \to W$ is a morphism of varieties we have a commutative square

$$\begin{array}{ccc} T(V) & \xrightarrow{T(\alpha)} & T(W) \\ {\scriptstyle p_V}\downarrow & & \downarrow{\scriptstyle p_W} \\ V & \xrightarrow{\alpha} & W \end{array}$$

It also commutes if we replace the p's by i's.

Recall that

$$V(K[\delta]) = \mathrm{Mor}_{K\text{-sch}}(\mathrm{spec}_K(K[\delta]), V).$$

It is clear that the scheme $\mathrm{spec}_K(K[\delta])$ consists of a single point, with local ring $K[\delta]$. Hence a point of $V(K[\delta])$ corresponds to a point $x \in V$ and a comorphism $\mathcal{O}_x \to K[\delta]$. The latter can be written in the form $e_x + \delta X$ for some K-linear map $X: \mathcal{O}_x \to K$. This sends $f \in \mathcal{O}_x$ to $f(x) + \delta X(f)$ in $K[\delta]$. According to (AG.15.9) the X's so obtained vary precisely over $\mathrm{Der}_K(\mathcal{O}_x, K(x)) = T(V)_x$. We shall denote the element $e_x + \delta X$ also by

$$e_x^{\delta X},$$

and view it both as a homomorphism $\mathcal{O}_x \to K[\delta]$ and as a point of $T(V)$. From the latter point of view we see that the projection p_V is given by

$$p_V : e_x^{\delta X} \mapsto x.$$

(Moreover i_V sends x to $e_x = e_x^{\delta o}$.) Thus we can reformulate the conclusion above as follows: *There is a natural bijection*

$$T(V)_x \to p_V^{-1}(x)$$

given by

$$X \mapsto e_x^{\delta X}$$

Suppose $\alpha: V \to W$ is a morphism. Then $T(\alpha)(e_x^{\delta X}) = e_x^{\beta X} \circ \alpha^o$, where $\alpha^o: \mathcal{O}_{\alpha(x)} \to \mathcal{O}_x$. Expanding the right side we obtain $(e_x + \delta X) \circ \alpha^o = e_x \circ \alpha^o + \delta X \circ \alpha^o = e_{\alpha(x)} + \delta (d\alpha)_x X$. Thus

$$T(\alpha)(e_x^{\delta X}) = e_{\alpha(x)}^{\delta (d\alpha)_x X}$$

In other words, the map that $T(\alpha)$ induces on the fibre over x corresponds to the differential $(d\alpha)_x$.

16.3 *$T(V)$ "is" a K-scheme.* To give $T(V)$ the structure of a K-scheme it suffices to do so when V is affine and to verify that the construction in that case is suitably functorial. Before doing this we recall some properties of symmetric algebras.

Let M be a module over a (commutative) ring A. The *symmetric algebra*, $S_A(M)$, is the largest commutative quotient of the tensor algebra of M. Both of these A-algebras are graded, with A in degree zero, and M in degree one. The universal property of the symmetric algebra is expressed by the identification

$$\mathrm{Hom}_{A\text{-alg}}(S_A(M), B) = \mathrm{Hom}_{A\text{-mod}}(M, B)$$

for all (commutative) A-algebras B. In other words, a module homomorphism $M \to B$ extends uniquely to an A-algebra homomorphism $S_A(M) \to B$.

The following facts are easily verified:

(a) If M is free with basis $t_1, \ldots, t_n$, then $S_A(M) = A[t_1, \ldots, t_n]$, the polynomial ring.

(b) $S_A(M \oplus N) = S_A(M) \bigotimes_A S_A(N)$.

(c) If $A \to A'$ is any base change then $S_{A'}\left(A' \bigotimes_A M\right) = A' \bigotimes_A S_A(M)$.

Now let $V = \mathrm{spec}_K(A)$ be an affine K-scheme, and put $\Omega = \Omega_{A/K}$, the A-module of K-differentials (see AG.15.1). We propose to construct a bijection

$$\varphi: T(V) \to \mathrm{spec}_K(S_A(\Omega)),$$

which is functorial in A, and so that p_V and i_V on the left correspond, on the right, to the inclusion $A \to S_A(\Omega)$ and the projection $S_A(\Omega) \to A$ sending Ω to 0, respectively. Moreover, if V has a k-structure given by $A_k \subset A$ then φ will be compatible with the k-structure on the right given by $S_{A_k}(\Omega_k)$, where $\Omega_k = \Omega_{A_k/k}$ (see AG.15.8 and (c) above).

We define the K-algebra homomorphism

$$\varphi(e_x^{\delta X}): S_A(\Omega) \to K$$

as follows: Viewing $e_x: A \to K(x)$ as a base change, it induces

$$e_x: S_A(\Omega) \to S_K(\Omega(x)),$$

where $\Omega(x) = K(x) \bigotimes_A \Omega$. We define $\varphi(e_x^{\delta X})$ to be the composite of this with some

$$h: S_K(\Omega(x)) \to K$$

to be explained now. We have

$$\mathrm{Hom}_{K\text{-alg}}(S_K(\Omega(x)), K) = \mathrm{Hom}_{K\text{-mod}}(\Omega(x), K).$$

If Ω_x is the localization of Ω at $\mathcal{O}_x$ then $\Omega(x) = K(x) \bigotimes_A \Omega = K(x) \bigotimes_{\mathcal{O}_x} \Omega_x = \Omega_x / \mathfrak{m}_x \Omega_x$. Moreover, with the aid of AG.15.5 and AG.15.3 we see that

$$\mathrm{Hom}_{K\text{-mod}}(\Omega(x), K) = \mathrm{Der}_K(\mathcal{O}_x, K(x)) = T(V)_x.$$

Combining these identifications, we can now choose $h \in \mathrm{Hom}_{K\text{-alg}}(S_K(\Omega(x)), K)$ to correspond to $X \in T(V)_x$.

The properties of φ claimed above are all easily verified, in particular, the fact that φ is bijective.

Suppose now that V is a variety. It does not then follow from the construction above that $T(V)$ is a variety, because $S_A(\Omega)$ may not be reduced. However, if Ω is free then (see (a) above) $S_A(\Omega)$ is a polynomial ring over A, so $T(V)$ is a variety of the form $V \times K^n$ for some n. More generally, then, we conclude that:

If V is a variety and if Ω is locally free then $T(V)$ is a variety locally isomorphic to the product of V with an affine space.

§17. Simple Points

17.1 A point x on a variety V is said to be *simple on* V if $\mathcal{O}_x$ is regular local ring (see (AG.3.9)). If all points of V are simple we say that V is *smooth.*

In the next theorem, Ω_x denotes the module of differentials $\Omega_{\mathcal{O}_x/K}$ (cf. AG.15.5).

Theorem. *The following conditions are equivalent*:

(1) *x is simple on V.*
(2) $\dim_K T(V)_x = \dim_x V$.
(3) *x lies on a unique irreducible component of V, and Ω_x is a free $\mathcal{O}_x$-module.*

Using (AG.15.4) we see that

$$T(V)_x = \mathrm{Der}_K(\mathcal{O}_x, K(x)) = \mathrm{Hom}_{K\text{-mod}}(\Omega_x/\mathfrak{m}_x\Omega_x, K)$$

and

$$\Omega_x/\mathfrak{m}_x\Omega_x \cong \mathfrak{m}_x/\mathfrak{m}_x^2.$$

Moreover (see (AG.3.9)) we have

$$\dim_K(\mathfrak{m}_x/\mathfrak{m}_x^2) \geqq \dim \mathcal{O}_x (= \dim_x V)$$

with equality if and only if $\mathcal{O}_x$ is regular. These remarks already show the equivalence of (1) and (2).

The point x lies on a unique irreducible component if and only if $\mathcal{O}_x$ is an integral domain. Since regular local rings are integral domains it suffices, for the rest of the proof, to assume V is irreducible. If not, pass to an irreducible open neighborhood of x.

Let S be a minimal set of generators of Ω_x as an $\mathcal{O}_x$-module. It follows from AG.3.2 that card $S = \dim_K(\Omega_x/\mathfrak{m}_x\Omega_x)$, and Ω_x is free if and only if S is a basis. The latter is equivalent to $1 \otimes S$ being a basis over $K(V)$, the field of fractions of $\mathcal{O}_x$, of $K(V) \bigotimes_{\mathcal{O}_x} \Omega_x$. Since $1 \otimes S$ spans the latter we conclude that Ω_x is $\mathcal{O}_x$-free if and only if $\operatorname{card} S = \dim_{K(V)}\left(K(V) \bigotimes_{\mathcal{O}_x} \Omega_x\right)$.

From the fact that $\Omega_x/\mathfrak{m}_x\Omega_x \cong \mathfrak{m}_x/\mathfrak{m}_x^2$ we see that

$$\operatorname{card} S = \dim T(V)_x \geqq \dim_x V.$$

On the other hand it follows from AG.15.5 and AG.15.6, using the separability of $K(V)$ over K, that $K(V) \bigotimes_{\mathcal{O}_x} \Omega_x = \Omega_{K(V)/K}$, and

$$\dim_{K(V)} \Omega_{K(V)/K} = \mathrm{tr{\cdot}deg{\cdot}}_K K(V) = \dim_x V.$$

Combining these remarks we have:

$$\Omega_x \text{ is } \mathcal{O}_x\text{-free} \Leftrightarrow \operatorname{card} S = \dim_x V \Leftrightarrow \dim_K T(V)_x = \dim_x V.$$

This proves (2)$\Leftrightarrow$(3), thus concluding the proof of the theorem.

17.2 Corollary. *Let V be a variety. The set U of simple points on V is an open dense subvariety whose irreducible and connected components coincide.*

It follows from AG.1.2 that the set U_o of points of V lying on a unique irreducible component is open and dense, and the irreducible and connected components of U_o coincide. Since $U \subset U_o$ we can therefore reduce to the case when V is irreducible. If Ω is the coherent sheaf of differentials on V (see AG.15.5) then it follows from criterion (3) above that $U = \{x \in V \mid \Omega_x$ is a free $\mathcal{O}_x$-module$\}$. To show this is open dense we can assume V is affine, say $\mathrm{spec}_K(A)$, and that Ω is the A-module $\Omega_{A/K}$. In this case it follows from (AG.3.5) that $U' = \{x \in \mathrm{spec}(A) \mid \Omega_x$ is a free A_x-module$\}$ is open in $\mathrm{spec}(A)$. Taking for x the zero prime ideal, in which case A_x is a field, we see that U' is not empty. Since $\mathrm{spec}(A)$ is irreducible, U' is dense, and hence likewise for $U = U' \cap \mathrm{spec}_K(A)$.

17.3 Theorem. *The following conditions on a morphism $\alpha: V \to W$ of varieties are equivalent:*

(1) *α is (dominant and) separable.*
(2) *There is a dense open subvariety V_o of V such that $(d\alpha)_x$ is surjective for all $x \in V_o$.*
(3) *In each irreducible component of V there is a simple point x (of V) such that $\alpha(x)$ is simple on W and such that $(d\alpha)_x$ is surjective.*

Suppose $V' \subset V$ and $W' \subset W$ are dense open subvarieties such that α induces a morphism $\alpha': V' \to W'$. Then clearly the theorem for α will follow once we prove it for α', thanks to the density of simple points. In this way one can easily reduce to the case where V and W are each irreducible, affine, and smooth. The latter condition implies that the modules $\Omega_V = \Omega_{K[V]/K}$ and $\Omega_W = \Omega_{K[W]/K}$ are locally free. By shrinking V and W still further we can assume they are (globally) free.

The comorphism $\alpha_0: K[W] \to K[V]$ induces $\Omega_W \to \Omega_V$, and $(d\alpha)_x$ then corresponds to the induced homomorphism from

$$\mathrm{Hom}_{K[V]\text{-mod}}(\Omega_V, K(x))$$

to

$$\mathrm{Hom}_{K[V]\text{-mod}}(K[V] \bigotimes_{K[W]} \Omega_W, K(x)).$$

Write $d: M \to N$ for the homomorphism $K[V] \bigotimes_{K[W]} \Omega_W \to \Omega_V$. The modules M and N are free of ranks $\dim W$ and $\dim V$, respectively, and d is represented by a matrix (f_{ij}) over $K[V]$. The description of $(d\alpha)_x$ above shows that it is represented by the matrix $(f_{ji}(x))$ over K. Thus $(d\alpha)_x$ is surjective if and only if the rank of $(f_{ji}(x))$ is $\dim W$. The set of such x is therefore open, and it is non-empty if and only if (f_{ji}) has rank $\dim W$ as a matrix over $K(V)$. The latter, in turn, is equivalent to the injectivity of $\Omega_W \to \Omega_V$. This is equivalent to (i) the injectivity of α_o (i.e. the dominance of α), and (ii) the surjectivity of $\mathrm{Der}_K(K(V),\ K(V)) \to \mathrm{Der}_K(K(W), K(V))$. The last condition means that K-derivations of $K(W)$ into $K(V)$ extend to $K(V)$, and this condition (see AG.15.6) characterizes separability of $K(V)$ over $K(W)$.

17.4 Corollary. *If $\alpha_i: V_i \to W_i$ $(i = 1, 2)$ are two separable morphisms then $\alpha_1 \times \alpha_2: V_1 \times V_2 \to W_1 \times W_2$ is separable.*

This follows easily from criterion (2).

§18. Normal Varieties

This section contains the main results needed in Chapter II, §6 for the construction of homogeneous spaces.

18.1 Definition. A point x on a variety V is said to be *normal on V* if the local ring $\mathcal{O}_x$ is normal, i.e. if $\mathcal{O}_x$ is an integral domain integrally closed in its field of fractions. In particular such an x lies on a unique irreducible component of V, i.e. it has an irreducible open neighborhood. Consequently most questions involving normality can be easily reduced to the case of irreducible varieties.

If every point of V is normal on V then V is called a *normal variety*.

As an example, every simple point of V is normal on V (i.e. a regular local ring is normal). It follows (see AG.17.2) that the set of normal points on V contains a dense open set; in fact, it is itself open.

Moreover, *a product of two normal varieties is normal* [Fond., Ch. V, I, Prop. 3].

18.2 *Normalization.* Let V be an irreducible algebraic variety, and let L be a finite (algebraic) extension of $K(V)$. Then there is a normal irreducible variety V' and a surjective morphism $\alpha: V' \to V$ with finite fibres, and a $K(V)$-algebra isomorphism $K(V') \to L$. Moreover these data are essentially unique. We usually identify $K(V')$ with L, and call $\alpha: V' \to V$ the *normalization of V in L*. It is determined by the following property: If U is open affine in V, then $U' = \alpha^{-1}(U)$ is $\mathrm{spec}_K[K[U]')$, where $K[U']$ is the integral closure of $K[U]$ in L, and α is induced by $K[U] \subset K[U]' = K[U']$. If $L = K(V)$ we just call $\alpha: V' \to V$ *the normalization of V.*

Note that *a normalization of an affine variety is affine.* Moreover, *a normalization of a projective (resp., complete) variety is projective (resp., complete).* For projectiveness see [M, Ch. III, §8, Thm. 4]. For completeness, we must show that $V' \times X \to X$ is closed for all X, knowing the analogous assertion for V. It clearly suffices to verify that $V' \times X \to V \times X$ is closed. This is a local property which need only be verified when V and X are affine, in which case it follows from the fact (see AG.3.6) that $\mathrm{spec}_K(B) \to \mathrm{spec}_K(A)$ is surjective and closed whenever $A \subset B$ is a finite integral extension.

The next theorem is taken from [Class., exp. 5, no. 2], on Zariski's Main Theorem.

Theorem. *Let $\alpha: V \to W$ be a dominant morphism of irreducible normal varieties. Assume the fibres of α have finite constant cardinality n. Then α is the normalization of W in $K(V)$, and n is the separable degree of $K(V)$ over $K(W)$. In particular, if α is birational then α is an isomorphism, and if α is bijective then $K(V)$ is purely inseparable over $K(W)$.*

Using this theorem the following result can be deduced [Class., exp. 8, Prop. 1].

Proposition. *Let $\alpha: V \to W$ be a dominant morphism of irreducible varieties, and suppose that $f \in K[V]$ is constant along the fibres of α. Then f is purely inseparable over $K(W)$.*

18.3. Proposition. *Let $\alpha: V \to W$ be a bijective morphism of varieties. Assume that V is irreducible and W normal:*

(1) *If W is complete, then V is complete.*
(2) *If V is affine, then W is affine.*

It is clear that W is irreducible, too.

Suppose W' is open in W, and set $V' = \alpha^{-1}(W')$. We claim that the inclusion $\alpha^o K[W'] \subset K[V'] \cap \alpha^o K(W)$, is an equality. Since $\alpha': V' \to W'$ inherits all of our hypotheses it suffices to treat the case $W' = W$. So suppose $f \in K[V]$ and that $f = \alpha^o h$ for some $h \in K(W)$. We must show that $h \in K[W]$, i.e. that h is everywhere defined on W. We use the following Lemma [Class., exp. 8, Lemma 1]:

Lemma. *Let x be a normal point on an irreducible variety W, and suppose $h \in K(W)$ is not defined at x. Then there is a $y \in W$ at which $1/h$ is defined and vanishes.*

Returning to the argument above, if h is not defined at $x \in W$ then choose y as in the lemma. writing $y = \alpha(z)$ we see that $1/f = \alpha^o(1/h)$ is defined and vanishes at $z \in V$, contrary to the assumption that $f \in K[V]$. Thus we have shown that

$$\alpha^o K[W'] = K[V'] \cap \alpha^o K(W),$$

for all open W' in W, where $V' = \alpha^{-1}(W')$.

The Proposition of 18.2 implies that $K(V)$ is purely inseparable over $\alpha^o K(W)$. This, together with the result just proved, implies that $K[V']$ is integral over $\alpha^o K[W']$. Therefore, if $\beta: \tilde{W} \to W$ is the normalization of W in $K(V)$, it follows that β factors as $\tilde{W} \xrightarrow{\gamma} V \xrightarrow{\alpha} W$, and γ is surjective.

Now if W is complete then, by 18.2, $\tilde{W}$ is complete, so it follows that V is also complete.

Next suppose V is affine. Since $\alpha^o K[W]$ contains $K[V]^{p^n}$ for some n ($p = \text{char}(K)$) it follows easily that $K[W]$ is an affine K-algebra. Therefore we have a morphism $\delta: W \to \text{spec}_K(K[W])$, and we claim δ is an isomorphism. Since W is normal so also is $\text{spec}_K(K(W])$. Hence, by 18.2, it suffices to see that δ is birational. But this follows easily from the fact, proved above, that $\alpha^o K[W]$ contains $K[V] \cap \alpha^o K(W)$, and the fact that, since V is affine, $K(V)$ is the field of fractions of $K[V]$.

18.4 Proposition [Fond., Ch. V, V, Prop. 3]. *Let $\alpha: V \to W$ be a dominant morphism of irreducible varieties, and put $r = \dim V - \dim W$. Let x be a point of V such that $y = \alpha(x)$ is normal on W. Suppose further that each irreducible component passing through x of the fibre of α over y has dimension r. Then if U is a neighborhood of x in V, $\alpha(U)$ is a neighborhood of y in W.*

Corollary. *Let $\alpha: V \to W$* be a dominant morphism of varieties, *where W is normal. Assume the dimensions of the irreducible components of the fibres of α are constant. Then α is an open map.*

18.5 *Algebraic curves* (cf. [M, Ch. III, §8, Cor. to Prop. 1, and Thm. 5]). An *algebraic curve* is an algebraic variety of dimension 1. In the discussion to follow we shall assume that all algebraic varieties are irreducible.

(a) *An algebraic curve is smooth if and only if it is normal.*

(b) *Let L be a finitely generated field extension of K of transcendence degree 1. Then there is an essentially unique complete smooth curve C whose function field is isomorphic (as K-algebra) to L. Moreover C is a projective variety.*

(c) *If V is any smooth algebraic curve then the dominant morphisms $\alpha: V \to C$ correspond bijectively to the K-algebra homomorphisms $\alpha_o: K(C) \to K(V)$. If α_o is an isomorphism then α is an open immersion.*

If we apply (b) to $K(V)$ and (c) to the identity map of $K(V)$ we obtain:

(d) *A smooth curve V is an open subset of a unique complete smooth curve $\bar{V}$.*

Another corollary of (c) is:

(e) *If C is a complete smooth curve then the anti-homomorphism*

$$\text{Aut}_{\text{alg var}}(C) \to \text{Aut}_{K\text{-alg}}(K(C))$$

is bijective.

Finally, we record:

(f) *Let $\alpha: V \to W$ be a morphism from a smooth curve V into a complete variety W. Then α extends to a morphism $\bar{\alpha}: \bar{V} \to W$.*

To see this we can first replace W by $\overline{\alpha(V)}$ and thus assume α is dominant. Forgetting the trivial case when W is a point we may then assume W is a (complete) curve. Let $\pi: \tilde{W} \to W$ be its normalization (in $K(W)$). Then (see 18.2, and (a) above) $\tilde{W}$ is a complete smooth curve. Since V is smooth (hence normal), α factors through π via $\beta: V \to \tilde{W}$. According to (c), the comorphism $\beta^o: K(\tilde{W}) \to K(V) = K(\bar{V})$ is induced by a morphism $\bar{\beta}: \bar{V} \to \tilde{W}$. Now $\pi \circ \bar{\beta} = \bar{\alpha}$ is the desired extension of α.

References

[N.B.] N. Bourbaki, (a) Algèbre, Chapitre 5, Corps commutatifs (1959), (b) Algèbre commutative, Hermann, éd. Paris.

[C.-C.] Séminaire Cartan-Chevalley, Géométrie algébrique, Paris (1955/56).

[Class.] Séminaire C. Chevalley, Classification des groupes de Lie algébriques, Paris (1956–58).

[Fond.] C. Chevalley, Fondements de la géométrie algébrique, Paris (1958).

[EGA] A. Grothendieck et J. Dieudonné, Eléments de géométrie algébrique, Publ. IHES.

[Ha] R. Hartshorne, Algebraic Geometry, GTM 52, Springer-Verlag 1977.

[M] D., Mumford, The red book of varieties and schemes, Springer LNM 1358.

[Z.-S.] O. Zariski and P. Samuel, Commutative algebra, van Nostrand, Princeton (1958); GTM, 28, 29, Springer-Verlag.

Chapter I

General Notions Associated With Algebraic Groups

§1. The Notion of an Algebraic Group

1.1 *Algebraic groups.* An algebraic group is an algebraic variety G together with:

(id): an element $e \in G$;
(mult): a morphism $\mu: G \times G \to G$, denoted $(x, y) \mapsto xy$;
(inv): a morphism $i: G \to G$, denoted $x \mapsto x^{-1}$,

with respect to which (the set) G is a group. We call G a *k-group* if G is a k-variety and if μ and i are defined over k (see AG.12). It follows then that $e \in G(k)$, because $\{e\}$ is the image of the k-morphism $\alpha \circ \delta$, where $\delta(x) = (x, x)$ and $\alpha(x, y) = xy^{-1}$ (see AG.14.5).

A *morphism* of algebraic groups is a morphism of varieties which is also a homomorphism of groups. The expression "$\alpha: G \to G'$ is a k-morphism of k-groups" means G and G' are k-groups and α is a morphism defined over k.

1.2 The *connected component of e* in an algebraic group G will be denoted

$$G^o.$$

Proposition. *Let G be an algebraic group.*

(a) *G is smooth (as a variety).*
(b) *G^o is a normal subgroup of finite index in G whose cosets are the connected, as well as irreducible, components of G. If G is defined over k, so is G^o.*
(c) *Every closed subgroup of finite index contains G^o.*

Proof. (a) G is "homogeneous," i.e. it has (as a variety) a transitive group of automorphisms. (Namely, the translations $x \mapsto xy$.) Since G has some simple points (AG.17.2) it follows that all points are simple. Moreover it now follows from AG.17.2 that the irreducible and connected components of G coincide.

(b) If $x \in G^o$ then $x^{-1}G^o$ is a connected component of G containing e, and hence equal to G^o. Thus $x^{-1}G^o = G^o$ for all $x \in G^o$. It follows that $G^o = (G^o)^{-1}$

and $G^o G^o = G^o$, so G^o is a group. Its cosets (say left) are each connected components of G, clearly, so they must be finite in number (the space G is noetherian). Finally, if $y \in G$ then $yG^o y^{-1}$ is a connected component of G containing e, and hence equal to G^o. Thus G^o is normal in G. Let G be defined over k. Then G^o and its cosets are defined over k_s (AG.12.3). They are permuted by the Galois group Γ of k_s over k, acting as in AG.14.3. Since $e \in G(k)$ (1.1), it follows that $G^o(k_s)$ is stable under Γ, hence G^o is defined over k (AG.14.4).

(c) If H is a closed subgroup of finite index in G, then the complement of H, being a *finite* union of the non identity left cosets, is also closed. Thus H is open and closed so it must contain G^o.

The proposition implies that the notions "connected" and "irreducible" coincide for algebraic groups. The term "connected" is preferred because the word "irreducible" has a different use in the representation theory of G.

1.3 Proposition. *Let G be a k-group and let H be a not necessarily closed subgroup. Let U and V be dense open sets in G.*

(a) $U \cdot V = G$.

(b) *$\bar{H}$ is a subgroup of G. If $H \subset G(k_s)$ and if H is stable under* $\mathrm{Gal}(k_s/k)$, *then $\bar{H}$ is defined over k.*

(c) *If H is constructible, then $H = \bar{H}$.*

Proof. (a) Given $x \in G$, the dense open sets U and xV^{-1} have a common point, say $u = xv^{-1}$, so $x = uv \in U \cdot V$.

(b) Since $x \mapsto x^{-1}$ is a homeomorphism we have $\bar{H}^{-1} = \overline{H^{-1}} = \bar{H}$. If $x \in H$ then $x\bar{H} = \overline{xH} = \bar{H}$, so $H\bar{H} = \bar{H}$. If $y \in \bar{H}$ then $Hy \subset \bar{H}$ so $\bar{H}y = \overline{Hy} \subset \bar{H}$. Thus $\bar{H}\bar{H} = \bar{H}$, so $\bar{H}$ is a group.

The assertions concerning rationality over k follows from AG.14.4.

(c) If H is constructible then it follows from AG.10.2 that H contains a dense open subset of $\bar{H}$. By part (b) $\bar{H}$ is a closed subgroup, so part (a) implies $\bar{H} = H \cdot H = H$.

1.4 Corollary. *Let G' be a k-group and $\alpha: G \to G'$ a morphism.*

(a) *$\alpha(G)$ is a closed subgroup of G; and it is defined over k if α is defined over k.*

(b) $\alpha(G^o) = \alpha(G)^o$.

(c) $\dim G = \dim \ker(\alpha) + \dim \alpha(G)$.

Proof. (a) According to AG.10.2 the subgroup $\alpha(G)$ is constructible, so 1.3(c) implies that it is closed. Moreover AG.14.5 implies that it is defined over k if α is so.

(b) By part (a), $\alpha(G^o)$ is closed. Since it is also connected and of finite index in $\alpha(G)$ it follows from 1.2 (c) that $\alpha(G^o) = \alpha(G)^o$.

(c) It follows from AG.10.1 that for all x in some dense open set in $\alpha(G)$, $\dim G - \dim \alpha(G) = \dim \alpha^{-1}(x)$. But $\dim \alpha^{-1}(x) = \dim \ker(\alpha)$ for all x, clearly.

1.5 *Affine groups.* Let $G = \mathrm{spec}_K(A)$ be an affine algebraic group, $A = K[G]$. We shall translate the elements of structure of G in terms of A.

$$e \in G: \quad e: A \to K, \quad e(f) = f(e).$$

(The latter homomorphism, evaluation at e, was formerly denoted "e_{e}.")

$$\mu: G \times G \to G: \quad \mu_0: A \to A \underset{K}{\otimes} A.$$

If $\mu^o f = \sum g_i \otimes h_i$ then $f(xy) = \sum g_i(x) h_i(y)$.
Next we have the inverse.

$$i: G \to G: \quad i^o: A \to A.$$
$$(i^o f)(x) = f(x^{-1}).$$

In order to formulate the group axioms we introduce

$$p: G \to G, \quad p^o: A \to A$$
$$x \mapsto e \quad (p^o f)(x) = f(e).$$

Now the group axioms are expressed by the commutativity of the following diagrams:

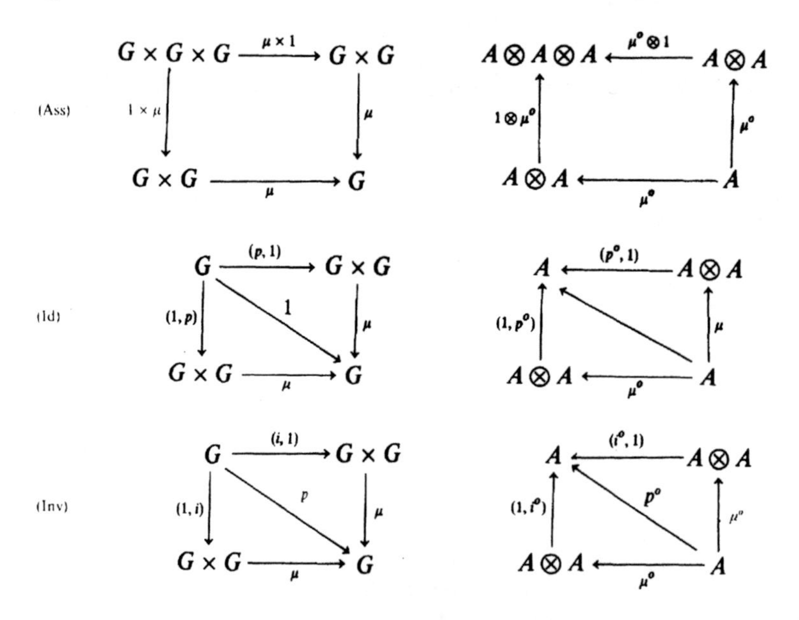

Note that p^o is just the composite of the augmentation $e: A \to K$ with the inclusion $K \subset A$. Thus, in terms of A, G is determined by the data (A, e, μ^o, i^o) subject to the above three axioms.

The data (A, e, μ^o) subject to (Ass) and (Id) are sometimes called an associative Hopf algebra with identity and μ^o is referred to as its *diagonal map*.

If C is any K-algebra we can describe the group structure on

$$G(C) = \mathrm{Hom}_{K\text{-}alg.}(A, C)$$

as follows: If $x, y \in G(C)$ then the product in $G(C)$ of x and y is the composite:

$$A \xrightarrow{\mu^o} A \otimes A \xrightarrow{x \otimes y} C \otimes C \xrightarrow{m} C,$$

where m is the multiplication in C ($m(a \otimes b) = ab$). If $C \to C'$ is an algebra homomorphism then $G(C) \to G(C')$ is a group homomorphism. Quite generally, for any (not necessarily affine) algebraic group its functor of points, $C \mapsto G(C)$, is a group valued functor.

1.6 Examples

(1) The *additive group* $\mathbf{G}_a$. Its affine ring is $k[\mathbf{G}_a] = k[T]$, a polynomial ring in one variable;

$$\mu^o(T) = (T \otimes 1) + (1 \otimes T);\ i^o(T) = -T;\ e(T) = 0.$$

(2) *The general linear group* $\mathbf{GL}_n$. The affine ring is

$$k[\mathbf{GL}_n] = k[T_{11}, T_{12}, \ldots, T_{nn}, D^{-1}],$$

where $D = \det(T_{ij})$. Thus $\mathbf{GL}_n$ is the principal open set $(K^{n^2})_D$, in affine n^2-space. We have

$$e(T_{ij}) = \delta_{ij}$$
$$\mu^o(T_{ij}) = \Sigma_h T_{ih} \otimes T_{hj}$$

and

$$i^o(T_{ij}) = (-1)^{i+j} D^{-1} \det(T_{rs})_{r \neq j, s \neq i}.$$

(3) *The multiplicative group* $\mathbf{GL}_1$ is sometimes denoted $\mathbf{G}_m$ in the literature. As a special case of the above formulas we have

$$k[\mathbf{GL}_1] = k[T, T^{-1}]$$
$$e(T) = 1, \quad \mu^o(T) = T \otimes T; \quad i^o(T) = T^{-1}.$$

(4) *The special linear group*, $\mathbf{SL}_n$, is the kernel of the morphism

$$\det : \mathbf{GL}_n \to \mathbf{GL}_1.$$

Thus $k[\mathbf{SL}_n] = k[T_{11}, T_{12}, \ldots, T_{nn}]/(\det(T_{ij}) - 1)$. The maps μ^o and i^o are induced, on passing to the quotient, by those of $k[\mathbf{GL}_n]$. The same remark applies to any closed subgroup of $\mathbf{GL}_n$, such as the following examples.

(5) The group of *upper triangular matrices*

$$\mathbf{T}_n = \{g \in \mathbf{GL}_n \mid g_{ij} = 0 \text{ for } j < i\}$$

and the *upper triangular unipotent group*

$$\mathbf{U}_n = \{g \in \mathbf{T}_n \mid g_{ii} = 1 (1 \leq i \leq n)\}.$$

$\mathbf{T}_n$ is the semi-direct product of $\mathbf{U}_n$ and of the *diagonal group*

$$D_n = \{g \in \mathbf{GL}_n | g_{ij} = 0 \text{ for } i \neq j\}.$$

(6) *The symplectic group*

$$\mathbf{Sp}_{2n} = \{g \in \mathbf{GL}_{2n} | {}^t g J g = J\}$$

where ${}^t g$ denotes the transpose of g and

$$J = \begin{pmatrix} O & I_n \\ -I_n & O \end{pmatrix}.$$

(7) If S is a non-singular symmetric n by n matrix then

$$O(S) = \{g \in \mathbf{GL}_n | {}^t g S g = S\}$$

is called the *orthogonal group* of S.

(8) Let V be a finite dimensional vector space, and let $S_K(V^*)$ be the symmetric algebra of its dual space. Then we can identify V with the affine variety $\operatorname{spec}_K(S_K(V^*))$. Indeed, for any K-algebra B we have $\operatorname{Hom}_{K\text{-alg}}(S_K(V^*), B) = \operatorname{Hom}_{K\text{-mod}}(V^*, B) = B \bigotimes_K V$. In case $B = K$ this gives the bijection $V \to \operatorname{spec}_K(S_K(V^*))$ making V a variety, and it shows that the points of V in B are just

$$V(B) = B \bigotimes_K V,$$

the B-module obtained by base change $K \to B$. We can make V an algebraic group using the addition $V \times V \to V$, and this is compatible with the natural addition in $B \bigotimes_K V$.

If V has a k-structure V_k as vector space, then it has a corresponding k-structure as variety given by $S_k(V_k^*)$ in $S_K(V^*)$. In case B is a k-algebra we then obtain, just as above, $V(B) = B \bigotimes_k V_k$.

The vector space $E = \operatorname{End}_{K\text{-mod}}(V)$ can also be made into a variety, and we then have $E(B) = B \bigotimes_K E = B \bigotimes_K \operatorname{End}_{K\text{-mod}}(V) = \operatorname{End}_{B\text{-mod}}(B \otimes V) = \operatorname{End}_{B\text{-mod}}(V(B))$. In this way the natural action of E on V extends naturally to the functor of points.

Relative to any basis for V, the determinant, det, is a polynomial with integer coefficients in the matrix coordinates of E. Thus if V has a k-structure then $E_k = \operatorname{End}_{k\text{-mod}}(V_k)$ is a k-structure on E, and we see that $\det \in S_K(E^*)$ is defined over k. The principal k-open set $E_{\det} = \{g \in E | \det(g) \neq 0\}$ is denoted

$$\mathrm{GL}(V) \quad \text{or} \quad \mathrm{GL}_V.$$

It inherits a multiplication from E making it a group. Since the inverse of matrix is a polynomial in the matrix coefficients and $\det^{-1}$ it follows that

GL_V is an algebraic group. Moreover, if B is any K-algebra we have

$$\mathrm{GL}_V(B) = \{g \in E(B) | \det(g) \text{ is invertible}\}$$

where we identify $E(B)$ with B-module endomorphisms of the free B-module $V(B)$. Thus

$$\mathrm{GL}_V(B) = \mathrm{Aut}_{B\text{-mod}}\left(B \bigotimes_K V\right).$$

A closed subgroup of GL_V is called a *linear algebraic group.* A morphism $\alpha: G \to \mathrm{GL}_V$ of algebraic group is called a *rational (linear) representation* of G. If G is a k-group we say α is defined over k, or that α is k-rational, if it is a k-morphism with respect to the k-structure on GL_V induced as above by one given on V. Relative to a k-rational basis of V, this just means that the corresponding matrix coefficients $\alpha(g)_{ij}$ are k-rational functions $G \to K$. Since these functions are all of the form $g \mapsto h(\alpha(g)(v))$ with $v \in V_k$ and $h \in V_k^*$ it follows that $\alpha: G \to \mathrm{GL}_V$ *is a* k-rational representation if and only if the corresponding map $G \times V \to V$ *is a k-morphism of varieties.*

A representation $\alpha: G \to \mathrm{GL}_V$ will be called *immersive* if it induces an isomorphism of G with the closed subgroup $\alpha(G)$ of GL_V, in other words, if it is a closed immersion.

(9) *The multiplicative group of an algebra.* Let Λ be a finite dimensional associative (not necessarily commutative) K-algebra, and let N be the norm, $N_{\Lambda/K}: \Lambda \to K$ (the determinant of the regular representation). Viewing Λ as an affine space, we see that the group $\mathbf{GL}_1(\Lambda)$ of invertible elements in Λ is the principal open set defined by N. Hence $\mathbf{GL}_1(\Lambda)$ is an affine algebraic group which is a "rational variety." The latter means that $\mathbf{GL}_1(\Lambda)$ is irreducible and that its function field, $K(\mathbf{GL}_1(\Lambda))$, is a field of rational functions (in $\dim_K \Lambda$ variables).

If Λ has a k-structure given by a k-subalgebra Λ_k then the norm N is defined over k, and $\mathbf{GL}_1(\Lambda)$ becomes a k-group (see AG.12.1). In this case $k(\mathbf{GL}_1(\Lambda))$ is already a purely transcendental extension of k, i.e. $\mathbf{GL}_1(\Lambda)$ is "k-rational." For any k-algebra k' the points, $\mathbf{GL}_1(\Lambda)(k')$, form the multiplicative group $\mathbf{GL}_1(\Lambda(k'))$ of $\Lambda(k') = \Lambda_k \bigotimes_k k'$.

1.7 *Actions of groups on varieties. An algebraic transformation space* is a triple (G, V, α) where G is an algebraic group, V is a variety, and $\alpha: G \times V \to V$, $(g, x) \mapsto gx = \alpha(g, x)$, is a morphism satisfying

$$ex = x \quad \text{and} \quad g(hx) = (gh)x$$

for all $x \in V$ and all g, $h \in G$. We sometimes refer to this situation by saying that "G acts morphically on the variety V." If G and V are given with k-structures we say G acts "k-morphically" if α is defined over k. Because of the notation gx or $g \cdot x$ or $g(x)$, a symbol for α is superfluous, and is usually omitted.

For subsets M and N of V we have the *transporter*

$$\mathrm{Tran}_G(M,N)=\{g\in G\,|\,gM\subset N\},$$

sometimes also denoted $\mathrm{Tr}_G(M,N)$. One calls

$$\mathscr{N}_G(M)=\mathrm{Tran}_G(M,M)$$

the *normalizer* of M in G. For example

$$G_x=\mathscr{N}_G(\{x\})$$

is the *stability group* or the *isotropy group* of $x\in V$, and

$$G(x)=\{gx\,|\,g\in G\}$$

is called the *orbit* of x. The set of fixed points of G on V, i.e. of points $x\in V$ for which $G_x=G$, is denoted V^G. If $V=G$, acted upon by inner automorphisms, and $M\subset G$, then

$$G^M=\bigcap_{x\in M}G_x$$

is called the *centralizer* of M in G and is denoted $\mathscr{Z}_G(M)$ or $\mathscr{Z}(M)$. In that case, the normalizer $\mathscr{N}_G(M)$ or $\mathscr{N}(M)$ of M can also be defined as

$$\mathscr{N}_G(M)=\{g\in G\,|\,gM=Mg\}$$

whereas

$$\mathscr{Z}_G(M)=\{g\in G\,|\,g.m=m.g \text{ for all } m\in M\}.$$

Proposition. *Let G be a k-group acting k-morphically on a k-variety V, and let M and N be subsets of V.*

(a) *We have* $\mathrm{Tran}_G(M,N)\subset\mathrm{Tran}_G(\bar M,\bar N)$, *with equality if N is closed.*
(b) *If N is k-closed and if $M\subset V(k)$, then* $\mathrm{Tran}_G(M,N)$ *is k-closed.*
(c) *If $\bar M\subset V(k)$ then $\mathscr{Z}_G(M)$ and $\mathscr{N}_G(\bar M)$ are k-closed.*

Proof. (a) If $gM\subset N$ then $g\bar M=\overline{gM}\subset\bar N$. If $N=\bar N$ then $g\bar M\subset\bar N$ implies $gM\subset N$.

(b) Define $\alpha_x\colon G\to V$ by $\alpha_x(g)=gx$ for $x\in V$. Then if $x\in V(k)$, α_x is defined over k, so $\alpha_x^{-1}(N)=\mathrm{Tran}_G(\{x\},N)$ is k-closed. Since $M\subset V(k)$ it follows that $\mathrm{Tran}_G(M,N)=\bigcap_{x\in M}\alpha_x^{-1}(N)$ is k-closed.

(c) The fixed points in V of any $g\in G$ are closed (because varieties are separated) so it follows that $\mathscr{Z}_G(M)=\mathscr{Z}_G(\bar M)$. Part (b) implies G_x is k-closed for $x\in V(k)$, so $\mathscr{Z}_G(M)=\bigcap_{x\in M}G_x$ is k-closed.

Moreover $\mathscr{N}_G(\bar M)=\mathrm{Tran}_G(M,\bar M)$ (part (a)), and the latter is k-closed by part (b).

Remarks. (1) It is not true in part (b) that $\mathrm{Tran}_G(M,N)$ need be defined over k even if N is defined over k and $M\subset V(k)$.

(2) The proposition applies notably to the action of G on itself by inner automorphisms.

1.8 *Closed orbit lemma.* The following simple result is a basic technical tool for the theory of algebraic groups.

Proposition. *Let G be an algebraic group acting morphically on a non-empty variety V. Then each orbit is a smooth variety which is open in its closure in V. Its boundary is a union of orbits of strictly lower dimension. In particular, the orbits of minimal dimension are closed.*

Proof. Let $M = G(x)$ be the orbit of $x \in V$. Since M is the image of the morphism $g \mapsto gx$ it follows from AG.10.2 that M contains a dense open set in $\bar{M}$. Now G operates transitively on M, and it evidently leaves $\bar{M}$ stable. Since M contains an $\bar{M}$-neighborhood of one of its points it follows from homogeneity that M is open in $\bar{M}$. Hence $\bar{M} - M$ is closed and of lower dimension, as well as being G-stable. Finally, the smoothness of M follows from homogeneity.

Corollary. *Closed orbits exist.*

A morphic action of G on V is said to be *closed* if all orbits are closed, *free* if only the identity of G has fixed points (i.e. $g.v. = v$ for some $v \in V$ implies $g = 1$). If it is free, then all orbits have the same dimension (that of G), hence are closed by the Proposition.

The graph F of an action is the image of the morphism $\alpha: G \times V \to V \times V$ defined by $(g, v) \mapsto (g \cdot v, v)$. If the action is free, then $\alpha: G \times V \to F$ is bijective. If it is an isomorphism of varieties, then the (free) action is called *principal*. This amounts to require that for $(z, y) \in F$, the unique $g = g(x, y) \in G$ such that $y = g.x$ is a morphic function on F. As an example, if V is itself a k-group, containing G as a closed subgroup acting by left (resp. right) translations, then the action is principal: Indeed it is obviously free and then the unique g bringing x into a point y of the orbit of x is $y.x^{-1}$ (resp. $x^{-1} \cdot y$).

1.9 *Translations.* Let G be an affine k-group acting k-morphically on an affine k-variety V, via $\alpha: G \times V \to V$. Thus α is defined by the comorphism

$$\alpha^o: k[V] \to k[G] \bigotimes_k k[V]$$

of affine rings over k.

If $g \in G$ we denote by λ_g the comorphism of $x \mapsto g^{-1}x$. Then

$$f \mapsto \lambda_g f, \quad (\lambda_g f)(x) = f(g^{-1}x),$$

is a linear automorphism of $K[V]$ which we call *left translation* of functions by g. The reason for the inverse is to make $g \mapsto \lambda_g$ a homomorphism: $\lambda_g \cdot \lambda_h = \lambda_{gh}$.

Proposition. *Let F be any finite dimensional vector subspace of $K[V]$. Then there is a finite dimensional subspace E which* (i) *contains F,* (ii) *is defined over k, and* (iii) *is stable under left translation by G. Moreover a necessary and sufficient condition that F be invariant under left translation is that*

$$\alpha^o F \subset K[G] \bigotimes_K F.$$

Proof. We begin with the first assertion. By enlarging F we may assume F is defined over k. We may further assume that F is spanned by a single function $f \in k[V]$, for the general case will then follow by taking the sum of the E's obtained for each element of a k-basis of F.

Write $\alpha^o f = \sum_{i=1}^{n} f_i \otimes h_i \in k[G] \bigotimes_k k[V]$, so that n is minimal. Then for $g \in G$ we have $(\lambda_g f)(x) = f(g^{-1}x) = \sum f_i(g^{-1})h_i(x)$, so that $\lambda_g f = \sum f_i(g^{-1})h_i$. There exists therefore a finite dimensional subspace of $k[V]$, defined over k, containing all $\lambda_g f (g \in G)$. The intersection of all such subspaces of $K[V]$ clearly satisfies the three required conditions.

To prove the last assertion, let F be a subspace of $K[V]$ and let $\{f_i\} \cup \{h_j\}$ be a basis for $K[V]$ such that $\{f_i\}$ spans F. If $f \in F$ and $g \in G$ we have $\lambda_g f = \sum r_i(g^{-1})f_i + \sum s_j(g^{-1})h_j$, where $\alpha^o f = \sum r_i \otimes f_i + \sum s_j \otimes h_j$. Hence $\lambda_g f \in F \Leftrightarrow s_j(g^{-1}) = 0$ for all j. Varying $g \in G$ and $f \in F$ we see that $\lambda_g F \subset F$ for all $g \in G \Leftrightarrow \alpha^o F \subset K[G] \bigotimes_K F$. Q.E.D.

Consider the special case where $V = G$ and G acts on itself by both left and right translations. More precisely we let $(g, h) \in G \times G$ act on $x \in G$ by $x \mapsto gxh^{-1}$. In this way we obtain two actions of G on functions $f \in K[G]$:

left translation: $(\lambda_g f)(x) = f(g^{-1}x)$
right translation: $(\rho_g f)(x) = f(xg)$.

They are both homomorphisms of G:

$$\lambda_{gh} = \lambda_g \lambda_h, \quad \rho_{gh} = \rho_g \rho_h;$$

and they commute:

$$\lambda_g \rho_h = \rho_h \lambda_g \quad \text{for all } g, h \in G.$$

Applying the proposition above we obtain the

Corollary. *Every finite dimensional subspace F of $K[G]$ is contained in a finite dimensional subspace E defined over k which is stable under both left and right translation by G.*

1.10 Proposition. *Let G be an affine k-group. Then G is k-isomorphic to a closed subgroup, defined over k, of some* $\mathbf{GL}_n$.

Proof. Write $k[G] = k[f_1, \ldots, f_n]$. Using 1.9 we can even do this so that $f_1, \ldots, f_n$ is a basis for a subspace E of $K[G]$ stable under right translation, i.e. so that $\mu^o E \subset E \bigotimes_K K[G]$ (see 1.9). Thus, for each i, we have

$$\mu^o f_i = \sum_i f_j \otimes m_{ji}$$

for some $m_{ji} \in k[G]$. If $g \in G$ then $(\rho_g f_i)(x) = f_i(xg) = \sum f_i(x) m_{ji}(g)$, i.e.

$$\rho_g f_i = \sum_j m_{ji}(g) f_j.$$

It follows that

$$\alpha : G \to \mathbf{GL}_n, \quad \alpha(g) = (m_{ji}(g))$$

is a morphism of algebraic groups, and it is evidently defined over k, because the m_{ji} are. Indeed, the comorphism

$$\alpha^o : k[\mathbf{GL}_n] = k[T_{ll}, \ldots, T_{nn}, D^{-1}] \to k[G]$$

is defined by $\alpha^o(T_{ji}) = m_{ji}$. Since $f_i(x) = f_i(ex) = \sum_j f_j(e) m_{ji}(x)$ we have $f_i = \sum_j f_j(e) m_{ji} \in \mathrm{im}(\alpha^o)$ for each i. Hence α^o is surjective, so α is a closed immersion. We know from 1.4 that $G' = \alpha(G)$ is defined over k, so a induces the desired k-isomorphism $G \to G'$.

Remark. It follows easily from an argument like the one above that, if E is any finite dimensional right invariant subspace of $K[G]$, the homomorphism $\alpha : G \to \mathrm{GL}(E)$ induced by right translation is a rational representation of G.

1.11 *Actions of groups on groups; semi-direct products.* Let G and H be k-groups, and let $\alpha : G \times H \to H$ be an action of G on H. (This means that elements of G act as group automorphisms of H.) The basic example of this occurs when G and H are subgroups of a larger group in which G normalizes H, and the action is induced by conjugation: $\alpha(g, h) = ghg^{-1}$. In fact this is essentially the most general case, as we see now by constructing the *semi-direct product*

$$H \cdot G,$$

as follows: as a variety it is $H \times G$, and the multiplication is defined by

$$(h_1, g_1)(h_2, g_2) = (h_1 \alpha(g_1, h_2), g_1 g_2).$$

It is easy to check that this makes $H \cdot G$ a group. For example

$$(h, g)^{-1} = (\alpha(g^{-1}, h)^{-1}, g^{-1})$$

Moreover we have the exact sequence of morphisms

$$l \to H \xrightarrow{i} H \cdot G \xrightarrow{p} G \to l,$$

and a section $s : G \to H \cdot G$ of p, defined by

$$i(h) = (h, e), \quad p(h, g) = g, \quad s(g) = (e, g)$$

If α is defined over k then it is clear that $H \cdot G$ has a natural k-structure so that i, p, and s are k-morphisms. The morphism i is an isomorphism of H with a normal subgroup of $H \cdot G$, and α is induced, via s, by conjugation of iH by sG:

$$(e, g)(h, e)(e, g)^{-1} = (\alpha(g, h), e).$$

Suppose G' is an algebraic group and G and H are closed subgroups with H normalized by G. Then we shall say G' is the *semi-direct product of the subgroups* G and H if the multiplication map

$$H \times G \to G', \quad (h, g) \mapsto hg$$

is an isomorphism of varieties. Then $\alpha(g, h) = ghg^{-1}$ defines an action of G on H so that G' is isomorphic to the group $H \cdot G$ constructed above.

1.12. Proposition. (Existence of equivariant embeddings). *Let G be a k-group operating k-morphically on a affine k-variety V. Then there exist a finite dimensional vector space E defined over k, a closed k-embedding $\phi: V \to E$ and a k-morphism $\mu: G \to GL(E)$ such that $\phi(g \cdot v) = \mu(g) \cdot \phi(v)$ for all $g \in G$ and $v \in V$.*

Proof. We may write $k[V] = k[f_1, \ldots, f_n]$, where the f_i generate $K[V]$ as a K-algebra and span over K a G-invariant subspace F (1.9). Let $\alpha: G \times V \to V$ be the map defining the action of G on V. By 1.9 we have $\alpha^o F \subset F \bigotimes_K K[G]$, and there exist uniquely defined $m_{ij} \in k[G]$ such that $\alpha^o f_i = \sum m_{ij} \otimes f^j$, whence

$$f_i(g \cdot v) = \sum m_{ij}(g) f_j(v) \quad (g \in G; v \in V). \tag{1}$$

Let $E = K^n$, with coordinates x_i. Define $\phi: V \to K^n$ by assigning to $v \in V$ the point with coordinates $x_i = f_i(v)$. It is a k-morphism. We have $\phi^o x_i = f_i$, hence ϕ^o is surjective, and ϕ is a closed embedding. The relation (1) can be written

$$\phi(g \cdot v) = M(g) \cdot \phi(v), \tag{2}$$

where $M(g) = (m_{ij}(g))$. It follows immediately that $M(gh) = M(g) \cdot M(h)$ $(g, h \in G)$, hence $\mu: g \mapsto M(g)$ is a k-morphism of G to $\mathbf{GL}_n$, which, in view of (2) satisfies our condition.

§2. Group Closure; Solvable and Nilpotent Groups

2.1 *Group closure.* Let M be a subset of a k-group G. We write

$$\mathscr{A}(M)$$

for the intersection of all closed subgroups of G containing M; thus $\mathscr{A}(M)$ is one of them, the smallest one. From 1.3(b) we have:

(a) *If M is a subgroup of G then $\mathscr{A}(M) = \bar{M}$.*

Put $N = M \cup \{e\} \cup M^{-1}$ and let N_m denote the image of the product map $\alpha_m: N \times \cdots \times N \to G$. Then $H = \bigcup_m N_m$ is the subgroup generated by M, so (a) implies that $\mathscr{A}(M) = \bar{H}$.

If M is a subvariety defined over k then so also is N. Since each α_m is a morphism defined over k it follows from AG.14.5 that each $\bar{N}_m$ is defined over k. Now AG.14.6 further implies that $\mathscr{A}(M) = \bar{H}$, being the closure of $\bigcup_m \bar{N}_m$, is also defined over k. Thus we have proved:

(b) *If M is a subvariety defined over k then $\mathscr{A}(M)$ is defined over k.*

Next we treat products:

(c) *If M_i is a subset of an algebraic group $G_i (i = 1, 2)$ then $\mathscr{A}(M_1 \times M_2) = \mathscr{A}(M_1) \times \mathscr{A}(M_2)$.*

The right-hand side is a closed subgroup containing $M_1 \times M_2$, and hence contains the left-hand side. On the other hand $\mathscr{A}(M_1 \times M_2)$ contains $M_1 \times \{e\}$ and hence also $\mathscr{A}(M_1 \times \{e\})$, which is clearly equal to $\mathscr{A}(M_1) \times \{e\}$. Similarly it contains $\{e\} \times \mathscr{A}(M_2)$, and hence also the right-hand side.

(d) *Let M and N be subsets of G such that N normalizes (resp. centralizes) M. Then $\mathscr{A}(N)$ normalizes (resp., centralizes) $\mathscr{A}(M)$.*

Let $C(X)$ denote the normalizer (resp. centralizer) of a subset X of G. By hypothesis $N \subset C(M)$, and evidently $C(M) \subset C(\mathscr{A}(M))$. It follows from (1.7) that $C(\mathscr{A}(M))$ is closed, and hence $\mathscr{A}(N) \subset C(\mathscr{A}(M))$.

(e) *If M and N are subgroups of G then the commutator groups (M, N) and $(\bar{M}, \bar{N})$ have the same closure.*

Let $c: G \times G \to G$, $c(x, y) = xyx^{-1}y^{-1}$. Since $M \times N$ is dense in $\bar{M} \times \bar{N}$ the same is true of $c(M \times N)$ in $c(\bar{M} \times \bar{N})$, so $\mathscr{A}(c(M \times N)) = \mathscr{A}(c(\bar{M} \times \bar{N}))$. But it follows from part (a) that these groups are the closures of (M, N) and of $(\bar{M}, \bar{N})$, respectively.

(f) *If $\alpha: G \to G'$ is a morphism of algebraic groups then*

$$\alpha(\mathscr{A}(M)) = \mathscr{A}(\alpha(M)).$$

$\alpha(\mathscr{A}(M))$ contains $\alpha(M)$ and, according to 1.4, it is closed. Hence it contains $\mathscr{A}(\alpha(M))$. On the other hand $\alpha^{-1}\mathscr{A}(\alpha(M))$ is closed (α is continuous) and contains M, so it contains $\mathscr{A}(M)$. Applying α we obtain $\mathscr{A}(\alpha(M)) \supset \alpha(\alpha^{-1}\mathscr{A}(\alpha(M))) \supset \alpha(\mathscr{A}(M))$, thus reversing the inclusion proved above.

2.2 Proposition. *Let $f_i: V_i \to G (i \in I)$ be a family of k-morphisms from irreducible k-varieties V_i into a k-group G, and assume $e \in f_i V_i = W_i$ for each $i \in I$. Put $M = \bigcup W_i (i \in I)$. Then $\mathscr{A}(M)$ is a connected subgroup of G defined over k. Moreover, there is a finite sequence $(\alpha(1), \ldots, \alpha(n))$ in I such that $\mathscr{A}(M) = W_{\alpha(1)}^{e_1} \cdots W_{\alpha(n)}^{e_n}$, where each $e_i = \pm 1$.*

Proof. By enlarging I if necessary we can assume the morphisms $x \mapsto f_i(x)^{-1}$ are also among the f_i's. If $\alpha = (\alpha(1), \ldots, \alpha(n))$ is a finite sequence in I put

$W_\alpha = W_{\alpha(1)} \cdots W_{\alpha(n)}$. The set W_α is the image of the k-morphism,

$$V_{\alpha(1)} \times \cdots \times V_{\alpha(n)} \xrightarrow{f_{\alpha(1)} \times \cdots \times f_{\alpha(n)}} G \times \cdots \times G \xrightarrow{\text{mult}} G;$$

it follows therefore from the hypotheses that W_α is constructible, and that $\bar{W}_\alpha$ is an irreducible k-variety (see AG. 10.2). As a consequence, for dimension reasons, there is an α such that $\bar{W}_\alpha$ is maximal.

If β and γ are two finite sequences, then

$$(1) \qquad \bar{W}_\beta \cdot \bar{W}_\gamma \subset \bar{W}_{(\beta,\gamma)}.$$

In fact, for $x \in W_\gamma$, the map $y \mapsto y \cdot x$ sends W_β into $W_{(\beta,\gamma)}$, hence $\bar{W}_\beta$ into $\bar{W}_{(\beta,\gamma)}$, whence $\bar{W}_\beta \cdot W_\gamma \subset \bar{W}_{(\beta,\gamma)}$. As a consequence, $x \cdot \bar{W}_\gamma \subset \bar{W}_{(\beta,\gamma)}$ for every $x \in \bar{W}_\beta$, from which (1) follows. Since $\bar{W}_\alpha$ is maximal, this yields in particular, for any β:

$$\bar{W}_\alpha \subset \bar{W}_\alpha \cdot \bar{W}_\beta \subset \bar{W}_{(\alpha,\beta)} = \bar{W}_\alpha.$$

Thus, $\bar{W}_\alpha$ is stable under products, and, taking β such that $W_\beta = W_\alpha^{-1}$, we also see that $\bar{W}_\alpha = \bar{W}_\alpha^{-1}$. Therefore, $\bar{W}_\alpha$ is a closed subgroup containing W_β for all β. Then, clearly, $\bar{W}_\alpha = \mathscr{A}(M)$. Since W_α contains a dense open subset of $\bar{W}_\alpha$, we have $\mathscr{A}(M) = W_\alpha \cdot W_\alpha = W_{(\alpha,\alpha)}$ by 1.3.

Remark. The proof shows that the n in the statement of the proposition can be taken to be $\leqq 2 \cdot \dim G$.

2.3 *Group closure of a commutator group.*

Corollary. *Let G' be a k-group and let G and H be closed subgroups defined over k, with G connected. Then the commutator group (G, H) is a closed connected subgroup defined over k.*

Proof. If $h \in H$, define $f_h: G \to G'$ by $f_h(g) = (g, h) = ghg^{-1}h^{-1}$. These are morphisms of the connected variety G into G', which all map e onto e, so 2.2 implies that the group generated by all $f_h(G)(h \in H)$, which is just (G, H), is closed.

It follows that $(G, H) = \mathscr{A}(M)$ where M is the image of the commutator map $G \times H \to G'$. The latter is a k-morphism, so $\bar{M}$ is defined over k, and 2.1(b) implies that $\mathscr{A}(\bar{M})$, which equals $\mathscr{A}(M)$, is defined over k.

If neither G nor H is connected then (G, H) need not be closed. One need only consider an infinite group generated by two finite subgroups G and H (for example the modular group $\mathbf{SL}_2(\mathbb{Z})/\{\pm 1\}$ in $\mathbf{PGL}_2$). However this cannot happen if G or H is normal.

Proposition. *Let G be a k-group and let H and N be closed subgroups defined over k such that N is normalized by H. Then (H, N) is a closed subgroup of G defined over k and normal in HN.*

Corollary. *The smallest normal subgroup of G containing H is a closed subgroup defined over k.*

For, since (H, G) is normal in G, that subgroup is $H(H, G)$, which is closed.

Proof (of the Proposition). There is no loss in assuming that $G = HN$. Once we show that (H, N) is closed the fact that it is defined over k follows just as in the proof of the first corollary above. That corollary further implies that (H^o, N) and (H, N^o) are closed and connected. Hence so also is the group L generated by them together with all their conjugates in G.

We shall now invoke the theorem of Baer in the appendix at the end of §2. We have first that (H, N) is normal, so that L, the least normal subgroup containing (H^o, N) and (H, N^o), is contained in (H, N). Since L is closed it suffices to show that L has finite index in (H, N); the latter will then be a finite union of cosets of L.

Pass to the group $G' = G/L$, and denote the image in G' of a subgroup $M \subset G$ by M'. Then H' and N' are such that $H^{o\prime}$ centralizes N' and $N^{o\prime}$ centralizes H' (by definition of L). Hence the set of commutators of elements of H' with elements of N' is a quotient of the finite set $(H'/H^{o\prime}) \times (N'/N^{o\prime})$. Now the desired finiteness of (H', N') follows from Baer's theorem (appendix).

2.4 *Solvable and nilpotent groups.* Let G be an abstract group. The *derived series* $(\mathscr{D}^n G)$ $(n \geqq 0)$, and the *descending central series* $(\mathscr{C}^n G)$ $(n \geqq 0)$, are defined inductively by:

$$\mathscr{D}^0 G = G, \quad \mathscr{D}^{n+1} G = (\mathscr{D}^n G, \mathscr{D}^n G), \quad (n \geqq 0)$$
$$\mathscr{C}^0 G = G, \quad \mathscr{C}^{n+1} G = (G, \mathscr{C}^n G), \quad (n \geqq 0).$$

We sometimes write $\mathscr{D}^\infty G = \cap \mathscr{D}^n G$ and $\mathscr{C}^\infty G = \cap \mathscr{C}^n G$. All these are characteristic subgroups (i.e. stable under all automorphisms) of G, evidently. One says that G is *solvable* (resp. *nilpotent*) if, for some n, we have $\mathscr{D}^n G = \{e\}$ (resp., $\mathscr{C}^n G = \{e\}$).

The center of G is denoted $\mathscr{C}G$.

Now suppose G is an algebraic group. Then it would be natural to introduce notions of "algebraic solvability" and "nilpotence" for G, using the series $\mathscr{A}(\mathscr{D}^n G)$ and $\mathscr{A}(\mathscr{C}^n G)$, respectively. However, it follows from the results of 2.3 that the groups $\mathscr{D}^n G$ and $\mathscr{C}^n G$ are closed, so these notions coincide with the abstract group notions of solvability and nilpotence.

Proposition. *Let G be an algebraic group, and let M and N be not necessarily closed subgroups such that M normalizes N. Then $\bar{M}$ normalizes $\bar{N}$ and $\overline{(M, N)} = (\bar{M}, \bar{N})$.*

Proof. It is clear that $\bar{M}$ normalizes $\bar{N}$ (cf. 1.7). Part (e) of 2.1 says (M, N) and $(\bar{M}, \bar{N})$ have the same closure, and 2.3 says $(\bar{M}, \bar{N})$ is closed (because $\bar{N}$ is normal in $\bar{M}\bar{N}$).

By a simple induction on n this implies:

Corollary 1. *For all $n \geqq 0$ we have*

$$\overline{\mathscr{D}^n(M)} = \mathscr{D}^n(\bar{M}) \text{ and } \overline{\mathscr{C}^n(M)} = \mathscr{C}^n(\bar{M}).$$

In particular, if M is closed, then so also are the groups in its derived and descending central series.

Corollary 2. *If N is a normal subgroup of M such that M/N is abelian (resp., nilpotent, resp. solvable) then the same is true of $\bar{M}/\bar{N}$.*

Corollary 3. *The following conditions on a k-group G are equivalent:*

(1) *G is solvable.*
(2) *There is a chain $G = G_o \supset G_1 \supset \cdots \supset G_n = \{e\}$ of closed subgroups defined over k such that $(G_i, G_i) \subset G_{i+1} (0 \leqq i < n)$.*

Proof. $(2) \Rightarrow \mathscr{D}^i G \subset G_i$ so G is solvable. Taking $G_i = \mathscr{D}^i G$ we see that $(1) \Rightarrow (2)$ by applying Corollary 1, plus (2.3) to get the G_i defined over k.

Corollary 4. *The following conditions on a k-group G are equivalent:*

(1) *G is nilpotent.*
(2) *There is a chain $G = G_o \supset G_1 \supset \cdots \supset G_n = \{e\}$ of closed subgroups defined over k such that $(G, G_i) \subset G_{i+1} (0 \leqq i < n)$.*

Proof. $(2) \Rightarrow \mathscr{C}^i G \subset G_i$ so G is nilpotent. Conversely if G is nilpotent then Corollary 1 and 2.3 imply that the $\mathscr{C}^i G$ satisfy the conditions in (2).

Appendix. We present here a proof, due to M. Rosenlicht, of the following result of R. Baer. (See M. Rosenlicht, Proc. A. M. S. **13** (1962), 99–101.)

Proposition. *Let H and N be subgroups of a group G such that H normalizes N. Then the commutator group (H, N) is normal in HN. If the set of commutators*

$$\{hnh^{-1}n^{-1} \mid h \in H, n \in N\}$$

is finite then (H, N) is finite.

We begin with a special case:

If $\mathscr{C}(G)$ has finite index in G then (G, G) is finite.

It suffices to show that any product of commutators of elements of G can be written as such a product with at most n^3 factors, n being the index of the center of G. Noting that there are at most n^2 distinct commutators, and that in any product of commutators any two factors may be brought together by replacing the intermediate factors by conjugates, also commutators, it suffices to show that the $(n+1)$th power of a commutator is the product of n commutators. But if $a, b \in G$, then $(aba^{-1}b^{-1})^n$ is central, so

$$(aba^{-1}b^{-1})^{n+1} = b^{-1}(aba^{-1}b^{-1})^n b(aba^{-1}b^{-1}),$$

which may be written

$$b^{-1}((aba^{-1}b^{-1})^{n-1}(ab^2a^{-1}b^{-2}))b,$$

a product of n commutators.

We proceed to prove the general result. It is worth remarking that if one is only interested in the case where both H and N are normal, the trickiest points below collapse to trivialities.

Assume, as we may, that $G = HN$, and consider the set S of all commutators of conjugates of elements of H by elements of N. Any conjugate of an element of H is of the form nhn^{-1}, with $n \in N$, $h \in H$, so each element of S is of the form

$$(nhn^{-1})n_1(nhn^{-1})^{-1}n_1^{-1} = (hnh^{-1}n^{-1})^{-1}(h(n_1n)h^{-1}(n_1n)^{-1}),$$

with $n_1 \in N$, which shows that S is a finite subset of (H, N). But S clearly generates (H, N) and each inner automorphism of G permutes the elements of S. We deduce that (H, N) *is normal in* G, and also that there exists a normal subgroup G_0 of G of finite index that centralizes S, hence also (H, N). Now $G_0 \cap (H, N)$ is a central subgroup of (H, N) of finite index, so $((H, N), (H, N))$ is finite. Since the latter subgroup is normal in G, we may divide by it to *suppose that* (H, N) *is commutative.*

We now claim that the subgroup $(H, (H, N))$ of (H, N) is normal in G. Conjugation by elements of H clearly leaves it invariant, so we must show that if $n \in N$, $h \in H$, $m \in (H, N)$, then $n(hmh^{-1}m^{-1})n^{-1} \in (H, (H, N))$. But the latter element can also be written

$$hn(n^{-1}h^{-1}nh)mh^{-1}m^{-1}n^{-1},$$

which, by the commutativity of (H, N), is equal to

$$hnm(n^{-1}h^{-1}nh)h^{-1}m^{-1}n^{-1} = h(nmn^{-1})h^{-1}(nmn^{-1})^{-1} \in (H, (H, N)).$$

Note also that any commutator of H and (H, N) is one of H and N, so there are only a finite number of such and they all commute. Furthermore, if we square any such commutator, say $hmh^{-1}m^{-1}$, we get $(hmh^{-1}m^{-1})^2 = (hmh^{-1})^2m^{-2} = hm^2h^{-1}m^{-2}$, which is also a commutator. Thus $(H, (H, N))$ is finite. Dividing G by this subgroup, we see that we may *suppose that* H *centralizes* (H, N).

To finish the proof, recall that (H, N) is commutative and generated by a finite number of commutators $hnh^{-1}n^{-1}$, and note that here too the square of such a commutator is also a commutator:

$$(hnh^{-1}n^{-1})^2 = (hnh^{-1}n^{-1})(nh^{-1}n^{-1}h) = hnh^{-2}n^{-1}h = h^2nh^{-2}n^{-1}.$$

Thus (H, N) is finite.

§3. The Lie Algebra of an Algebraic Group

In this section G is a k-group and $A = K[G]$.

3.1 *Restricted Lie Algebras.* Let p denote the characteristic exponent of k($p = \text{char}(k)$ if $\text{char}(k) > 0$, and $p = 1$ if $\text{char}(k) = 0$). A restricted Lie algebra over k is a Lie algebra $\mathfrak{g}$ together with a "p operation," $X \mapsto X^{[p]}$, such that:

If $p = 1$ then $X^{[p]} = X$ and if $p > 1$ the p-operation satisfies

(i) $\text{ad}(X^{[p]}) = \text{ad}(X)^p$ $(X \in \mathfrak{g})$

(ii) $(tX)^{[p]} = t^p X^{[p]}$ $(t \in k, X \in \mathfrak{g})$

(iii) $(X + Y)^{[p]} = X^{[p]} + Y^{[p]} + \sum_{i=1}^{p-1} i^{-1} s_i(X, Y)$, where $s_i(X, Y)$ is the coefficient of t^i in $\text{ad}(tX + Y)^{p-1}(X) \cdot (X, Y \in \mathfrak{g})$.

Here, as usual, we write

$$\text{ad}(X)(Y) = [X, Y].$$

We shall have no occasion to use formula (iii) except in the special case

(iii′) If $[X, Y] = 0$ then $(X + Y)^{[p]} = X^{[p]} + Y^{[p]}$.

For a general discussion of restricted Lie algebras, and, in particular, of the following examples, the reader can consult Jacobson, pp. 185 ff. in [10].

Examples. (1) An associative k-algebra A gives rise to a restricted Lie algebra with underlying k-module A, where

$$[X, Y] = XY - YX, \quad X^{[p]} = X^p.$$

(2) In case $A = \text{End}_k(E)$, where E is a vector space over k, we write $\mathfrak{gl}(E)$ for the corresponding restricted Lie algebra. If E is identified to K^n, hence A to $\mathbf{M}_n(K)$, we write $\mathfrak{gl}_n(K)$ for $\mathfrak{gl}(E)$.

(3) Suppose E itself is a not necessarily associative k-algebra. Then

$$\text{Der}_k(E, E) = \{X \in \mathfrak{gl}(E) \mid X(f \cdot g) = (Xf) \cdot g + f \cdot (Xg) \text{ for all } f, g \in E\}$$

is a restricted Lie subalgebra of $\mathfrak{gl}(E)$.

Let F be a set of k-automorphisms of E. Then

$$L = \{X \in \text{Der}_k(E, E) \mid Xs = sX, (s \in F)\}$$

is a restricted Lie subalgebra of $\text{Der}_k(E, E)$.

(4) Let $\mathfrak{g}$ be a restricted Lie algebra, and let $\mathfrak{h}$ and S be a subalgebra and subset, respectively, of $\mathfrak{g}$. Then

$$\mathfrak{h}^S = \{X \in \mathfrak{h} \mid [X, Y] = 0 \text{ for all } Y \in S\}$$

is a restricted Lie subalgebra of $\mathfrak{h}$, called the centralizer of S in $\mathfrak{h}$.

3.2 *Derivatives of products.* Let G be an algebraic group, let $\alpha_i: V_i \to G$ be a morphism of varieties, and let $v_i \in V_i$ be a point such that $\alpha_i(v_i) = e(1 \leqq i \leqq n)$.

Put

$$v = (v_1, \ldots, v_n) \in V = V_1 \times \cdots \times V_n$$

and define $\alpha: V \to G$ to be the product map

$$\alpha(x_1, \ldots, x_n) = \alpha_1(x_1) \cdots \alpha_n(x_n).$$

Define $\beta_i: V_i \to V$ by

$$\beta_i(x) = (v_1, \ldots, v_{i-1}, x, v_{i+1}, \ldots, v_n).$$

Since $\alpha_i(v_i) = e$, we have $\alpha_i = \alpha \circ \beta_i$ $(1 \leqq i \leqq n)$. By AG.16.1 there is a canonical isomorphism

$$T(V)_v \cong T(V_1)_{v_1} \oplus \cdots \oplus T(V_n)_{v_n},$$

so that

$$(d\alpha)_v(X_1, \ldots, X_n) = \sum_i (d\alpha \circ d\beta_i)_{v_i} X_i = \sum_i (d\alpha_i)_{v_i} X_i.$$

Applying this to $\mu: G \times G \to G$ we obtain

$$T(G \times G)_{(e,e)} = T(G)_e \oplus T(G)_e,$$

and

$$(d\mu)_{(e,e)}(X, Y) = X + Y.$$

(The α_i $(i = 1, 2)$ both correspond, in this case, to the identity morphism $G \to G$.) Next consider the composite

$$G \xrightarrow{(id, i)} G \times G \xrightarrow{\mu} G$$

sending x to $xx^{-1} = e$. Its derivative is zero, clearly, so we have

$$\begin{aligned} 0 &= d(\mu \circ (id, i))_e(X) \\ &= (d\mu)_{(e,e)}(d(id, i)_e X) \\ &= (d\mu)_{(e,e)}(X, (di)_e X) \\ &= X + (di)_e X. \end{aligned}$$

Thus

$$(di)_e X = -X.$$

3.3 *Left invariant derivations.* From AG.15.5, 16.1 we have

$$T(G)_x = \mathrm{Der}_K(A, K(x)) \tag{1}$$

and see that

$$T(G)_{x,k} = \mathrm{Der}_k(A_k, k(x)) \quad (x \in G) \tag{2}$$

is a k-form of $T(G)_x$. Recall also that $e_x: A \to K(x)$ is the evaluation map at x, defined by $e_x(f) = f(x)$ $(f \in A, x \in G)$. Given $D \in \mathrm{Der}_K(A, A)$, let $D_x = e_x \circ D$. In view of (1), it belongs to $T(G)_x$. Therefore D can be viewed as a "vector field", assigning to each $x \in X$ an element of $T(G)_x$. Let

(3) $$\mathrm{Lie}(G) = \{D \in \mathrm{Der}_K(A, A) \mid \lambda_x \circ D = D \circ \lambda_x, \text{ for all } x \in G\}$$

(4) $$\mathrm{Lie}(G)_k = \mathrm{Lie}\, G \cap \mathrm{Der}_k(A_k, A_k) = \{D \in \mathrm{Lie}(G) \mid D(A_k) \subset A_k\}.$$

Let $f \in A$ and $y \in G$. It follows from the definitions that

$$(\lambda_x \circ D) f(y) = Df(x^{-1} \cdot y) = D_{x^{-1} \cdot y} f,$$

and from AG 16.1 that $(D \circ \lambda_x) f(y)$ is the image of D_y under the differential at y of the translation $gl \to x^{-1} \cdot g$, to be denoted by $x^{-1} \cdot D_y$.

Therefore, the condition on D in (3) can also be written

(5) $$x \cdot D_y = D_{x \cdot y} \quad (x, y \in G)$$

i.e., the vector field $y \mapsto D_y$ is invariant under left translations.

3.4 Theorem. *The map* $v: D \mapsto D_1 = e_1 \circ D$ *is an isomorphism of vector space of* $\mathrm{Lie}(G)$ *onto* $T(G)_1$, *which maps* $\mathrm{Lie}(G)_k$ *onto* $T(G)_k$. *In particular* $\mathrm{Lie}(G)_k$ *is a* k*-form of* $\mathrm{Lie}(G)$ *and* v *is defined over* k.

v is injective: Let $D \in \mathrm{Lie}(G)$ and assume that $D_1 = 0$. Then 3.3(5) shows that $D_x = 0$ for all x's, hence $D = 0$.

v is surjective. Let $X \in T(G)_1$. We have to find $D \in \mathrm{Lie}(G)$ such that $D_1 = X$.

Since D is to be a left invariant vector field, we try to define Df $(f \in A)$ by

(1) $$Df(x) = (x \cdot X) f = X(\lambda_{x^{-1}} f).$$

We prove first that $Df \in A$. There is a finite set I and elements $u_i, v_i \in A$ $(i \in I)$ such that

(2) $$\mu^o f = \sum_i u_i \otimes v_i$$

(where μ^o is the comorphism of the product in G, as usual), i.e. such that

(3) $$f(y \cdot x) = \sum_i u_i(y) \cdot v_i(x), \quad (x, y \in G).$$

We have then

$$\lambda_y f = \sum_i u_i(y^{-1}) \cdot v_i,$$

hence

$$Df(y) = X(\lambda_{y^{-1}} f) = \sum_i u_i(y) \cdot X v_i.$$

This shows that

(4) $$Df = \sum_i u_i \cdot X v_i$$

is indeed a regular function. There remains to see that $D \in \mathrm{Lie}(G)$. Let $u, v \in A$. It follows from the definitions that we have

$$D(u \cdot v)(x) = x \cdot X(u \cdot v) = x \cdot X(u) \cdot v(x) + u(x) \cdot x \cdot Xv(y)$$
$$= Du(x) \cdot v(x) + u(x) \cdot Dv(x),$$

hence D is a derivation. Fix $z \in G$. With the notation of (3) we have

$$\lambda_z f(x \cdot y) = f(z^{-1} \cdot x \cdot y) = \sum_i u_i(z^{-1} \cdot x) \cdot v_i(y) = \sum_i (\lambda_z u_i)(x) v_i(y),$$

therefore

$$D(\lambda_z f)(x) = \sum_i (\lambda_z u_i)(x) \cdot X v_i.$$

But the right-hand side is equal to $\lambda_z\left(\sum_i u_i \cdot X v_i\right)(x)$ hence to $(\lambda_z \circ D) f(x)$, which shows that $D \circ \lambda_z = \lambda_z \circ D$, hence that $D \in \mathrm{Lie}(G)$. Assume now $f \in A_k$. Then we can choose the u_i and v_i in (2) to be rational over k. If moreover $X \in A_k$ then $X v_i \in k$ $(i \in I)$ hence Df, as defined by (4), is defined over k, and $D \in \mathrm{Lie}(G)_k$. Conversely, if we start from $D \in \mathrm{Lie}(G)_k$, then it follows directly from the definitions that $e_1 \circ D \in T(G)_k$. Therefore ν is an isomorphism of $\mathrm{Lie}(G)_k$ onto $T(G)_k$. This completes the proof of 3.4.

3.5 *The Lie algebra of* G. We define the Lie algebra $L(G)$ of G to be $T(G)_1$ endowed with restricted Lie algebra structure of $\mathrm{Lie}(G)$, carried over by means of ν. By 3.3(5), we may then also identify $L(G)$ to the Lie algebra of left invariant vector fields on G. If $\alpha: G \to G'$ is a morphism of affine algebraic groups, then we write also $L(\alpha)$ instead of $(d\alpha)_1$. We shall see in 3.19 that it is a morphism $L(G) \to L(G')$ of restricted Lie algebras, which is defined over k if α is so.

We shall then view $G \mapsto L(G) = T(G)_1$ as a functor from (affine) algebraic groups to restricted Lie algebras. The Lie algebra of a k-group $G, H, M, \ldots$ will also often we denoted by the corresponding German letter $\mathfrak{g}, \mathfrak{h}, \mathfrak{m}, \ldots$. Note that the equality $T(G)_1 = T(G^o)_1$ implies

3.6 Corollary. *We have* $L(G) = L(G^o)$ *and* $\dim_K L(G) = \dim G$.

3.7 *Definition of* $*X$ *and* $X*$. The inverse map to ν associates to $X \in T(G)_1$ the left invariant vector field on G which is equal to X at 1. We shall often denote it by $*X$ and write $f \mapsto f * X$ for its action on A. [This notation is suggested by the convolution on a Lie group, see 3.19.] As above, given $f \in A$, let $u_i, v_i \in A$ be such that

$$f(x \cdot y) = \sum_i u_i(x) \cdot v_i(y), \quad (x, y \in G). \tag{1}$$

We have then, by 3.3(5)

$$f * X(x) = \sum_i u_i(x) \cdot X v_i \tag{2}$$

$$f * X = \sum_i u_i \cdot X v_i. \tag{3}$$

Similarly, one can define the right invariant vector field $X*$, equal to X at 1, and its action on regular functions. It is a derivation which commutes with right translations and is given on f by

$$X * f = \sum_i X u_i \cdot v_i. \tag{4}$$

3.8 Proposition. *Let H be a closed subgroup of G and $I \subset A$ be the ideal of H. Then*

(i) $L(H) = \{X \in L(G) | XI = 0\}$
(ii) $\mathrm{Lie}\, H = \{X \in \mathrm{Lie}(G) | I * X \subset I\}$.

Proof.

(i) Let $j: H \to G$ be the inclusion map. Then $L(j): T(H)_1 \to T(G)_1$ is the map

$$\mathrm{Der}_K(A/I, K(1)) \to \mathrm{Der}_K(A, K(1))$$

associated to $A \to A/I$. Therefore X belongs to the image of $L(j)$ if and only if $XI = 0$.

(ii) Let $X \in \mathrm{Lie}(G)$ be such that $I * X \subset I$ and let $P \in I$. Then $XP = (P * X)(1) = 0$, therefore $XI = 0$ and $X \in \mathrm{Lie}(H)$ by 3.4 and (i). Assume now that $X \in \mathrm{Lie}(H)$. If $P \in I$, and $h \in H$ then $\lambda_h P \in I$, obviously, therefore $(P * X)(h) = X(\lambda_{h^{-1}} P)(1) = 0$, hence $P * X \in I$.

Corollary. *Let $G \subset \mathbf{GL}_n$ be a closed subgroup, and let J be the ideal of all polynomials in $B = K[T_{11}, T_{12}, \ldots, T_{nn}]$ vanishing on G. Then*

$$G = \{g \in \mathbf{GL}_n | \rho_g J = J\}$$

and

$$\mathfrak{g} = \{X \in \mathfrak{gl}_n | J * X \subset J\}.$$

Proof. Put $A' = K[\mathbf{GL}_n] = B[D^{-1}]$, where $D = \det(T_{ij})$, and let J' be the ideal of functions in A' vanishing on G. The proposition above asserts that $\mathfrak{g} = \{X \in \mathfrak{gl}_n | J' * X \subset J'\}$, and the fact that $G = \{g \in \mathbf{GL}_n | \rho_g J' = J'\}$ is obvious.

Now it is easy to see that $J' = A'J$ and $J' \cap B = J$. Suppose $f \in J$ and $f' \in A'$. Then $\rho_g(ff') = \rho_g(f)_g(f')$ and $(ff') * X = (f * X)f' + f(f' * X)$. Hence $\rho_g J = J \Rightarrow \rho_g J' = J'$ and $J * X \subset J \Rightarrow J' * X \subset J'$.

For the converse it suffices to show that each ρ_g and each $*X$ leave $B \subset A'$ stable. For then they leave $J = J' \cap B$ stable as soon as they leave J' stable.

Since $*X = (I \otimes X) \circ \mu^o$ (see 3.7(3)) we have

$$T_{ij} * X = (I \otimes X)\left(\sum_i T_{ih} \otimes T_{hj}\right) = \sum_i T_{ih} X(T_{hj}) \in B.$$

Similarly, if $g, h \in \mathbf{GL}_n$ then

$$(\rho_g T_{ij})(h) = T_{ij}(hg) = \sum_i T_{im}(h) T_{mj}(g),$$

so $\rho_g T_{ij} = \sum_i T_{im} T_{mj}(g) \in B$. Thus $B * X \subset B$ and $\rho_g B \subset B$, as required.

Remark. This corollary is the basis of one of the classical approaches to the Lie algebra of a matrix group, using only polynomial functions in the coordinates of the matrices.

3.9 Examples

(a) Let $G = \mathbf{G}_a$, the additive group, so that $G(K) = K$ and $K[G] = K[T]$, a polynomial ring. If $D \in \mathrm{Lie}(G)$ then D is determined completely by $f(T) = DT$. Left invariance requires that, for all $x \in G$ we have $\lambda_{-x} DT = f(T + x)$ equal to

$$D(\lambda_{-x}(T)) = D(T + x) = DT + Dx = f(T).$$

But $f(T + x) = f(T)$ for all x means that f is constant, so $\mathrm{Lie}(G)$ consists of all K-multiplies of $D = d/dT$. Since $D^{[p]} T^n = n(n-1) \cdots (n-(p-1)) T^{n-p}$ (or zero if $n < p$) it follows that, if $\mathrm{char}(k) > 0$, the p-operation is zero in $\mathrm{Lie}(\mathbf{G}_a)$ (because the product of p consecutive integers is divisible by p).

(b) Consider $G = \mathbf{GL}_1$, so that $G(K) = K^*$ and $K[G] = K[T, T^{-1}]$. If $D \in \mathrm{Lie}(G)$, then D is determined by the Laurent polynomial $DT = f(T)$. This time left invariance requires that, for all $x \in G$ we have $f(xT) = xf(T)$. It is easy to see that this implies $f(T) = aT$ for some $a \in K$. It follows that $D^{[p]} T = a^p T$ in this case. Thus $\mathrm{Lie}(G)$ is isomorphic to the one dimensional Lie algebra K with p-operation $a \mapsto a^p$. If $\mathrm{char}(k) > 0$, therefore the p-operations distinguish the Lie algebras of the additive and multiplicative groups.

(c) $G = \mathbf{GL}_n, \mathbf{SL}_n$. We use the notation of 1.6(2). To $X \in T(G)_1$ we associate the matrix (X_{ij}), where $X_{ij} = X \cdot T_{ij}$ $(1 \leqq i, j \leqq n)$. This yields an injective linear map of $T(G)_1$ into $\mathbf{M}_n(K)$. It is surjective since any derivation of the polynomial algebra in the T_{ij}'s extends uniquely to one of A (AG.15.5) and a derivation of a polynomial algebra is determined by its values on the generators, which are arbitrary (AG.15.2). We claim that

$$L(G) = \mathbf{gl}_n(K), \tag{1}$$

(notation of 3.1, Examples). To see this, consider the matrix $T = (T_{ij}) \in \mathbf{M}_n(A)$ and let $T * X \in \mathbf{M}_n(A)$ be the matrix with coefficients $(T * X)_{ij} = (T_{ij} * X)$. Since

$\mu^o(T_{ij}) = \sum_i T_{im} \otimes T_{mj}$ (see 1.6), we have

$$T_{ij} * X = \sum_i T_{im} \cdot X T_{mj} = \sum_i T_{im} \cdot X_{mj}$$

hence

$$T * X = T \cdot X \quad \text{(matrix product).}$$

As a consequence, $T * X * Y = T \cdot X \cdot Y (X, Y \in L(G))$ and therefore

(2) $$T * [X, Y] = T \cdot [X, Y], T * X * \cdots * X = T \cdot X^p$$

(p factors on the left hand side of the second equality). This proves (1).

The value at 1 of (Det T)$* X$ is the sum of the determinants of the n matrices obtained from the identity matrix by replacing the i-th column by the entries X_{ij} $(1 \leqq i, j \leqq n)$. Therefore

(3) $$((\mathrm{Det}\ T) * X)_1 = \mathrm{Tr}\ X = \sum_i X_{ii}$$

If X is in the Lie algebra of $\mathbf{SL}_n$, then the left hand side must be zero, hence $\mathrm{Tr}\ X = 0$. Since $\mathbf{SL}_n$ has codimension one in $\mathbf{GL}_n$, this characterizes the Lie algebra $\mathbf{sl}_n$, i.e.

(4) $$L(\mathbf{SL}_n)(k) = \{X \in \mathbf{M}_n(k) \mid \mathrm{Tr} X = 0\}.$$

(d) Stated more intrinsically, the above shows that if V is a finite dimensional vector space over K with a k-structure, then

$$L(\mathrm{GL}(V))(k) = \mathfrak{gl}(V)(k) \quad L(\mathrm{SL}(V))(k) = \mathfrak{sl}(V),$$

where $\mathfrak{sl}(V)$ denotes the Lie algebra of endomorphisms of V with trace zero.

The relation $X_{ij} = X \cdot T_{ij}$ yields

(5) $$\langle \alpha, Xv \rangle = X(g \mapsto \langle \alpha, g \cdot v \rangle) \quad (v \in V, \alpha \in V^*, g \in GL(V)).$$

3.10. Let $\pi: G \to \mathrm{GL}(V)$ be a rational representation of V. We know that $d\pi(X)$ is an endomorphism of V $(X \in \mathfrak{g})$. We claim that its effect on $v \in V$ can be described in the following way:

Let $o_v: G \to V$ be the orbit map $g \mapsto \pi(g) \cdot v$. It is a morphism of varieties. Then

(1) $$d\pi(X)(v) = (do_v)_1(X),$$

where the right hand side is viewed as an element of V via the canonical identification of $T(V)_v$ with V. To show this, we have to prove that

(2) $$\langle \alpha, d\pi(X) \cdot (v) \rangle = \langle \alpha, (do_v)_1(X) \rangle \quad (\alpha \in V^*).$$

By 3.9(5) we have $d\pi(X) \cdot v = d\pi(X)(h \mapsto h \cdot v), (h \in \mathrm{GL}(V))$, hence

(3) $$\langle \alpha, d\pi(X)(v) \rangle = X(g \mapsto \langle \alpha, \pi(g) v \rangle) \quad (g \in G).$$

But the right hand side of (3) is by definition equal to the right hand side of (2).

3.11 Proposition. *Let $V \subset A$ be a finite dimensional vector space of A which is invariant under right (resp. left) translations. Let $\rho: G \to \mathrm{GL}(V)$ (resp. $\lambda: G \to \mathrm{GL}(V)$) be the rational representation $g \mapsto \rho_g | V$ (resp. $g \mapsto \lambda_g | V$). Then for $X \in \mathfrak{g}$, and $f \in V$ we have*

(1) $$(d\rho)(X)(f) = f * X, \quad (\text{resp.}\, d\lambda(X)(f) = X * f).$$

We give the proof for ρ. The other case is of course quite similar. By 3.10, $d\rho(X) \cdot f = X(g \mapsto \rho_g f)$. In the notation of 3.7, we have $\rho_g f = \sum_i u_i \cdot v_i(g)$, and therefore, in view of 3.7(3)

$$X(g \mapsto \rho_g f) = \sum_i u_i \cdot X v_i = f * X.$$

3.12 Corollary

(i) *Under right (resp. left) convolution, $\mathfrak{g}$ leaves stable every subspace of A which is right (resp. left) invariant under G.*

(ii) *Let $G = \mathbf{GL}_n$ and D be the determinant function on $\mathbf{GL}_n$. Then $D * X = D \cdot \mathrm{Tr}(X)(X \in \mathfrak{g})$.*

The first assertion follows obviously from 3.11.

We have $\rho_g \cdot D = D \cdot \lambda_{g^{-1}} = D \cdot \det g$ $(g \in G)$, hence D spans a one-dimensional subspace of A which is left and right invariant. We see also that $D * X = D \cdot (D * X)_1$, and we already have seen that $(D * X)_1 = \mathrm{Tr}\, X$ (3.9(3)).

3.13 *The adjoint representation.* The group G operates on itself by inner automorphisms

(1) $$\mathrm{Int}\, x : y \mapsto x \cdot y \cdot x^{-1} = {}^x y \quad (x, y \in G).$$

The differential $d(\mathrm{Int}\, x)$ will be denoted $\mathrm{Ad}\, x$. It is an automorphism of $\mathfrak{g}$. We claim that $\mathrm{Ad}: G \to GL(\mathfrak{g})$ is a k-morphism. Since G may be identified to a k-subgroup of some $\mathbf{GL}_n$, it suffices to show that when $G = \mathbf{GL}_n$. In that case $\mathfrak{g}$ may be identified with $\mathbf{gl}_n(K)$ (3.9(c)). We claim that

(2) $$\mathrm{Ad}\, x(X) = x \cdot X \cdot x^{-1} \quad (x \in G, X \in \mathfrak{g}),$$

where the product on the right is just matrix multiplication. We have

$$(\mathrm{Int}\, x)^o(T_{ij})(y) = T_{ij}(x \cdot y \cdot x^{-1}) = (x \cdot y \cdot x^{-1})_{ij} = \sum_{l,m} x_{il} y_{lm} (x^{-1})_{mj}.$$

But $y_{lm} = T_{lm}(y)$ by definition, hence $(\mathrm{Int}\, x)^o(T_{ij})(y) = (x \cdot T \cdot x^{-1})_{ij}(y)$, i.e.

(3) $$(\mathrm{Int}\, x)^o(T_{ij}) = (x \cdot T \cdot x^{-1})_{ij} \quad (x \in G, 1 \leqq i, j \leqq n).$$

We now have

$$((\mathrm{Ad}\, x)X)(T) = X(\mathrm{Int}\, x)^o T) = X(x \cdot T \cdot x^{-1}) = x \cdot X T \cdot x^{-1}$$
$$= x \cdot X \cdot x^{-1} \quad (x \in G, X \in \mathfrak{g}).$$

This shows that Ad is a morphism of G into $\mathrm{GL}(\mathfrak{g})$. We can identify $\mathrm{GL}(\mathfrak{g})$ to an open subset of $\mathrm{End}\,\mathfrak{g} = \mathfrak{g} \otimes \mathfrak{g}^*$. The ring $k[\mathrm{End}\,\mathfrak{g}]$ is the polynomial ring with coefficients in k over the $T_{ij} \otimes T_{lm}$. In this presentation $\mathrm{Ad}\,x$ is given by $x \otimes {}^t x^{-1}$. Therefore $\mathrm{Ad}^o(T_{ij} \otimes T_{lm})$ is the function $x \mapsto x_{ij} \cdot ({}^t x^{-1})_{lm}$. It belongs to A_k, hence Ad is defined over k.

3.14 *The differential of the adjoint representation.* The differential dAd_1 of Ad at the identity is a k-morphism of restricted Lie algebras of $\mathfrak{g}$ into $\mathfrak{gl}(\mathfrak{g})$. We claim that

$$(\mathrm{dAd})(X) = \mathrm{ad}\,X \quad (X \in \mathfrak{g}) \tag{1}$$

i.e. $\mathrm{dAd}(X)$ is the endomorphism $Y \mapsto [X, Y]$ of $\mathfrak{g}$. We may assume $G = \mathbf{GL}_n$. By definition

$$\mathrm{dAd}(X)_{ij} = (\mathrm{dAd}(X))T_{ij} \quad (1 \leqq i,j \leqq n).$$

For $Y \in \mathfrak{g}$, let $u_Y : G \to \mathfrak{g}$ be defined by $g \mapsto g \cdot Y \cdot g^{-1}$. It is a morphism of varieties. Its differential $du_{Y,1}$ at the identity maps $\mathfrak{g}$ into the tangent space to $\mathfrak{g}$ at the origin, i.e. into $\mathfrak{g}$ itself. By definition

$$(\mathrm{dAd}(X)(Y))(T_{ij}) = (du_Y(X))(T_{ij}) = X(u_Y^o T_{ij}) \tag{2}$$

If $g \in G$, we have $u_Y^o(T_{ij})(g) = T_{ij}(g \cdot Y \cdot g^{-1}) = (g \cdot Y \cdot g^{-1})_{ij}$, which can be written

$$u_Y^o(T_{ij}) = (T \cdot Y \cdot T^{-1})_{ij}.$$

Here we have denoted by T^{-1} the $n \times n$ matrix over A whose (i,j)-coefficient is the function $g \mapsto (g^{-1})_{ij}$. We have now

$$X(u_Y^o(T)) = X\left(\sum_{lm} T_{il} Y_{lm} T_{mj}^{-1}\right).$$

X is a derivation and, at 1, T and T^{-1} are the identity matrix, therefore

$$X(u_Y^o T_{ij}) = \sum_l X_{il} Y_{lj} + \sum_m Y_{im} X T_{mj}^{-1} \tag{3}$$

We have $T_{ij}^{-1} = i^o(T)_{ij}$, where i is, as usual, the inversion map $x \mapsto x$ on G, therefore

$$X(T_{ij}^{-1}) = X(i^o(T)_{ij}) = di(X)(T_{ij}) = -T_{ij},$$

whence

$$X(u_Y^o T_{ij}) = (X \cdot Y)_{ij} - (Y \cdot X)_{ij}$$

which, in view of (2) proves (1).

3.15 Ker(Ad) *can be larger than* $\mathscr{C}(G)$. Clearly one has $\mathscr{C}(G) \subset \ker(\mathrm{Ad})$. This is even an equality if $\mathrm{char}(k) = 0$ or if G is semi-simple. The following example of Chevalley shows that the inclusion can be proper in general.

Assume $\mathrm{char}(k) = p > 0$ and let $G = \{g(a,b) | a \in K^*, b \in K\}$ where

$$g(a,b) = \begin{bmatrix} a & 0 & 0 \\ 0 & a^p & b \\ 0 & 0 & 1 \end{bmatrix}.$$

Then $g(a,b)g(a',b') = g(aa', a^p b' + b)$, so that G is a closed subgroup of $\mathbf{GL}_3$. It is, as a group, the semi-direct product of the normal subgroup $H = \{g(1,b) | b \in K\} \cong \mathbf{G}_a$ with the group $L = \{g(a,0) | a \in K^*\} \cong \mathbf{GL}_1$, the action of L on H being given by the Frobenius homomorphism. In particular G is not commutative; indeed $\mathscr{C}(G) = \{e\}$. The Lie algebra $\mathfrak{h}$ is spanned by $g(0,1)$. We claim that $\mathfrak{l}$ is spanned by the diagonal matrix with diagonal entries $(1,0,0)$. In fact, if θ is the isomorphism $a \mapsto g(a,0)$ of $\mathbf{GL}_1$ onto H, and X is the standard generator of the Lie algebra of $\mathbf{GL}_1$, then $d\theta(X)(T_{22}) = X(T_{22}^p) = 0$, while obviously $d\theta(X)(T_{11}) = 1$ and $d\theta(X)T_{ij} = 0$ for other values of i,j. From 3.13 we see immediately that $\mathrm{Ad}\, g(a,b) = 1$ if and only if $a = 1$, i.e. $\ker \mathrm{Ad} = H$, which is not central.

We note also that $[\mathfrak{h}, \mathfrak{l}] = 0$, therefore $\mathfrak{g}$ is commutative, although G is not, and in fact $(G,G) = H$. This cannot happen in characteristic zero (see 7.8).

3.16 Some applications. (a) If $a \in G$ then $g \mapsto ag^{-1}a^{-1}$ is the composite of $\mathrm{Int}(a)$ with i, so its differential is $-\mathrm{Ad}(a)$. Multiplying by the identity map we obtain the commutator map $c_a: g \mapsto (g,a) = gag^{-1}a^{-1}$, so $dc_a = Id - \mathrm{Ad}(a)$. Following c_a by right multiplication by a, we get a formula for the differential of $g \mapsto gag^{-1}$, the map of G onto the conjugacy class of a. This differential is $Id - \mathrm{Ad}(a)$, followed by the differential $(d\rho_a)_e: \mathfrak{g} \to T(G)_a$ where $\rho_a(g) = ga$ as usual.

(b) Fix $X \in \mathfrak{g}$. Define $\alpha_X: G \to \mathfrak{g}$ by $\alpha_X(x) = \mathrm{Ad}(x)\cdot X - X (x \in G)$. Then we have

(1) $$(d\alpha_X)_1 = -\mathrm{ad}\, X.$$

Proof. Let $f \in K[\mathfrak{g}]$, and $Y \in \mathfrak{g}$. Then, by definition

$$(d\alpha_X)_1(Y)(f) = Y(\alpha_X^o f).$$

$\alpha_X^o f$ is the function $x \mapsto f(\mathrm{Ad}\, x(X) - X)$. Assume first $f \in \mathfrak{g}^*$. Then $\alpha_X^o f(x) = f(\mathrm{Ad}\, x(X)) - f(X)$, hence, by 3.13, 3.14

$$Y(\alpha_X^o f) = Y(x \mapsto f(\mathrm{Ad}\, x(X))) = (\mathrm{dAd})(Y)(X)f = \mathrm{ad}\, Y(X)(f) = [Y,X]f.$$

This proves that the two sides of (1) are equal on $\mathfrak{g}^*$. Since $K[\mathfrak{g}]$ is the symmetric algebra on $\mathfrak{g}^*$ and both sides of (1) are derivations, (1) follows.

3.17 Proposition. *Let M and N be closed subgroups of an affine algebraic group G, and let H be the closure of the commutator group (M,N). Then $\mathfrak{h}$ contains all elements of the forms*

$$[X,Y] \quad (X \in \mathfrak{m}, Y \in \mathfrak{n}),$$
$$\mathrm{Ad}(m)(Y) - Y \quad (m \in M), \quad \mathrm{Ad}(n)(X) - X \quad (n \in N).$$

Proof. For $m \in M$ define $\alpha_m : N \to H$ by $\alpha_m(n) = mnm^{-1}n^{-1}$. Then by 3.16

$$(d\alpha_m)_1 = (\mathrm{Ad}(m) - Id) : \mathfrak{n} \to \mathfrak{h}.$$

This secures all elements of the second form, and those of the third form are obtained similarly.

If $Y \in \mathfrak{n}$ define $\alpha_Y : M \to \mathfrak{h}$ by $\alpha_Y(m) = \mathrm{Ad}(m)(Y) - Y$. This lies in $\mathfrak{h}$ thanks to the conclusion established above. From (3.16)(b) we have $(d\alpha_Y)_e = -\mathrm{ad}(Y)$, thus securing all elements $[X, Y](X \in \mathfrak{m}, Y \in \mathfrak{n})$ in $\mathfrak{h}$.

Remark. The elements in the proposition span $\mathfrak{h}$ if $\mathrm{char}(k) = 0$, (see §7) but not in general (see (3.15)).

3.18 Corollary

(i) *If* $N \subset \mathrm{Norm}\, M$, *then* $[L(N), L(M)] \subset L(M)$ *i.e.* $L(N) \subset \mathrm{Norm}(L(M))$.
(ii) *If* $N \subset \mathscr{Z}_G(M)$, *then* $[L(N), L(M)] = 0$.
(iii) *If G is solvable, of length m, then $L(G)$ is solvable, of length $\leqq m$.*
(iv) *If G nilpotent with a central series of length m, then $L(G)$ is nilpotent with a central series of length $\leqq m$, in particular $L(G)$ is commutative if G is so.*

We have $H \subset M$ in case (i) and $H = (1)$ in case (ii), this implies the first two assertions. Then (i) and (ii) imply (iii) and (iv) by a trivial induction.

Remark. In characteristic zero, there are converses to the assertions in 3.18 (see 7.8), but not in general in positive characteristic. We already saw in 3.15 a counterexample to the converse to the last assertion in (iv), hence to a converse to (ii). It may also happen that $L(G)$ is solvable, even nilpotent while G is simple, in particular equal to its derived group. Examples are provided by $\mathbf{SL}_2$ and $\mathbf{PSL}_2$ in characteristic two (see 17.5(2)).

3.19 *Convolution.* In a real or complex Lie group, it is often convenient to view translations by elements as convolutions by point measures and the action of a tangent vector as a convolution with a distribution whose support is a point. In this way, operations by group elements and Lie algebra elements are all special cases of convolutions by distributions, and one can avail oneself of a rather efficient formalism. In this section, we outline an analogue in our context of this point of view. We keep the notation $\alpha = G \to G'$ and $\alpha^o : A' \to A$. If V is a vector space over K we shall write

$$A(V) = \mathrm{Hom}_{K-\mathrm{mod}}(A, V),$$

If W is another vector space, we define a K-bilinear pairing

$$(X, Y) \mapsto X \cdot Y,$$

$$A(V) \times A(W) \to A\left(V \bigotimes_k W\right)$$

where

$$X \cdot Y = (X \otimes Y) \circ \mu^0.$$

Example. *If* $g, h \in G$ *then*

$$e_g \circ e_h = e_{gh} \quad (\text{in } A(K)).$$

More generally, if B is any K-algebra, then $G(B)$, the group of points of G in B, corresponds (under the map $g \mapsto e_g$) to $\operatorname{Hom}_{K\text{-alg}}(A, B) \subset A(B)$, and the above formula becomes

$$e_{gh} = m \cdot (e_g \cdot e_h)$$

where $m: B \otimes_K B \to B \otimes_B B = B$ is the canonical map. We shall freely identify $K \underset{K}{\otimes} V$ and $V \underset{K}{\otimes} K$ with V.

Lemma 1. *Let* U, V, *and* W *be vector spaces, and let* $X \in A(U)$, $Y \in A(V)$, *and* $Z \in A(W)$.

(a) $e \cdot X = X = X \cdot e$.
(b) $(X \cdot Y) \cdot Z = X \cdot (Y \cdot Z)$.
(c) $A(K)$ *is an associative* K*-algebra with identity* e, *and* $A(V)$ *is an* $A(K)$*-bimodule. The map* $g \mapsto e_g$ *is a monomorphism from* G *to the group of invertible elements of* $A(K)$.
(d) *We have*

$$(X \cdot Y) \circ \alpha^o = (X \circ \alpha^o) \cdot (Y \circ \alpha^o),$$

where the product on the right is defined with respect to $\mu'^o: A' \to A' \underset{K}{\otimes} A'$. *In particular* $\circ\, \alpha^o: A(K) \to A'(K)$ *is an algebra homomorphism inducing* $\alpha: G \to G'$ *via the embedding defined in* (c).
(e) *If* U *and* V *have* k*-structures such that* X *and* Y *are defined over* k, *then* $X \cdot Y$ *is defined over* k.

Proof. Let $f \in A$ and write $\mu^o f = \sum_i f_i \otimes h_i$. Then

$$f(x) = f(ex) = \sum_i f_i(e) h_i(x) = f(xe) = \sum_i f_i(x) h_i(e).$$

Hence

$$f = \sum_i f_i(e) h_i = \sum_i f_i h_i(e).$$

(a) We have, since X is K-linear,

$$(e \cdot X)(f) = (e \otimes X)\mu^o f = \sum f_i(e) X(h_i) = X\left(\sum f_i(e) h_i\right) = X(f).$$

Thus $e \cdot X = X$, and similarly $X \cdot e = X$.

(b) With $I:A \to A$ standing for the identity map, the associativity of μ is expressed by

$$(I \otimes \mu^o) \circ \mu^o = (\mu^o \otimes I) \circ \mu^o$$

(or $I \cdot \mu^o = \mu^o \cdot I$ in the present notation). Now

$$(X \otimes Y \otimes Z) \circ (I \otimes \mu^o) \circ \mu^o = (X \otimes ((Y \otimes Z) \circ \mu^o)) \circ \mu^o = X \cdot (Y \cdot Z)$$

and, similarly,

$$(X \otimes Y \otimes Z) \circ (\mu^o \otimes I) \circ \mu^o = (X \cdot Y) \cdot Z.$$

Part (c) is an immediate consequence of parts (a) and (b), together with the example above.

(d) The fact that $\circ\alpha^o$ is a homomorphism is expressed by the equation

$$\mu^o \circ \alpha^o = (\alpha^o \otimes \alpha^o) \circ \mu'^o.$$

Now

$$(X \cdot Y) \circ \alpha^o = (X \otimes Y) \circ \mu^o \circ \alpha^o = (X \otimes Y) \circ (\alpha^o \otimes \alpha^o) \circ \mu'^o = (X \circ \alpha^o) \cdot (Y \circ \alpha^o).$$

The remaining assertions of (d) are clear.

(e) follows from the formula $X \cdot Y = (X \otimes Y) \circ \mu^o$ and the fact that μ^o is defined over k. This completes the proof of Lemma 1.

As above let $I \in A(A)$ denote the identity map. If $X \in A(K)$, we define *right convolution by* X,

$$*X = I \cdot X : A \to A$$

and *left convolution by* X,

$$X* = X \cdot I : A \to A.$$

If $f \in A$ and $\mu^o f = \sum f_i \otimes h_i$ then

$$f * X = \sum f_i X(h_i), \quad X * f = \sum X(f_i) h_i.$$

We have

$$(f * X)(g) = X(\lambda_{g^{-1}} f) \quad \text{and} \quad (X * f)(g) = X(\rho_g f).$$

The first equation follows because

$$(f * X)(g) = \sum f_i(g) X(h_i) = X(\sum f_i(g) h_i) = X(\lambda_{g^{-1}} f),$$

and similarly for the second. These are the formulas of 3.7, but in a more general situation.

We shall now establish several identities for these operation. Let $g \in G$, let $X, Y \in A(K)$, and let $\mu^o f$ be as above.

(1) $$*e_g = \rho_g, \quad \text{and} \quad e_g * = \lambda_{g^{-1}}.$$

For $f * e_g = \sum f_i h_i(g) = \rho_g f$, and similarly for $e_g *$.

(2) $$X \circ (*Y) = X \cdot Y \quad \text{and} \quad X \circ (Y*) = Y \cdot X.$$

We have $X\circ(I\otimes Y)\circ\mu^o=(X\otimes Y)\circ\mu^o$, and similarly $X\circ(Y\otimes I)\circ\mu^o=(Y\otimes X)\circ\mu^o$. Using part (a) of Lemma 1 we thus obtain:

$$e\circ(*X)=X=e\circ(X*). \tag{3}$$

Combining (2) and (3) we find that

$$e\circ((*X)\circ(*Y))=X\cdot Y. \tag{4}$$

A simple check shows that

$$(I\otimes X)\circ\mu^o\circ(I\otimes Y)=((I\otimes X)\circ\mu^o)\otimes Y,$$

so that $(*X)\circ(*Y)=(*X)\cdot Y$. The latter is $(I\cdot X)\cdot Y$ which, by part (b) of Lemma 1, equals $I\cdot(X\cdot Y)$. Thus,

$$(*X)\circ(*Y)=*(X\cdot Y). \tag{5}$$

Similar computations, starting from

$$(I\otimes X)\circ\mu^o\circ(Y\otimes I)=Y\otimes((I\otimes X)\circ\mu^o)$$
$$(Y\otimes I)\circ\mu^o\circ(I\otimes X)=((Y\otimes I)\circ\mu^o)\otimes X$$

show that

$$(*X)\circ(Y*)=Y\cdot(*X)=Y\cdot(I\cdot X), \quad \text{and} \quad (Y*)\circ(*X)=(Y\cdot I)\cdot X,$$

whence

$$(*X)\circ(Y*)=(Y*)\circ(*X) \tag{6}$$

Lemma 2. *The composite $X\mapsto e\circ(I\cdot X)=e\circ(*X)$ of the K-linear maps*

$$A(K)\xrightarrow{I\cdot}A(A)\xrightarrow{e\circ}A(K)$$

is the identity. Moreover $I\cdot$ is a K-algebra monomorphism onto the K-algebra of elements in $A(A)$ commuting with all left translations $\lambda_g(g\in G)$. In particular $I\cdot$ maps $T(G)_1$ isomorphically onto Lie(G). *Finally, both $e\circ$ and $I\cdot$ preserve the property that an element is defined over k.*

Proof. The first assertion is just (3), and it implies that $I\cdot$ is an isomorphism onto its image, whose inverse is induced by $e\circ$. From (4) we see moreover that $e\circ$ is an algebra homomorphism. Formula (6) says that left and right convolutions commute. Hence all $*X$ (i.e. $m(I\cdot)$) commute with all $\lambda_{g^{-1}}=e_g{}^*$ (see (1)). To show that $I\cdot$ maps $A(K)$ *onto* the set of left invariant elements of $A(A)$ it suffices to show that $e\circ$ is injective on left invariant elements. So let $D\in A(A)$ be left invariant and suppose $e\circ D=0$. Then $(Df)(g)=(\lambda_{g^{-1}}(Df))(e)=(D(\lambda_{g^{-1}}f))(e)=(e\circ D)(\lambda_{g^{-1}}f)=0$ for all g, so $Df=0$ for all f.

We know that e carries derivations to derivations so it remains to check that $I\cdot$does so also. If $X\in A(K)$ is a derivation $A\to K(e)$ then $I\otimes X: A\bigotimes_K A\to A\bigotimes_K K(e)$ is the derivation obtained by the base change $K\to A$, so $*X=(I\otimes X)\circ\mu^o$ is also a derivation because μ^o is an algebra homomorphism.

Finally it is clear that $e\circ$ preserves elements defined over k (because e is defined over k). That $I\cdot$ does also follow from part (e) of Lemma 1.

This completes the proof of Lemma 2.

Lemma 3. *$L(G)$ is a restricted Lie subalgebra of $A(K)$. That is, if $X, Y\in T(G)_1$, we have*

$$[X, Y] = X\cdot Y - Y\cdot X, X^{[p]} = X\cdot X\cdot \ldots \cdot X \ (p \text{ factors}).$$

Since we have seen, in part (d) of Lemma 1, that $\alpha^o: A' \to A$ induces an algebra homomorphism $\circ\alpha^o: A(K) \to A'(K)$, it follows that its restriction, $L(\alpha): T(G)_1 \to T(G')_1$ is a restricted Lie algebra homomorphism. This provides another approach to 3.4.

3.20 *The tangent bundle as a split extension of G by* $\mathfrak{g}$. Recall from AG 16.2 that we have

$$\begin{array}{ccc} T(G) & = & G(K[\delta]) \\ p\downarrow\uparrow s & & p\downarrow\uparrow s \\ G & = & G(K) \end{array}$$

where $K[\delta]$ is the algebra of dual numbers ($\delta^2 = 0$) and p and s are induced respectively by the homomorphism $K[\delta] \to K$ sending δ to 0 and by the inclusion of K in $K[\delta]$. A typical element of $T(G)$ is of the form

$$e_g^{\delta X} = e_g + \delta X \quad (g\in G; X\in T(G)_g)$$

(AG.16.2). It is the algebra homomorphism $K[G] \to K[\delta]$ sending f to $f(g) + \delta X(f)$. According to 3.4 the group multiplication in $T(G)$ is given by

$$(e_g + \delta X)(e_h + \delta Y) = m\circ((e_g + \delta X)\otimes(e_h + \delta Y))\circ\mu^o$$

where $m: K[\delta]\otimes K[\delta] \to K[\delta]$ is the multiplication in the k-algebra $K[\delta]$. Thus, with the notation $X\cdot Y = (X\otimes Y)\circ\mu^o$ introduced in 3.19, we have

$$\begin{aligned}(e_g + \delta X)(e_h + \delta Y) &= m\cdot(e_g\cdot e_h + (1\otimes\delta)e_g\cdot Y + (\delta\otimes 1)X\cdot e_h + (\delta\otimes\delta)X\cdot Y) \\ &= e_{gh} + \delta(e_g\cdot Y + X\cdot e_h),\end{aligned}$$

or

$$(1) \qquad e_g^{\delta X} e_h^{\delta Y} = e_{gh}^{\delta(e_g\cdot Y + X\cdot e_h)}.$$

The map p sends $e_g^{\delta X}$ to g and s sends g to $e_g (= e_g^{\delta 0})$. Since the composite $p\circ s$ is the identity on G it follows that the group $T(G)$ is the semi-direct product of sG with $\ker(p) = p^{-1}(e)$. Writing $e^{\delta X}$ in place of $e_e^{\delta X}$, we see from (AG.16.2) that $X \mapsto e^{\delta X}$ is a bijection from $T(G)_e = \mathfrak{g}$ to $\ker(p)$. Moreover (1) and Lemma 1 of 3.19 imply that it is a homomorphism of groups:

$$e^{\delta X} e^{\delta Y} = e^{\delta(X+Y)}$$

Thus we have a split group extension

$$0 \to \mathfrak{g} \xrightarrow{e^{\delta(\cdot)}} T(G) \xrightarrow{p} G \to 1$$

with everything defined over k.

If $\mathrm{Int}: G \times G \to G$ is the action of G on G by inner automorphisms, then the commutative diagram

$$\begin{array}{ccc} G \times G & \xrightarrow{\mathrm{Int}} & G \\ \downarrow & & \downarrow \\ T(G) \times T(G) & \xrightarrow{T(\mathrm{Int})} & T(G) \end{array}$$

shows that $T(\mathrm{Int}(g)) = \mathrm{Int}(e_g)$. The restriction of $T(\mathrm{Int}(g)): T(G) \to T(G)$ to $\mathfrak{g} = \ker(p)$ is (see AG.16.2) just $d(\mathrm{Int}(g)) = \mathrm{Ad}(g)$. Explicitly, this says that $\mathrm{Ad}(g)$ is defined by the formula

(2) $$\mathrm{Int}(e_g)(e^{\delta X}) = e^{\delta\,\mathrm{Ad}(g)X}.$$

Viewing G and $\mathfrak{g}$ as subgroups of $T(G)$, both defined over k, we see that Ad is just the action $T(\mathrm{Int}): G \times \mathfrak{g} \to \mathfrak{g}$.

3.21 *Some differentiation formulas.* As an application of this formalism, we now derive a few more differentiation formulas.

(a) Let $\mathscr{V}$ denote the category of finite dimensional K-modules, and let $F: \mathscr{V} \times \cdots \times \mathscr{V} \to \mathscr{V}$ be a functor of n variables which is K-multilinear on the Hom's. Let $\alpha_i: G \to GL(V_i)$ be rational representations of an algebraic group $G (1 \leqq i \leqq n)$. Then $\alpha = F(\alpha_1, \ldots, \alpha_n): G \to \mathrm{GL}(V)$ is a rational representation on $V = F(V_1, \ldots, V_n)$. Moreover we have

(1) $$d\alpha(X) = \sum_{i=1}^{n} F(1_{V_1}, \ldots, d\alpha_i X, \ldots, 1_{V_n}), \quad (X \in \mathfrak{g}).$$

This follows from $\alpha(e^{\delta H}) = e^{\delta d\alpha(X)}$, and

$$\begin{aligned} \alpha(e^{\delta X}) &= F(\alpha_1(e^{\delta X}), \ldots, \alpha_n(e^{\delta X})) = F(e^{\delta d\alpha_1 X}, \ldots, e^{\delta d\alpha_n X}) \\ &= F(1_{V_1} + \delta d\alpha_1 X, \ldots, 1_{V_n} + \delta d\alpha_n X) \\ &= F(1_{V_1}, \ldots, 1_{V_n}) + \delta\left(\sum_{i=1}^{n} F(1_{V_1}, \ldots, d\alpha_i X, \ldots, 1_{V_n})\right). \end{aligned}$$

(b) If $\alpha_i: G \to \mathrm{GL}(V_i) (i = 1, 2)$ are rational representations and if $\beta: V_1 \to V_2$ is a homomorphism of G-representations, i.e. if β is linear and $\beta(\alpha_1(g)v) = \alpha_2(g)\beta(v)$ for $g \in G$ and $v \in V$, then β is also a homomorphism of $\mathfrak{g}$-representations, i.e.

$$\beta(d\alpha_1(X)v) = d\alpha_2(X)\beta(v)$$

for $X \in \mathfrak{g}$ and $v \in V$. For the first formula, applied to the tangent bundles, gives

$$\beta(\alpha_1(e^{\delta X})v) = \alpha_2(e^{\delta X})\beta(v).$$

The left side is $\beta(e^{\delta d\alpha_1 X}(v)) = \beta(v) + \delta(d\alpha_1 X(v))$, while the right side is $\beta(v) + \delta d\alpha_2 X(\beta(v))$, thus establishing the formula above.

(c) In the setting of (b) we can apply (1) to $\alpha = \alpha_1 \otimes \alpha_2 : G \to \mathrm{GL}\left(V_1 \bigotimes_K V_2\right)$, to obtain

$$d(\alpha_1 \otimes \alpha_2)(X) = (d\alpha_1 X \otimes 1_{V_2}) + (1_{V_1} \otimes d\alpha_2 X).$$

(d) Let $\alpha: G \to GL(V)$ be a rational representation. Then we have $T^n(\alpha): G \to GL(T^n(V))$, where $T^n(V) = V \bigotimes_K \cdots \bigotimes_K V$ (n factors) and $T^n(\alpha) = \alpha \otimes \cdots \otimes \alpha$. From (1) we have

$$d(\alpha \otimes \cdots \otimes \alpha) = \sum_{i=1}^{n} 1_V \otimes \cdots \otimes d\alpha \otimes \cdots \otimes 1_V.$$

Thus, on the tensor algebra $\coprod_n T^n(V)$, we see that the differential of the action of G as algebra automorphisms (extending α in degree 1) is the action of $\mathfrak{g}$ as derivations (extending $d\alpha$ in degree 1).

The passage from the tensor algebra to the symmetric algebra $S(V)$ and the exterior algebra $\Lambda(V)$ can be viewed as epimorphisms of G-representations in each degree. Thus it follows from (b) that the differentials of the actions of G as algebra automorphisms of $S(V)$ and of $\Lambda(V)$ are again given by the actions of $\mathfrak{g}$ as derivations extending $d\alpha$ in degree 1. Explicitly, if $e_1, \ldots, e_n \in V$ then

$$dS^n(\alpha)(X)(e_1 \cdots e_n) = \sum_{i=1}^{n} e_1 \cdots d\alpha X(e_i) \cdots e_n$$

and

$$d\Lambda^n(\alpha)(X)(e_1 \wedge \cdots \wedge e_n) = \sum_{i=1}^{n} e_1 \wedge \cdots \wedge d\alpha X(e_i) \wedge \cdots \wedge e_n.$$

(e) If $\dim V = n$ then $\Lambda^n(\alpha) = \det \circ \alpha$ and it follows from the above remarks that

$$d(\det \circ \alpha) = \mathrm{Tr} \circ d\alpha,$$

which generalizes slightly 3.12(ii).

(f) Let $\alpha: G \to GL(V)$ be a rational representation, and suppose V is a not necessarily associative algebra. Then if G acts via α as algebra automorphisms of V it follows that $\mathfrak{g}$ acts, via $d\alpha$, as derivations. This follows by expanding the formula $\alpha(e^{\delta X})(uv) = \alpha(e^{\delta X})(u)\alpha(e^{\delta X})(v)$ for $X \in \mathfrak{g}$ and $u, v \in V$.

3.22. For the reader familiar with real or complex Lie groups, we relate the algebraically defined operations $*X$ and $*e_x$ with the usual convolution. If

$X \in T(G)_1$, then as usual $t \mapsto e^{tX}$ denotes the one-parameter subgroup spanned by X. For a smooth function f on G, we have then

$$Xf(x) = \frac{d}{dt}(f(x \cdot e^{tX}))|_{t=0}.$$

Given a distribution S on G, let us denote symbolically by $\int f(x)dS_x$ its value $S(f)$ on a test function $f \in C_c^\infty(G)$. If S and T are distributions, one with compact support, then their convolution $S * T$ is a distribution whose value on a test function f is given by

$$(S * T)(f) = \iint f(x \cdot y) dS_x \cdot dT_y.$$

We now identify elements of $\mathfrak{g}$ (and more generally of the universal enveloping algebra $U(\mathfrak{g})$ of $\mathfrak{g}$) to distributions supported by the identity. Then

$$(f * X)(x) = \frac{d}{dt} f(x \cdot e^{-tX})|_{t=0} \quad (X * f)(x) = \frac{d}{dt} f(e^{-tX} \cdot x)|_{t=0} \quad (x \in G; f \in C^\infty(G))$$

Moreover $e_x *$ and $* e_x$ (cf. 3.19) are now defined as convolutions with the Dirac measure at x.

§4. Jordan Decomposition

4.1 *Nilpotent, unipotent and semi-simple endomorphisms.* Let V be a finite dimensional vector space over K with a k-rational structure $V(k)$. Then $E = \mathrm{End}_K(V)$ also has a k-structure given by $E(k) = \mathrm{End}_k(V(k))$. An $a \in E$ is called *nilpotent* if $a^n = 0$ for some $n > 0$, and *unipotent* if $a - I$ is nilpotent, where I denotes the identity on V. Thus a is nilpotent (resp., unipotent) if and only if all eigenvalues of a are 0 (resp. 1).

(a) *If* $\mathrm{char}(k) = p > 0$ *then a is unipotent if and only if* $a^{p^r} = I$ *for some* $r \geqq 0$.

For if $a = I + n$, with n nilpotent, then $a^{p^r} = I + n^{p^r} = I$ for sufficiently large r. Conversely, $a^{p^r} = I$ implies the minimal polynomial of a divides $T^{p^r} - 1 = (T-1)^{p^r}$, so all eigenvalues of a are 1.

(b) *Let $a \in E(k)$. We call a semi-simple if it satisfies the following conditions, which are equivalent*:

(i) *$V(\bar{k})$ is spanned by eigenvectors of a; i.e. a is diagonalizable over $\bar{k}$.*
(ii) *The algebra $\bar{k}[a] \subset E(\bar{k})$ is semi-simple, i.e. it is a product of copies of $\bar{k}$.*

That (i)$\Rightarrow$(ii) is obvious once a is put in diagonal form. (Alternatively, $\bar{k}[a] \cong \bar{k}[T]/(P(T))$, where $P(T)$, the minimal polynomial of a, is a product of *distinct* linear factors.) Conversely, if $\bar{k}[a]$ is a product of copies of $\bar{k}$, then any module over it, e.g. $V(\bar{k})$, is a direct sum of one dimensional submodules.

(c) *If $a \in E(k)$ is semi-simple, then the eigenvalues of a are separable over k. Hence a is diagonalizable over k_s.*

For $k[a] \cong k[T]/(P(T))$, P the minimal polynomial of a. Since $a \in E(k)$ we have $\bar{k}[a] = \bar{k} \bigotimes_K k[a]$, and the absence of nilpotent elements in the latter implies that P has no multiple roots, i.e. that P is a separable polynomial.

(d) *Suppose $a, b \in E$ commute. Then:*

(i) *a, b nilpotent $\Rightarrow a + b$ is nilpotent.*
(ii) *a, b unipotent $\Rightarrow ab$ is unipotent.*
(iii) *a, b semi-simple $\Rightarrow ab$ and $a + b$ are semi-simple.*

If $a^n = b^m = 0$ then $(a + b)^{n+m} = 0$, thus proving (i). Since $ab - I = (a - I)b + (b - I)$, (ii) follows from (i). Part (iii) is left as an exercise (cf. 4.6).

Finally, we record the obvious remark:

(e) *If a is both semi-simple and nilpotent (resp. unipotent) then $a = 0$ (resp. $a = I$).*

4.2 Proposition. *We keep the previous notation. Let $a \in E$.*

(1) *There exist unique a_s and a_n in E such that a_s is semi-simple, a_n is nilpotent, $a_s a_n = a_n a_s$, and such that $a = a_s + a_n$. We call this the (additive) Jordan decomposition of a.*
(2) *There are polynomials $P(T)$ and $Q(T)$ in $K[T]$, with zero constant term, such that $a_s = P(a)$ and $a_n = Q(a)$.*
(3) *The centralizer of a in E centralizes a_s and a_n. If $A \subset B \subset V$ are subspaces such that $aB \subset A$, then $a_s B \subset A$ and $a_n B \subset A$.*
(4) *If $A \subset V$ is a subspace invariant under a then the Jordan decomposition of a induces those of $a|A$ and of $a_{V/A}$, the endomorphism a induces on V/A.*
(5) *If $a \in E(k)$ then $a_s, a_n \in E(k^{p^{-\infty}})$. Moreover, the polynomials P and Q in (2) can be chosen in $k^{p^{-\infty}}[T]$.*

Proof. (1) Write $\det(T - a) = \prod (T - \alpha_i)^{m_i}$ where the α_i are distinct, and put $V_i = \ker(a - \alpha_i I)^{m_i}$. Then it is easy to see that $V = \coprod V_i$. Suppose $a = b + c$ with b semi-simple, c nilpotent, and $bc = cb$. Then b commutes with a, hence with $(a - \alpha_i I)^{m_i}$, and so b leaves each V_i invariant. Since $a - b = c$ is nilpotent, a and b have the same eigenvalues on V_i. Since $a|V_i$ has only one, α_i, and since b is semi-simple, it follows that $b|V_i = \alpha_i I|V_i$. Therefore b is uniquely determined, and so also is $c = a - b$. On the other hand, if we define a_s by $a_s|V_i = \alpha_i I|V_i$, and $a_n = a - a_s$, then these data clearly satisfy our requirements.

(2) Choose $P(T)$ to solve the congruences

$$P(T) \equiv \alpha_i \bmod (T - \alpha_i)^{m_i} \quad \text{and} \quad P(T) \equiv 0 \bmod (T).$$

These are consistent in case some $\alpha_i = 0$, so there is a solution ("Chinese Remainder Theorem"). We take $Q(T) = T - P(T)$.

(3) is an immediate corollary of (2).

(4) Let a' and a'' denote the endomorphisms induced by a on A and V/A, respectively. Part (3) implies that a_s and a_n leave A invariant, so we can similarly

define a'_s, a''_s and a'_n, a''_n. The fact that $a' = a'_s + a'_n$ and $a'' = a''_s + a''_n$ are Jordan decompositions is obvious.

(5) If $a \in E(k)$ then each of the α_i above are in $\bar{k}$, so the construction in (1) shows that $a_s, a_n \in E(\bar{k})$. If $s \in \mathrm{Gal}(\bar{k}/k)$ then s operates on $E(\bar{k})$, and we have $a = s(a) = s(a_s) + s(a_n)$. Since s acts as an algebra automorphism we see that $s(a_n)$ is nilpotent and commutes with $s(a_s)$. Moreover, since $\bar{k}[a_s]$ is a semi-simple algebra (see 4.1) the same is true of $s(\bar{k}[a_s]) = \bar{k}[s(a_s)]$, so a_s is still semi-simple. Therefore the uniqueness of the Jordan decomposition implies $s(a_s) = a_s$ and $s(a_n) = a_n$. But the elements of $E(\bar{k})$ fixed by $\mathrm{Gal}(\bar{k}/k)$ are just $E(k^{p^{-\infty}})$. In particular $a_s, a_n \in \bar{k}[a] \cap E(k^{p^{-\infty}}) = k^{p^{-\infty}}[a]$, so the P and Q in (2) can be chosen with coefficients in $k^{p^{-\infty}}$

Corollary 1. *Let $g \in \mathrm{GL}(V)$, and put $g_u = I + g_s^{-1} g_n$.*

(1) *We have $g = g_s g_u = g_u g_s$ with g_s semi-simple and g_u unipotent, and this is the unique factorization of g of this type. (It is called the multiplicative Jordan decomposition of g.)*
(2) *If $A \subset V$ is a subspace invariant under g, then it is invariant under g_s and g_u, and the Jordan decomposition of g induces those of the automorphisms induced by g on A and on V/A.*
(3) *If g is rational over k, then g_s and g_u are rational over $k^{p^{-\infty}}$*

Proof. (1) Since g_s and g_n commute, $g_u = I + g_s^{-1} g_n$ is unipotent, and $g = g_s g_u = g_u g_s$. Suppose $g = bc = cb$ where b is semi-simple and $n = c - I$ is nilpotent. Then bn is nilpotent and commutes with b, so $g = b + bn$ is the additive Jordan decomposition of g. Hence $b = g_s$ and $bn = g_n$.

In view of the formula for g_u parts (2) and (3) follow immediately from parts (4) and (5), respectively, of the proposition.

Corollary 2. *If $a, b \in E$ commute then $a + b = (a_s + b_s) + (a_n + b_n)$ is the additive Jordan decomposition of $a + b$. If, moreover, they are invertible, then $ab = (a_s b_s)(a_u b_u)$ is the multiplicative Jordan decomposition. All elements appearing above commute.*

Proof. This follows from 4.1(d), 4.2(1) and Cor. 1.

Corollary 3. *If $g \in GL(V)$ and $h \in GL(W)$ then $g \otimes h = (g_s \otimes h_s)(g_u \otimes h_u)$ is the Jordan decomposition of $g \otimes h$.*

Proof. Apply Corollary 2 to $g \otimes 1_W$ and $1_V \otimes h$.

Convention. Suppose that V is not necessarily finite dimensional. We shall say that $a \in E$ is "locally finite" if V is spanned by finite dimensional subspaces stable under a. In this case one says that a is *locally nilpotent* (resp., *unipotent*, resp., *semi-simple*) if its restriction to each finite dimensional a-stable subspace has this property. The uniqueness of the above Jordan decompositions gives us, for a locally finite endomorphism a, a Jordan decomposition $a = a_s + a_n$, and, if a is invertible, $a = a_s a_u$, such that these induce the usual ones on finite dimensional a-stable subspaces. Thus, for example, a_s is locally semi-simple, a_n is locally nilpotent, and $a_s a_n = a_n a_s$. These properties characterize a_s and a_n, and similarly for the multiplicative decomposition.

By abuse of language we shall often drop the word "locally" in the above situation.

Let G be an affine algebraic group. If $g \in G$ and $X \in \mathfrak{g}$ then ρ_g and $*X$ are locally finite endomorphisms of $A = K[G]$ (see 1.9 and 3.11). Hence we have Jordan decompositions

$$\rho_g = (\rho_g)_s(\rho_g)_u$$

and

$$*X = (*X)_s + (*X)_n.$$

The main result of this section asserts that these decompositions can be realized already in G and in $\mathfrak{g}$, respectively.

4.3 We keep the notation and conventions of 4.2.

Proposition. *Let $g \in \mathrm{GL}(V)$ and $X \in \mathfrak{gl}(V)$, and let $A = K[\mathrm{GL}(V)]$.*

(1) *g is semi-simple (resp., unipotent) if and only ρ_g is semi-simple (resp., unipotent).*
(2) *X is semi-simple (resp., nilpotent) if and only if $*X$ is semi-simple (resp., nilpotent).*

Proof. We have $A = B[D^{-1}]$ where $B = K[\mathrm{End}(V)]$, and where $D:\mathrm{End}(V) \to K$ is the determinant. Since right translation by $GL(V)$ is defined on $\mathrm{End}(V)$, and since $*X$ is its differential (see 3.11) it follows that ρ_g and $*X$ leave B invariant, and their extensions to A are defined, for $f \in B$, by

$$\rho_g(fD^{-n}) = \rho_g(f)\rho_g(D)^{-n} = D(g)^{-n}\rho_g(f)D^{-n}$$

and

$$\begin{aligned}(fD^{-n})*X &= (f*X)D^{-n} - nfD^{-n-1}(D*X)\\ &= (f*X)D^{-n} - nTr(X)fD^{-n}.\end{aligned}$$

Here we have used the fact that $\rho_g(D) = D(g)D$ and $(D*X) = (XD)D = \mathrm{Tr}(X)D$ (see 3.12). These formulas show that, if f is an eigenvector for ρ_g (resp., $*X$) then so also is fD^{-n} for each $n \geqq 0$. This proves that ρ_g (resp., $*X$) is semi-simple if and only if its restriction to B is.

Suppose ρ_g on B is unipotent. Then since $\rho_g D = D(g)D$, we have $D(g) = 1$. Hence $(\rho_g - I)(fD^{-n}) = ((\rho_g - I)(f))D^{-n}$, and it follows that ρ_g is unipotent on A. The converse is obvious.

Similarly, if $*X$ on B is nilpotent then $(D*X) = \mathrm{Tr}(X)D$ implies $\mathrm{Tr}(X) = 0$, and hence $(fD^{-n})*X = (f*X)D^{-n}$. This proves $*X$ is nilpotent on A, and the converse is obvious.

These remarks show that it suffices to prove the analogue of the proposition with $B = K[\mathrm{End}(V)]$ in place of $A = K[\mathrm{GL}(V)]$. The algebra B is the symmetric algebra, $S(E^*)$, on the dual $E^* = \mathrm{Hom}_K(E, K)$ of $E = \mathrm{End}\, V$. Moreover, ρ_g and $*X$ are just the automorphism and derivation, respectively, of the algebra $S(E^*)$ induced by (the transposes of) right multiplication on E by $g \in GL(V)$ and by $X \in \mathfrak{gl}(V) = E$, respectively.

If we identify End V with $V^* \otimes V (f \otimes v : x \mapsto f(x)v)$, then right multiplication by $a \in E$ corresponds to $a^* \otimes I$, where a^* is the transpose of a. It suffices to check this for a of the form $g \otimes w$, in which case

$$(f \otimes v)(g \otimes w) : x \mapsto f(w)g(x)v = (f(w)g \otimes v)(x),$$

and $(a^* \otimes I)(f \otimes v) = a^*(f) \otimes v = f(w)g \otimes v$. Since a is nilpotent (resp., unipotent, resp., semi-simple) if and only if $a^* \otimes I$ is, the proof of the proposition is completed by the next lemma, in which we let E^* play the role of V.

Lemma. *Let $g \in GL(V)$ and $X \in \mathfrak{gl}(V)$.*

(1) *g is semi-simple (resp., unipotent) if and only if the automorphism $S(g)$ of $S(V)$ induced by g is semi-simple (resp., unipotent).*
(2) *X is semi-simple (resp., nilpotent) if and only if the derivation $s(X)$ of $S(V)$ induced by X is semi-simple (resp., nilpotent).*

Proof. On $S^1(V) = V$ the restrictions of $S(g)$ and $s(X)$ are g and X, respectively, so the "if's" are clear.

Since $S^n(g)$ is induced by $T^n(g)$ on $T^n(V) = V \otimes \cdots \otimes V$ by passing to the quotient, the "only if" in part (1) follows from corollary 3 in 4.2.

Similarly, $s(X)$ is induced, on passing to the quotient $S^n(V)$ of $T^n(V)$, by

$$\sum_i I \otimes \cdots \otimes X \otimes \cdots \otimes I;$$

(where X is in the ith place in the ith summand). These summands commute, and are semi-simple (resp., nilpotent) if X is, so the same is true of their sum.

4.4 *Jordan decomposition in affine groups.* Let G be an affine k-group with coordinate ring $A = K[G]$. If $g \in G$ and $X \in \mathfrak{g}$ then ρ_g and $*X$ on A have Jordan decompositions in the sense of the convention of 4.2.

Theorem. *Let $g \in G$ and $X \in \mathfrak{g}$.*

(1) *There is a unique factorization $g = g_s g_u$ in G such that $\rho_g = \rho_{g_s}\rho_{g_u}$ is the (multiplicative) Jordan decomposition of ρ_g. If $g \in G(k)$ then $g_s, g_u \in G(k^{p^{-\infty}})$.*

(2) *There is a unique decomposition* $X = X_s + X_n$ *in* $\mathfrak{g}$ *such that* $*X = (*X_s) + (*X_n)$ *is the (additive) Jordan decomposition of* $*X$. *If* $X \in \mathfrak{g}(k)$ *then* $X_s, X_n \in \mathfrak{g}(k^{p^{-\infty}})$.

(We refer to the above as the *Jordan decompositions* of g in G and of X in $\mathfrak{g}$, respectively.)

(3) *In case* $G = GL(V)$, *and so* $\mathfrak{g} = \mathfrak{gl}(V)$, *the Jordan decompositions above coincide with those defined in* 4.2.

(4) *If* $\alpha: G \to G'$ *is a morphism of affine groups then* α *and* $d\alpha$ *preserve Jordan decompositions in the groups and Lie algebras, respectively.*

Proof. *Case* 1. $G = GL(V)$ and $\mathfrak{g} = \mathfrak{gl}(V)$. Let $g = g_s g_u$ and $X = X_s + X_n$ be the Jordan decompositions of 4.2. Then Proposition 4.3(1) implies ρ_{g_s} is semi-simple and ρ_{g_u} is unipotent. Since $g \mapsto \rho_g$ is a group homomorphism, ρ_{g_s} and ρ_{g_u} commute, so $\rho_g = \rho_{g_s}\rho_{g_u}$ is the Jordan decomposition. Similarly, 4.3(2) implies $*X_s$ is semi-simple and $*X_n$ is nilpotent. Since $X \mapsto *X$ is a Lie algebra homomorphism, $*X_s$ and $*X_n$ commute, so $*X = (*X_s) + (*X_n)$ is the Jordan decomposition of $*X$. Both $g \mapsto \rho_g$ and $X \mapsto *X$ are compatible with k-structures, so the rationality assertions follow from those of 4.2.

The uniqueness in this, as in the general, case follows from the faithfulness of $g \mapsto \rho_g$ (see 1.10) and of $X \mapsto *X$ (see 3.4, 3.7).

General case. Choose a k-rational embedding $G \subset \mathrm{GL}(V)$ for some V (see 1.10), so that $\mathfrak{g} \subset \mathfrak{gl}(V)$. Then ρ_g and $*X$ are induced, on passing to the quotient A of $B = K[GL(V)]$, by the corresponding actions on B. Hence we have $g = g_s g_u$ in $GL(V)$ and $X = X_s + X_n$ in $\mathfrak{gl}(V)$, from case 1, and if we show that $g_s, g_u \in G$ and $X_s, X_n \in \mathfrak{g}$, then they will give the required decompositions of g and X. Moreover the uniqueness and rationality properties will follow just as in case 1.

Let J be the ideal in B defining G. According to 3.8, Corollary

$$G = \{g \in GL(V) \mid \rho_g J = J\}$$

and

$$\mathfrak{g} = \{X \in \mathfrak{gl}(V) \mid J * X \subset J\}.$$

But 4.2 implies that for $g \in G$ and $X \in \mathfrak{g}$, J is invariant under $(\rho_g)_s, (\rho_g)_u, (*X)_s$, and $(*X)_n$, and case 1 implies these are $\rho_{g_s}, \rho_{g_u}, *X_s$, and $*X_n$, respectively. This completes the proof of (1), (2), and (3).

Proof of (4). By factoring α through $\alpha(G)$ it suffices to treat the two cases

(i) α is the inclusion of a closed subgroup, and
(ii) α is surjective.

In case (i) we have $G \subset G'$ and the compatibility of Jordan decompositions follows from (3) after embedding G' in a linear group.

In case (ii) the comorphism $\alpha^o: A' \to A$ is injective, so we can view A' as a subring of A. Then, for $g \in G$ and $X \in \mathfrak{g}$ we have $\rho_{\alpha(g)} = \rho_g | A'$ and

$*d\alpha(X) = *X|A'$. Hence, we have the corresponding relationships between the Jordan decompositions according to 4.2.

Corollary

(1) *If $g, h \in G$ commute then $gh = (g_s h_s)(g_u h_u)$ is the Jordan decomposition of gh, and all elements appearing commute.*
(2) *If $X, Y \in \mathfrak{g}$ commute (i.e. $[X, Y] = 0$) then $X + Y = (X_s + Y_s) + (X_n + Y_n)$ is the Jordan decomposition of $X + Y$, and all elements appearing commute.*

Proof. After embedding G in a $GL(V)$ this follows from 4.2, Corollary 2.

4.5 *Semi-simple and unipotent elements in affine groups.* For an affine k-group G they are the elements of

$$G_s = \{g \in G | g = g_s\}$$

and of

$$G_u = \{g \in G | g = g_u\},$$

respectively. We define the analogous sets,

$$\mathfrak{g}_s = \{X \in \mathfrak{g} | X = X_s\} \quad \text{and} \quad \mathfrak{g}_n = \{X \in \mathfrak{g} | X = X_n\}$$

of semi-simple and nilpotent elements, respectively, in $\mathfrak{g}$.

It follows from part (4) of Theorem 4.4 that, if $\alpha: G \to G'$ is a morphism of affine k-groups, then

$$\alpha(G_s) \subset G'_s, \alpha(G_u) \subset G'_u$$
$$(d\alpha)(\mathfrak{g}_s) \subset \mathfrak{g}'_s, (d\alpha)(\mathfrak{g}_n) \subset \mathfrak{g}'_n.$$

In fact we have $\alpha(G_s) = \alpha(G)_s$ and $\alpha(G_u) = \alpha(G)_u$, and similarly for $\mathfrak{g}$. Moreover, the corollary to Theorem 4.4 implies that a product of two commuting elements in G_s (resp., G_u) is again in G_s (resp., G_u). Similarly for sums of commuting elements in $\mathfrak{g}_s$ (resp., $\mathfrak{g}_n$). In particular, *if G is commutative then G_s and G_u are subgroups of G, and $\mathfrak{g}_s$ and $\mathfrak{g}_n$ are subspaces of* $\mathfrak{g}$. Moreover, in general, it follows from 4.1(e) that

$$G_s \cap G_u = \{e\} \quad \textit{and} \quad \mathfrak{g}_s \cap \mathfrak{g}_n = 0.$$

If we embed G in a $GL(V)$, then the elements of G_u (resp., $\mathfrak{g}_n$) are defined by the equation $(g - e)^n = 0$ (resp., $X^n = 0$) in $\text{End}(V)$ (for large enough n). These equations, with respect to a k-rational basis for V, have coefficients in $\mathbb{Z}$, so we conclude that:

G_u is a k-closed subset of G

and

$\mathfrak{g}_n$ is a k-closed subset of $\mathfrak{g}$.

4.6 *Trigonalization and diagonalization.* Let M be a subset of $\mathfrak{gl}_n$. We say that M is *trigonalizable* (over k) if there is a $g \in \mathbf{GL}_n$ (resp., $g \in \mathbf{GL}_n(k)$) such that gMg^{-1} is in upper triangular form (i.e. lies in $L(\mathbf{T}_n)$). We say M is *diagonalizable* (over k) if there is a $g \in \mathbf{GL}_n$ (resp., $g \in \mathbf{GL}_n(k)$) such that gMg^{-1} is in diagonal form (i.e. lies in $L(\mathbf{D}_n)$).

More generally, if V is any finite dimensional vector space with a k-structure $V(k)$ then we can speak of trigonalizing or diagonalizing (over k) a family of endomorphisms of V. This means they assume triangular or diagonal form, respectively, with respect to a suitable (k-rational) basis of V.

Proposition. *Let $M \subset \mathfrak{gl}_n(k)$ be a commuting family of endomorphisms, and let L be the field extension of k generated by the eigenvalues of elements of M.*

(a) *M is trigonalizable over L.*
(b) *If M consists of semi-simple endomorphisms then $L \subset k_s$ and M is diagonalizable over L.*

Proof. The fact that $L \subset k_s$ in case (b) follows from 4.1(c). For the rest of the proof therefore we can replace k by L and assume all eigenvalues of elements of M are in k.

If $X \in M$ and if $a \in k$ then $W = \ker(X - aI)$ is visibly defined over k, and it is stable under all Y commuting with X, in particular all Y in M.

If M does not consist of scalar matrices (otherwise there is nothing to prove) then we can choose X and a so that $0 \neq W \neq V$. Then, by induction on dimension, we can find an $e_1 \in W(k)$ which spans an M-stable line. Applying induction to V/Ke_1 we can complete e_1 to a k-rational basis $e_1, \ldots, e_n$ such that M leaves $Ke_1 + \cdots + Ke_i$ invariant for each $i = 1, \ldots, n$. This proves (a).

To prove (b) we can again assume there is a non-scalar X in M. Write $V = V_1 \oplus \cdots \oplus V_r$ where $V_i = \ker(X - a_i I)$ and $a_1, \ldots, a_r$ are the distinct eigenvalues of X. Then each V_i is defined over k and stable under M so, by induction on dim V, we can diagonalize over k the action of M on each V_i. This yields the desired diagonalization of M on V.

4.7 Theorem. *Let G be a commutative k-group. Then G_s and G_u are closed subgroups and the product morphism*

$$\alpha: G_s \times G_u \to G$$

is an isomorphism of algebraic groups.

Proof. We have already seen in 4.5 that G_u is a k-closed subgroup and that G_s is a subgroup meeting G_u in e. Hence α is an isomorphism of abstract groups.

Embed G in some $\mathbf{GL}_n$. Using 4.6(b) we can further arrange that $G_s = G \cap D_n$. In particular it follows that G_s is a closed subgroup, and clearly α is then a morphism of algebraic groups.

Write $K^n = V_1 \oplus \cdots \oplus V_r$ where the V_i are the distinct simultaneous eigenspaces for G_s. Then G_u leaves each V_i stable so we can, by 4.6(a), trigonalize the action of G_u in each V_i. Thus we can assume $G \subset \mathbf{T}_n$ and G_s still equals $G \cap \mathbf{D}_n$.

If $g \in G$ then $g_s \in G_s \subset \mathbf{D}_n$ so it follows easily (for example, from the fact that g is triangular and that g_s is a polynomial in g) that g_s is just the projection of g onto its diagonal component, $g \mapsto \operatorname{diag}(g_{11}, \ldots, g_{nn})$. This is clearly a morphism. Hence $g \mapsto g_u = g_s^{-1} g$ is likewise a morphism, so $g \mapsto (g_s, g_u)$ gives the required inverse to α.

Remark. We have seen in 4.5 that G_u is k-closed. It will further be shown in §10 that G_s is defined over k. If $\operatorname{char}(k) = 0$ this follows from the obvious invariance of $G_s(\bar{k})$ under $\operatorname{Gal}(\bar{k}/k)$. If $\operatorname{char}(k) = p > 0$ then, if $G \subset \mathbf{GL}_n$, we have $g_u^{p^n} = e$ for all $g \in G$. Hence $\beta: G \to G, \beta(g) = g^{p^n}$, is a k-morphism of k-groups, clearly, with image in G_s. In fact it will be seen in §8 that the pth power map in G_s is surjective. From this it follows that $G_s =: \beta(G)$ is defined over k.

4.8 *Trigonalizing unipotent groups.* Let Λ be the algebra of upper triangular $n \times n$ matrices, and let N be the ideal in Λ of matrices with zero diagonal. Then $N^n = 0$, so $\mathbf{U}_n = I + N$ is a *unipotent group*, i.e. one consisting of unipotent elements. It is easy to verify that the $I + N^i$ $(1 \leqq i \leqq n)$ are normal subgroups of U satisfying the commutator formula: $(I + N^i, I + N^j) \subset I + N^{i+j}$. In particular, taking $i = 1$ and varying j, we see that $\mathbf{U}_n$ is a nilpotent group. Its Lie algebra $\mathfrak{u}$ is the set of all upper triangular matrices with eigenvalues zero, hence it consists of nilpotent matrices.

Theorem. *Let G be a not necessarily closed unipotent subgroup of $\mathbf{GL}_n(k)$. Then G is conjugate over k to a subgroup of U. In particular, G is a nilpotent group.*

Proof. In view of the remarks made above, it suffices to prove the first assertion. For this, it suffices to show that there is a line L in V fixed by G. For then the set W of fixed points under G is a non zero subspace of V defined over k, and we can finish by applying induction to the induced action of G in V/W.

Henceforth, we may assume therefore that k is algebraically closed. Using induction on $\dim V$ we may further assume that V is an irreducible G-module. Then the vector space A spanned by G is a k-algebra acting irreducibly on V, so (by Wedderburn theory) it must be all of $\operatorname{End}(V)$.

Every $g = I + x \in G$ is unipotent, so $\operatorname{Tr}(g) = \operatorname{Tr}(I) = \dim V$ is independent of g. If also $g' \in G$ then $\operatorname{Tr}(xg') = \operatorname{Tr}((g - I)g') = \operatorname{Tr}(gg') - \operatorname{Tr}(g') = 0$, therefore. But we saw above that such g' span $\operatorname{End}(V)$, and hence $x = 0$, i.e. $g = I$. This means $G = \{I\}$ so $\dim V = 1$. Q.E.D.

Corollary. *Let G be a unipotent algebraic group (i.e. $G = G_u$). Then G is isomorphic to a closed subgroup of $\mathbf{U}_n \subset \mathbf{GL}_n$ for some n. Hence $L(G)$ consists of nilpotent elements.*

Proof. We apply the theorem to an immersive representation $\pi: G \to \mathbf{GL}_n$ (see 1.10) to embed G in $\mathbf{U}_n$. Then $L(G)$ is embedded in $L(\mathbf{U}_n)$ which consists of upper triangular matrices with zeroes on the diagonal.

4.9 Remark. It will be shown later that an element $X \in \mathfrak{g}$ is nilpotent (resp. semi-simple) if and only if it is tangent to a closed unipotent subgroup (resp. to a torus, cf. §8).

4.10 Proposition. *Let G be a unipotent affine group and V a quasi-affine variety on which G operates morphically. Then all orbits are closed.*

Proof. Every orbit of G is a finite union of orbits of G^o, hence we may assume G to be connected. Let $v \in V$. We have to prove that $G \cdot v = \overline{G \cdot v}$. Let $W = \overline{G \cdot v}$ and assume $F = W - G \cdot v$ is not empty. By 1.8, F is closed. Let J be the ideal of F in $K[W]$. We claim it is not the zero ideal: Consider the closure $\bar{V}$ of V in some affine embedding. Let $\bar{F}$ be the closure of F in $\bar{V}$. Since $\bar{F} \cap V = F$, any $v \in V - F$ does not belong to $\bar{F}$, hence there exists a regular function on $\bar{V}$ which vanishes on $\bar{F}$ and is equal to 1 at v. Its restriction to V is then a non-zero element of J. The ideal J is stable under G and is the union of G-invariant finite dimensional subspaces (1.9). By 4.8 any such space has non-zero elements fixed under G. But the invariants of G in $K[W]$ are the constant functions, contradiction. This proves the proposition.

Bibliographical Note

The Jordan decomposition in algebraic groups is discussed in [1]. However, its existence is equivalent to the theorem 4.7 on commutative algebraic groups, proved earlier by Kolchin [20]. The proof given here is different from the one of [1], and follows a suggestion made by Springer for the Lie algebra case. Jordan decomposition in the Lie algebra is introduced in [2]. However, the definition adopted there and in [3] is more stringent, and the existence proof is less elementary. Here, it becomes the theorem mentioned in 4.9 to be proved in 11.8 and 14.26. Proposition 4.10 is due to M. Rosenlicht [27].

Chapter II

Homogeneous Spaces

§5. Semi-Invariants

In this section *all algebraic groups are affine.* The results here prepare the way for the construction of quotients in §6.

5.1 Theorem. *Let G be a k-group and let H be a closed subgroup defined over k. Then there is an immersive representation $\alpha: G \to GL(E)$ defined over k, and a line $D \subset E$ defined over k, such that*

$$H = \{g \in G \mid \alpha(g)D = D\} \quad \textit{and} \quad \mathfrak{h} = \{X \in \mathfrak{g} \mid d\alpha(X)D \subset D\}.$$

Proof. Let I denote the ideal in $A = K[G]$ of functions vanishing on H; it is generated by $I_k = I \cap k[G]$, and even by a finite subset of I_k. Therefore, using 1.9, we can find a finite dimensional right G-invariant subspace V of A, defined over k, and such that, if $W = V \cap I$, the ideal I is generated by W_k.

Both V and I are right H-invariant and defined over k, so the same is true of W. We claim now that

$$H = \{g \in G \mid \rho_g W = W\} \quad \text{and} \quad \mathfrak{h} = \{X \in \mathfrak{g} \mid W * X \subset W\}.$$

We know from 3.8 that the analogous equations hold if we replace W by I.

We have already remarked that W is right H-invariant, so it is $\mathfrak{h}$-invariant also because convolution is the differential of right translation (3.11).

Conversely, suppose $g \in G$ and $\rho_g W = W$. Since ρ_g is an algebra automorphism we have $\rho_g I = \rho_g(WA) = \rho_g(W)A = WA = I$, so $g \in H$. Similarly, if $X \in \mathfrak{g}$ and $W * X \subset W$, then

$$I * X = (WA) * X \subset (W * X)A + W(A * X) \subset WA = I, \quad \text{so} \quad X \in \mathfrak{h}.$$

Now put $E = \Lambda^d(V)$, where $d = \dim W$, and let $D = \Lambda^d W \subset E$. The representation $\rho: G \to GL(V)$ induces $\alpha = \Lambda^d \rho: G \to GL(E)$, a k-rational representation. In case α is not immersive replace E by $E \oplus F$ using any k-rational immersive representation $G \to GL(F)$. Then all the conditions of the theorem are achieved thanks to the following lemma from linear algebra.

Lemma. *Let W be a d-dimensional subspace of a vector space V, and let $D = \Lambda^d W \subset E = \Lambda^d V$. Let $g \in GL(V)$ and let $X \in \mathfrak{gl}(V)$. Then*

$$(1) \qquad (\Lambda^d g)D = D \Leftrightarrow gW = W,$$

$$(2) \qquad (d\Lambda^d)(X)D \subset D \Leftrightarrow XW \subset W.$$

Proof. In both cases the implication $\Leftarrow$ is clear.

Let $(e_i)(1 \leqq i \leqq m)$ be a basis for V such that $e_1, \ldots, e_d$ span W. In case (1) we can further arrange that, for some $n \geqq 1$, $e_n, \ldots, e_{n+d-1}$ span gW. Then $(\Lambda^d g)(e_1 \wedge \cdots \wedge e_d)$ is a multiple of $e_n \wedge \cdots \wedge e_{n+d-1}$, so $(\Lambda^d g)D = D$ implies $n = 1$, i.e. $gW = W$.

In case (2) we can replace X by $X - Y$ for some Y leaving W stable, if necessary, to achieve the condition $W \cap XW = 0$. Then we can choose the basis above so that Xe_i is a multiple of $e_{d+i}(1 \leqq i \leqq d)$. In this case

$$(d\Lambda^d)_e(X)(e_1 \cdots e_d) = \sum_{1 \leqq i \leqq d} e_1 \wedge \cdots \wedge e_{i-1} \wedge Xe_i \wedge e_{i+1} \wedge \cdots \wedge e_d.$$

Since the vectors $e_1 \wedge \cdots \wedge e_{i-1} \wedge e_{d+i} \wedge e_{i+1} \wedge \cdots \wedge e_d$ are part of a basis of $\Lambda^d V$ that includes $e_1 \wedge \cdots \wedge e_d$, it follows that the sum above can be a multiple of $e_1 \wedge \cdots \wedge e_d$ only if each $Xe_i = 0$, i.e. only if $X = 0$.

5.2 *Characters and semi-invariants.* Let G and G' be k-groups. We shall write $\mathrm{Mor}(G, G')$ for the algebraic group morphisms from G to G', and $\mathrm{Mor}(G, G')_k$ for the set of those defined over k.

Recall from AG.14.3 that $\Gamma = \mathrm{Gal}(k_s/k)$ operates on $\mathrm{Mor}(G, G')_{k_s}$, and ${}^s\alpha$, for $s \in \Gamma$ and $\alpha \in \mathrm{Mor}(G, G')$ is characterized by:

$$({}^s\alpha)(g) = s(\alpha(s^{-1}g)) \quad (g \in G(k_s)).$$

Moreover we have

$$\mathrm{Mor}(G, G')_k = \mathrm{Mor}(G, G')_{k_s}^{\Gamma},$$

i.e. α is defined over k if and only if α is defined over k_s and is a Γ-equivariant homomorphism of $G(k_s)$ into $G'(k_s)$.

Note that when G' is commutative, $\mathrm{Mor}(G, G')$ is an abelian group and $\mathrm{Mor}(G, G')_{k_s}$ is a Γ-module, the product in $\mathrm{Mor}(G, G')$ being defined by: $aa'(g) = a(g) \cdot a'(g)$.

We shall write

$$X(G) = \mathrm{Mor}(G, \mathbf{GL}_1)$$

and call its elements *characters* of G. Thus $\chi \in X(G)$ means $\chi \in K[G]$, and $\chi(g) \neq 0$ and $\chi(gg') = \chi(g)\chi(g')$ for all $g, g' \in G$. The condition $\chi \in X(G)_k$ just means that moreover $\chi \in k[G]$.

Let $\alpha: G \to GL(V)$ be a k-rational representation. A *semi-invariant* of G in V is a non-zero vector $v \in V$ spanning a G-stable line in V. Thus we can write

$$\alpha(g)v = \chi(g)v$$

for some function $\chi: G \to K^*$, and evidently χ is a character, which is defined over k if $v \in V(k)$. This character is called the *weight* of the semi-invariant v.

We shall also use the term semi-invariant with respect to the action by translation of functions induced by an action of G on a variety.

With V as above and $\chi \in X(G)$ write

$$V_\chi = \{v \in V \mid \alpha(g)v = \chi(g)v \text{ for all } g \in G\}.$$

We can even restrict to $g \in G(k_s)$ here, because $G(k_s)$ is dense in G (see AG.13.3). If, further, χ is defined over k_s, then the equation $\alpha(g)v = \chi(g)v$ is a linear equation in v defined over k_s, so we see that V_χ is defined over k_s if $\chi \in X(G)_{k_s}$.

Suppose $\chi \in X(G)_{k_s}$, $g \in G(k_s)$, $v \in V_\chi(k_s)$, and $s \in \Gamma$. Then

$$({}^s\chi((g)(sv) = (s(\chi(s^{-1}g)))(sv) = s(\chi(s^{-1}g)(v))$$

since Γ acts semi-linearly

$$({}^s\chi)(g)(sv) = s(\alpha(s^{-1}g)(v)) = \alpha(g)(sv) \quad (v \in V_\chi)$$

since α is defined over k. This shows that $sV_\chi(k_s) \subset V_{({}^s\chi)}(k_s)$, and the reverse inclusion follows by applying s^{-1} to this. Thus we have

$$sV_\chi(k_s) = V_{({}^s\chi)}(k_s)$$

for $\chi \in X(G)_{k_s}$ and $s \in \Gamma$. In particular (see AG.14.1):

If χ is defined over k, then V_χ is defined over k.

A *weight* of G in V is a $\chi \in X(G)$ such that $V_\chi \neq 0$.

Lemma. *The subspaces V_χ $(\chi \in X(G))$ of V are linearly independent. In particular G has only finitely many weights in V.*

Proof. If not, choose n minimal such that there exist distinct χ_i $(1 \leqq i \leqq n)$ and non-zero $v_i \in V_{\chi_i}$ such that $v_1 + \cdots + v_n = 0$. Clearly $n > 1$, so there is a $g \in G$ such that $\chi_1(g) \neq \chi_2(g)$. Since $\sum \chi_i(g)v_i = 0$ we can subtract $\chi_1(g)^{-1}$ times the last equation from the first to obtain a non trivial dependence relation of length $< n$; contradiction.

5.3 Corollary. *In the setting of 5.1, there exist $\chi \in X(H)_k$, and functions $f_1, \ldots, f_n \in k[G]$, which are semi-invariants of the same weight, χ, for H under right translations, such that*

$$H = \{g \in G \mid \rho_g f_i \in K f_i,\ 1 \leqq i \leqq n\}, \tag{1}$$

$$\mathfrak{h} = \{X \in \mathfrak{g} \mid f_i * X \in K f_i,\ 1 \leqq i \leqq n\}. \tag{2}$$

Proof. With E and D as in 5.1, let $e_1, \ldots, e_n$ be a k-rational basis of E such that $D = Ke_1$, and let T_{ij} denote the (i, j) coordinate function on $\mathfrak{gl}(E) \cong \mathfrak{gl}_n$, the isomorphism being defined relative to the basis above.

In this coordinate system we can paraphrase Theorem 5.1 as follows:

$$(*) \qquad H = \{g \in G \mid T_{i1}(\alpha(g)) = 0 \text{ for } i > 1\}$$

$$(**) \qquad \mathfrak{h} = \{X \in \mathfrak{g} \mid T_{i1}((d\alpha)X) = 0 \text{ for } i > 1\}.$$

The first formula implies that $\chi = T_{11} \circ \alpha$ is a character on H, evidently defined over k. Put $f_i = T_{i1} \circ \alpha (l < i \leqq n)$. Then $f_i \in k[G]$, and, for $g \in G$ and $h \in H$, we have

$$(\rho_h f_i)(g) = f_i(gh) = T_{i1}(\alpha(gh)) = \sum_j T_{ij}(\alpha(g)) T_{j1}(\alpha(h))$$
$$= T_{i1}(\alpha(g)) T_{11}(\alpha(h)) = \chi(h) f_i(g).$$

Thus each f_i is a semi-invariant of weight χ for H. If $g \in G$ and $\rho_g f_i \in K f_i$ then $\rho_g f_i(e)$ is a multiple of $f_i(e) = T_{i1}(\alpha(e)) = 0$, for each $i > l$. Thus $g \in H$, thanks to $(*)$. This proves (1).

It remains to be shown that if $X \in \mathfrak{g}$ and if $f_i * X \in K f_i$ for all $i > 1$ then $X \in \mathfrak{h}$. The identification of the Lie algebra of $\mathbf{GL}_n$ with $\mathfrak{gl}_n$ assigns the tangent vector Y to the matrix $(Y(T_{ij}))$ (see 3.6). Viewing the T_{ij} as coordinate functions on the Lie algebra, we have $T_{ij}(Y) = Y(T_{ij})$. Applying this to the X above, we have $T_{i1}((d\alpha)(X)) = (d\alpha)(X)(T_{i1}) = X(T_{i1} \circ \alpha) = X f_1 = (f_i * X)(e) =$ (a multiple of $f_i(e)$) $=$ (a multiple of $T_{i1}(\alpha(e))$) $= 0$ for all $i > l$. Hence $(**)$ implies $X \in \mathfrak{h}$.

5.4 Corollary. *Let $G \subset \mathbf{GL}_n$ be an algebraic matric group defined over k. Then there exist $\chi \in X(G)_k$ and polynomials $f_1, \ldots, f_m \in k[T_{11}, \ldots, T_{nn}]$ which are semi-invariants of weight χ for G with respect to right translations, such that*

$$G = \{g \in \mathbf{GL}_n \mid \rho_g f_i \in K f_i,\ 1 \leqq i \leqq m\},$$
$$\mathfrak{g} = \{X \in \mathfrak{gl}_n \mid f_i * X \in K f_i,\ 1 \leqq i \leqq m\}.$$

Proof. $k[\mathbf{GL}_n] = k[T_{11}, T_{12}, \ldots, T_{nn}, D^{-1}]$ where $D = \det(T_{ij})$. From 5.3 we obtain functions $f'_i \in k[\mathbf{GL}_n] (1 \leqq i \leqq m)$ which are semi-invariants of some weight $\chi' \in X(G)_k$, and which satisfy conditions of the above type. For r large enough we can write $f'_i = D^{-r} f_i$ with f_i a polynomial $(l \leqq i \leqq m)$. Evidently D is a semi-invariant of weight D for $\mathbf{GL}_n$. Consequently the f_i are semi-invariants for G of weight $\chi = (D|G)^r \chi'$. One sees immediately that χ and $f_1, \ldots, f_m$ satisfy the conditions above.

5.5 *Invariants.* It is not true in general that Theorem 5.1 and its corollaries can be strengthened to give invariants (i.e. $\chi = 1$) instead of semi-invariants. However, there are two important cases when this can be done. One, clearly, is when $X(H)_k = \{1\}$. Another case is deduced as follows: Let $\alpha: G \to GL(E)$ be as in Theorem 5.1, and let χ be the character by means of which H acts on D. Suppose we can find a second representation $\alpha': G \to GL(E')$ and a $D' \subset E'$ with the analogous properties, but so that H acts on D' via χ^{-1}. Then $\alpha \otimes \alpha'$ gives a representation of G on $E \bigotimes_K E'$ so that H acts trivially on $D \bigotimes_K D'$. Moreover, if $D = Kv$ and $D' = Kv'$ then it is easy to see that H is exactly the isotropy group of $v \otimes v'$ and $\mathfrak{h}$ is the isotropy algebra of $v \otimes v'$.

How can we find such an E' and D'? We can try the contragredient representation $\alpha^*:G\to GL(E^*)$. Then the one dimensional H-invariant subspace D of weight χ leads to a one-dimensional *quotient* space D^* of E^* on which H acts via χ^{-1}. To lift D^* H-equivariantly back into E^* it would suffice to know that H acts completely reducibly on E. In characteristic zero this happens if H is a reductive group.

5.6 Theorem. *Let G be a k-group, and let N a normal k-subgroup. Then there is a linear representation $\alpha:G\to GL(V)$ defined over k such that $N=\ker(\alpha)$ and $\mathfrak{n}=\ker(d\alpha)$.*

Proof. Theorem 5.1 gives us an $\alpha:G\to GL(E)$ and a line $D\subset E$, all defined over k, such that N is the stability subgroup of D in G and such that $\mathfrak{n}$ is the stability subalgebra of D in $\mathfrak{g}$.

The action of N on D is via some $\chi\in X(N)_k$. Let F denote the sum of all the subspaces E_φ, φ ranging over $X(N)_{k_s}$. We saw above (5.2) that this sum is direct. If $x\in E_\varphi, g\in G(k_s)$, and $n\in N$, then

$$\alpha(n)\alpha(g)x=\alpha(g)\alpha(g^{-1}ng)x=\varphi(g^{-1}ng)\alpha(g)x.$$

Thus, if we define $(g\varphi)(n)=\varphi(g^{-1}ng)$, then $g\varphi\in X(N)_{k_s}$, and

$$\alpha(g)E_\varphi=E_{(g\varphi)},(g\in G(k_s),\varphi\in X(N)_{k_s}).$$

It follows that F is G-invariant. Moreover, F is defined over k_s and invariant under $\mathrm{Gal}(k_s/k)$ (see 5.2) so F is defined over k. Finally, since $D\subset F$, there is no loss in assuming that $E=F$; otherwise follow α by restriction to F.

This done, let $V\subset\mathfrak{gl}(E)$ be the set of endomorphisms of $E=\bigoplus E_\varphi$ $(\varphi\in X(N)_{k_s})$ which leave each of the E_φ stable. Evidently $V=\bigoplus\mathfrak{gl}(E_\varphi)$. If $g\in G(k_s)$ and if $v\in V$ then, for each $\varphi\in X(N)k_s$,

$$\alpha(g)v\alpha(g)^{-1}E_\varphi=\alpha(g)vE_{(g^{-1}\varphi)}\subset\alpha(g)E_{(g^{-1}\varphi)}=E_\varphi.$$

Thus $\alpha(G)$ normalizes V, so we can define $\beta:G\to GL(V)$ by $\beta(g)(v)=\alpha(g)v\alpha(g)^{-1}$. Since β is just the restriction to V of α followed by Ad on $GL(E)$, it is a morphism. To see that V, and hence also β, are defined over k, take $s\in\mathrm{Gal}(k_s/k)$ and $v\in V(k_s)$. Then $({}^sv)E_\varphi(k_s)=(s.v.s^{-1})E_\varphi(k_s)=svE_{(s^{-1}\varphi)}(k_s)=sE_{(s^{-1}\varphi)}(k_s)=E_\varphi(k_s)$. Thus V is a subspace defined over k_s, and $V(k_s)$ is $\mathrm{Gal}(k_s/k)$ stable, so V is defined over k (see AG.14.1).

If $n\in N$ then $\alpha(n)$ is a multiple of the identity on each E_φ, so $\alpha(n)$ centralizes V, and hence $\beta(n)=e$. Conversely, if $\beta(g)=e$ then $\alpha(g)$ must leave each E_φ stable and induce a scalar multiplication in each one. (This is a simple calculation in $\mathfrak{gl}(E)$.) Since $D\subset E_\chi$ it follows that $\alpha(g)$ leaves D stable, so $g\in N$. This shows that $N=\ker(\beta)$, and hence that $\mathfrak{n}\subset\ker(d\beta)$.

Since β is the restriction to V of $Ad_{GL(E)}\circ\alpha$ it follows that $d\beta$ is the restriction to V of $\mathrm{ad}\circ d\alpha$. Therefore $X\in\ker(d\beta)\Rightarrow ad((d\alpha)(X))V=0\Rightarrow(d\alpha)(X)$ centralizes $V\Rightarrow(d\alpha)(X)$ leaves each E_φ stable and induces a scalar multiplication in each one (same calculation as above) $\Rightarrow(d\alpha)(X)$ leaves $D\subset E_\chi$ stable $\Rightarrow X\in\mathfrak{n}$.

§6. Homogeneous Spaces

Given an algebraic group G and a closed subgroup H we want to give the coset space G/H the structure of a variety in a natural way, for example, so that the projection $\pi: G \to G/H$ is a morphism satisfying a suitable universal mapping property. We shall do this here for G affine. The method is to use the results of §5 to realize G/H as the orbit of a point with isotropy group H under a suitable action of G on a projective space. In order to verify that this construction of G/H has the required properties we shall have to invoke several results from algebraic geometry which are quoted in Chapter AG.

We hasten to point out that the use of the term "quotient" here is not the categorical one, and hence it should be regarded as a provisional terminology adjusted to our present needs.

6.1 *Quotient morphisms.* Let $\pi: V \to W$ be a k-morphism of k-varieties. We say π is a *quotient morphism* (over k) if

(1) π is surjective and open.
(2) If $U \subset V$ is open, then π^0 induces an isomorphism from $K[\pi(U)]$ onto the set of $f \in K[U]$ which are constant on the fibres of $\pi | U$.

Recall from AG.8.2 that (1) implies that π is dominant.

Universal Mapping Property: *Let $\pi: V \to W$ be a quotient morphism over k. If $\alpha: V \to Z$ is any morphism constant on the fibres of π then there is a unique morphism $\beta: W \to Z$ such that $\alpha = \beta \circ \pi$. If α is a k-morphism of k-varieties then so also is β.*

Proof. It is clear that β exists and is unique topologically because π is open. It remains to show that if U is open in Z, then $f \mapsto f \circ \beta$ carries $K[U]$ into $K[\beta^{-1}(U)]$. But π^0 identifies $K[\beta^{-1}(U)] = K[\pi(\alpha^{-1}U))]$ with the set of $h \in K[\alpha^{-1}(U)]$ which are constant on the fibres of $\pi | \alpha^{-1}(U)$. Since α_0 maps $K[U]$ into the ring of such functions in $K[\alpha^{-1}(U)]$ (because α is constant on the fibres of π) it follows indeed that $\beta^0 K[U] \subset K[\beta^{-1}(U)]$. Thus β is a morphism of varieties.

In the above argument β^0 was seen to be the unique map rendering the diagram

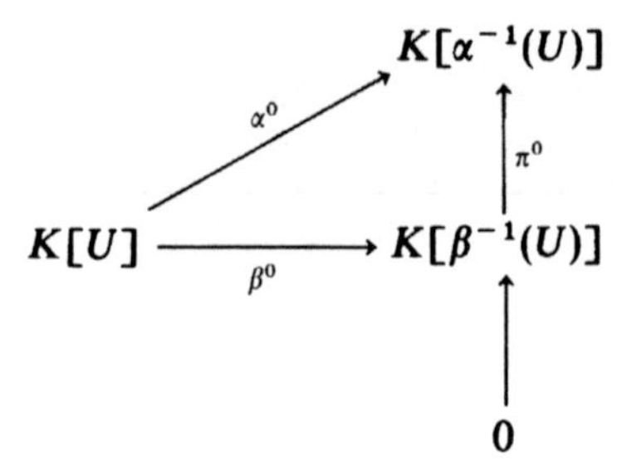

commutative. If U is k-open then so also are $\alpha^{-1}(U)$ and $\beta^{-1}(U) = \pi(\alpha^{-1}(U))$ (see AG.11.3). Moreover α^0 and π^0 are defined over k, so it follows that β^0 is also defined over k.

Corollary. *A bijective quotient morphism is an isomorphism.*

For if π is bijective we can apply the universal mapping property to $\alpha = 1_V$ to obtain π^{-1}.

6.2 Lemma. *Let $\pi: V \to W$ be a surjective open separable morphism of irreducible varieties, and assume W is normal. Then π is a quotient morphism.*

Proof. We must verify condition (2) of the definition of quotient morphism for each open U in V. Since $\pi | U: U \to \pi(U)$ inherits all of the hypotheses made on π it suffices to treat the case $U = V$. Then we must show that every $f \in K[V]$ constant on the fibres of π lies in the subring $\pi^0 K[W]$. According to (AG.18.2, Prop.) f is purely inseparable over $\pi^0 K(W)$, so the separability of π implies that $f = \pi^0 f'$ for some $f' \in K(W)$. It remains to be shown that f' is everywhere defined. If f' is not defined at $\pi(x)$, then, because W is normal, it follows from AG.18.3; Lem. that there is a point $\pi(y)$ where $1/f'$ is defined and vanishes. But then $1/f = \pi^0(1/f')$ is defined and vanishes at y, contrary to the fact that $f \in K[V]$.

6.3 *The quotient of V by G.* For the next few sections (until 6.7) we fix a k-group G acting k-morphically on a k-variety V. An *orbit map* is a surjective morphism $\pi: V \to W$ of varieties such that the fibres of π are the orbits of G in V. A *quotient* of V by G *over* k is an orbit map $\pi: V \to W$ which is a quotient morphism over k in the sense of 6.1. In particular such a π satisfies the following:

Universal Mapping Property: *If $\alpha: V \to Z$ is any morphism constant on the orbits of G there is a unique morphism $\beta: W \to Z$ such that $\alpha = \beta \circ \pi$. If α is a k-morphism of k-varieties so also is β.*

It follows that *the quotient*, if it exists, *is unique up to a unique k-isomorphism.* We are thus permitted to denote it by the symbol $G \backslash V$. If the action is defined so that G operates on the right on V, as with right translation in a larger group containing G, then we shall use the symbol V/G.

In general quotients do not exist. For example, the next proposition shows that the existence of a quotient implies that the dimensions of the orbits cannot vary. Moreover: *if an orbit map $\pi: V \to W$ exists, then the orbits of G in V are closed.* This is because they are inverse images of points under a morphism.

6.4 Proposition. *Let $\pi: V \to W$ be a dominant orbit map and assume that W is irreducible.*

(a) *G acts transitively on the set of irreducible components of V. In particular, if G is connected, then V is irreducible.*

Assume now that the irreducible components of V are open.

(b) *The orbits of G in V have constant dimension $d = \dim V - \dim W$.*

(c) *If W is normal then π is open.*

Proof. (a) Let F and F' be irreducible components of V. Since π is dominant and W is irreducible it follows that πF and $\pi F'$ contain dense open sets in W. Hence $\pi^{-1}(\pi F') = G \cdot F'$ contains a non-empty, hence dense open set in F. But $G \cdot F'$ is the union of those irreducible components of V into which G transforms F'. In particular it is closed, so it contains F, and hence $F = gF'$ for some $g \in G$ because F is irreducible. The stability group H of F in G is a closed subgroup of finite index, so H contains G^o (see 1.2). This proves (a).

To prove (b) and (c) we replace V by F and G by H, and thus reduce to the case when V is irreducible. This reduction is justifiable in view of (a) and of the disjointness of the irreducible components of V.

Now that V is irreducible part (c) is a consequence of (b), by virtue of AG.18.4, so it remains to prove (b). For this we use the results of AG.10.1 on the dimension of the fibres of a morphism. The orbits are homogeneous so all irreducible components of an orbit have the same dimension. Moreover, $\dim G(x) \geqq d$, with equality whenever $\pi(x) \in U$, where U is some dense open set in W.

Next consider the graph of the action of G on V: $\Gamma = \{(g, x, gx) \in G \times V \times V\}$. Let D be the diagonal in $V \times V$, put $Z = \Gamma \cap (G \times D)$, and let $p: Z \to D$ be the projection. If $x \in V$, then $p^{-1}(x, x) = \{(g, x, x) | g \in G,\ gx = x\} = G_x \times \{(x, x)\}$. Hence all irreducible components of the fibre of p over (x, x) have the same dimension. Let Z_o be an irreducible component of Z containing $\{e\} \times D$, and let $p_1: Z_0 \to D$ be the restriction of p. Then p_1 is surjective, and $p_1^{-1}(x, x)$ is a non-empty union of irreducible components of $p^{-1}(x, x)$. Thus by applying the theorems on fibres of a morphism (AG.10.1) to p_1 we see that

$$\dim G_x \geqq d' = \dim Z_o - \dim D$$

with equality whenever $x \in U'$, where U' is some open dense set in V. Combining this and the above, we have, for all $x \in V$,

$$d \leqq \dim G(x) = \dim G - \dim G_x \leqq \dim G - d'.$$

Choosing $x \in U' \cap \pi^{-1}(U)$, which is possible because the latter set is open dense, we see that the inequalities become equalities, so $d = \dim G - d'$. Hence $\dim G(x) = d$ for all x.

Remark. Given an orbit map $\pi: X \to Y$ which is a candidate for a quotient morphism, we shall have to check that it is open. If U is open in X, then so is $G(U)$, and $\pi(U) = \pi(G(U))$. From this it is elementary that the following three conditions are equivalent: (i) π is open, (ii) the image of every G-stable open subset of X is open; (iii) the image of every G-stable closed subset of X is closed.

In particular, if π is closed, then it is open.

6.5 *The function field of a quotient.* If U is a dense open set in V and if $g \in G$ then $g^{-1}U$ is also open dense, and we have the comorphism $\lambda_g: K[g^{-1}U] \to K[U]$ (where $(\lambda_g f)(x) = f(g^{-1}x)$). As U varies we obtain an automorphism of the direct system of $K[U]$'s, and hence of their direct limit $K(V)$. If $f \in K(V)$ has domain of definition U, then $\lambda_g f$ has domain of definition gU. In this way G acts, by left translation, as a group of K-algebra automorphisms of $K(V)$, and we denote the fixed ring by $K(V)^G$.

Proposition. *Suppose a quotient $\pi: V \to W$ of V by G exists. Then π is a separable morphism and π^o induces an isomorphism of $K(W)$ onto $K(V)^G$. If V is irreducible then, for each $x \in V$, π^o maps $\mathcal{O}_{W,\pi(x)}$ isomorphically onto $\mathcal{O}_{V,x} \cap K(V)^G$.*

Proof. Since π is dominant, π^o induces a monomorphism of $K(W)$ into $K(V)$ whose image clearly lies in $K(V)^G$. On the other hand, if $f \in K(V)^G$, then the domain of definition U of f is G-stable, and f is constant on the fibres of $\pi|U$. Hence the definition of quotient implies that f lies in the image of $\pi^o: K[\pi(U)] \to K[U]$.

To show that π is separable, therefore, if suffices to prove that if F is a finite product of fields, and if H is a group of automorphisms of F, then F is separable over $E = F^H$. To a decomposition of E as a product of fields corresponds a decomposition of F which is clearly H-stable, so we can reduce to the case when E is a field. Then H operates transitively on the factors of F; otherwise we could separate the latter into H-orbits and this would yield a product decomposition of F^H.

If L is one of the fields into which F factors we must show that L is separable over E. But the remarks above imply that $E = L^{H'}$ where H' is the stability group of L in H. Hence the desired separability follows from AG.2.4.

If V is irreducible, then the field $K(V)$ contains all the local rings $\mathcal{O}_{V,x} (x \in V)$. Identifying $K(W)$ with $K(V)^G$, we have $\mathcal{O}_{W,\pi(x)} \subset \mathcal{O}_{V,x} \cap K(V)^G$, and the map $\pi_o: \mathcal{O}_{W,\pi(x)} \to \mathcal{O}_{V,x}$ is just the inclusion. It remains to show that every $f \in \mathcal{O}_{V,x} \cap K(V)^G = \mathcal{O}_{V,x} \cap K(W)$ lies in $\mathcal{O}_{W,\pi(x)}$. If U is an open neighborhood of x on which f is defined, then the definition of a quotient implies that, as an element of $K[U]$, f lies in $\pi^0 K[\pi(U)]$. In particular, as a rational function on W, f is defined at $\pi(x)$, i.e. $f \in \mathcal{O}_{W,\pi(x)}$.

6.6 Proposition. *Suppose $\pi: V \to W$ is a separable orbit map, and assume that W is normal and that the irreducible components of V are open. Then (W, π) is the quotient of V by G.*

Proof. We can easily reduce to the case when W is connected, and hence (being normal) irreducible. Then it follows from 6.4(c) that π is open, and from 6.4(a), that G operates transitively on the components of V. Since these components are disjoint we can replace V by one of them and G by its stability group, and retain all of our hypotheses. Thus we see that it suffices to prove the proposition when V is also irreducible. But then the fact that π is a quotient morphism follows from Lemma 6.2.

Corollary. *Let G_1, G_2 be k-groups, and V_1, V_2 k-varieties. Assume that G_i operates k-morphically on V_i and that V_i/G_i exists and is normal $(i = 1, 2)$. Then $(V_1 \times V_2)/(G_1 \times G_2)$ exists and is canonically isomorphic to $(V_1/G_1) \times (V_2/G_2)$.*

The product $V_1/G_1 \times V_2/G_2$ is normal (AG.18.1). The projections $V_i \to V_i/G_i$ are separable 6.5, hence their product is AG.17.3, Cor. The fibres of the latter map are the orbits of $G_1 \times G_2$; we may then apply the proposition.

6.7 Proposition. *Suppose $x \in V(k)$, and let $\pi: G \to G(x)$ be the k-morphism $g \mapsto g \cdot x$. Then $G(x)$ is a smooth variety defined over k and locally closed in V. Moreover π is an orbit map for the action of G_x on G by right translation. The following conditions are equivalent:*

(a) *π is a quotient of G by G_x.*
(b) *π is separable, i.e. $(d\pi)_e: L(G) \to T(G(x))_x$ is surjective.*
(c) *The kernel of $(d\pi)_e$ is contained in $L(G_x)$. When these conditions hold G_x is defined over k, and hence π is a quotient of G by G_x over k.*

Proof. The first assertion follows from 1.8, and the second one is obvious.

In view of the homogeneity of G and of $G(x)$ the interpretation of separability given in (b) is justified by AG.17.3. We obtain (a)$\Rightarrow$(b) from 6.5 and (b)$\Rightarrow$(a) from 6.6. Since $\dim G = \dim G_x + \dim G(x)$, and since the tangent spaces to a smooth variety have the same dimension as the variety, the equivalence of (b) and (c) follows from the obvious inclusion $L(G_x) \subset \ker(d\pi)_e$.

If π is separable, then it follows from (AG.13.2) that there is a dense open set $W \subset G(x)$ such that, if $w \in W(k_s)$, the fibre $\pi^{-1}(w)$ has a dense set of separable points. Since W contains a separable point (AG.13.3) w, we can translate w to deduce the corresponding property for every separable point of $G(x)$. Since x is rational over k, it follows that $G_x = \pi^{-1}(x)$ has a dense and Galois stable set of separable points, so (AG.14.4) G_x is defined over k.

Remark. The above argument shows that the kernel of a separable k-morphism of k-groups is defined over k.

6.8 Theorem. *Let G be an affine k-group and let H be a closed subgroup defined over k. Then the quotient $\pi: G \to G/H$ exists over k, and G/H is a smooth quasi-projective variety. If H is a normal subgroup of G, then G/H is an affine k-group and π is a k-morphism of k-groups.*

Proof. Theorem 5.1 gives us a k-rational representation $\alpha: G \to GL(E)$ and a line D in E defined over k such that

$$H = \{g \in G \mid \alpha(g)D = D\}, \quad \mathfrak{h} = \{X \in \mathfrak{g} \mid d\alpha(X)D \subset D\}.$$

Let $q: E - \{0\} \to P$ denote the projection onto the projective space $P = P(E)$ of lines in E, and let $x = q(D - \{0\}) \in P(k)$. Via α, we have a k-morphic action of G on P, and we propose to construct G/H from the orbit map $\pi: G \to G(x)$, $\pi(g) = gx$. Since H is the isotropy group of x, clearly, it remains only to show, thanks to 6.7, that $\ker(d\pi)_e = \mathfrak{h}$.

Choose $v \neq 0$ in D and define $\beta: G \to E - \{0\}$ by $\beta(g) = \alpha(g)v$. Then $\pi = q \circ \beta$ and $(d\beta)_e(X) = (d\alpha)(X)v$, where, as usual, we identify $T(E - \{0\})_v = T(E)_v$ with E. Therefore, since $\mathfrak{h} = (d\beta)_e^{-1}(D)$, the fact that $\ker(d\pi)_e = \mathfrak{h}$ follows from the fact that the kernel of $(dq)_v: T(E - \{0\})_v \to T(P)_{q(v)}$ is just D.

Finally, if H is a normal subgroup of G, then Theorem 5.6 permits us to choose $\alpha: G \to GL(E)$ above so that $H = \ker(\alpha)$ and $\mathfrak{h} = \ker(d\alpha)$. It then follows from 1.4 that $G' = \alpha(G)$ is a closed subgroup of $GL(E)$ defined over k. Letting G act, via α, by left translation on $GL(E)$, we can view $\pi: G \to G'$, $\pi(g) = \alpha(g)$ $(= \alpha(g)e)$ as the orbit map onto the orbit of e. Since $\ker(d\pi)_e = \mathfrak{h}$ and since H is the stability group of e under the above action, it follows again from 6.7 that π is the quotient of G by H. This completes the proof of the theorem.

Caution. Even though $G \to G/H$ is a surjective k-morphism, it is not true in general that $G(k) \to (G/H)(k)$ is surjective. This is true if $k = k_s$, and a general study of this problem leads to questions in Galois cohomology which will not be discussed here, (see e.g. [30]).

6.9 Corollary. (*a*) *Let $\alpha: G \to G'$ be a morphism of algebraic groups. If G is affine, so is $\alpha(G)$.*

(*b*) *Assume G to be unipotent. Then every homogeneous space of G is an affine variety.*

Proof. (*a*) Let $N = \ker(\alpha)$. Then α induces a bijective morphism $\beta: G/N \to \alpha(G)$ and we know that G/N is affine. Then it follows from AG. 18.3 that $\alpha(G)$ is affine.

(*b*) Let V be a homogeneous space of G and H an isotropy group. Again, by AG.18.3, it suffices to show that G/H is affine. Let $G \to GL(E)$, $D \subset E$ be as in 5.1. Since H is unipotent, too, every one-dimensional representation of H is trivial, hence D is pointwise fixed under D. If $d \in D - \{0\}$, then the orbit map $g \mapsto g \cdot d$ yields an isomorphism of G/H onto $G \cdot d$. But the latter is closed by 4.10 applied to the action of G on E, hence is affine.

6.10 Corollary. *Let G be an affine k-group acting k-morphically on a k-variety V, and let N be a closed normal subgroup of G defined over k.*

(1) *If V/N exists over k and is a normal variety, then G/N acts* k-morphically *on V/N (in the natural way). In particular, if N acts trivially on V, then G/N acts k-morphically on V.*

(2) *If, moreover, V/G exists and is a normal variety then the quotient of V/N by G/N exists and is canonically isomorphic to V/G.*

Proof. Let $\alpha: G \times V \to V$ be the action and let $\pi: V \to V/N$ and $p: G \to G/N$ be the quotient morphisms. Using the corollary to 6.6, we see that the vertical arrows in the commutative diagram

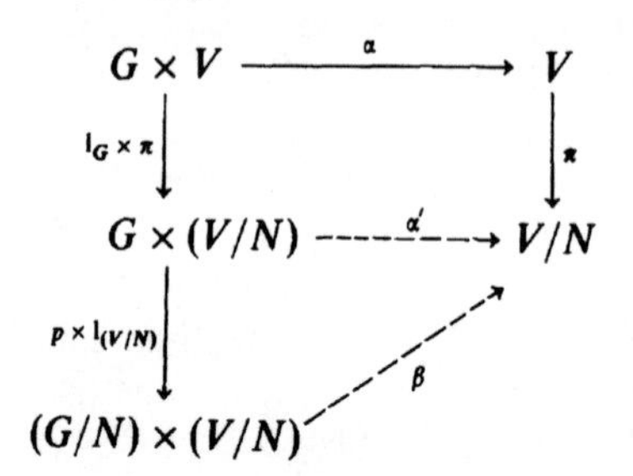

are quotient morphisms.

Now we can fill the diagram with α' and then β using the universal mapping property 6.1 for quotients. The k-action of G/N on V/N is then given by β. In case N acts trivially on V we have $V = V/N$, so this proves (1).

For part (2) let $\pi_G: V \to V/G$ be the quotient. The universal mapping property for π gives us a π' making the triangle

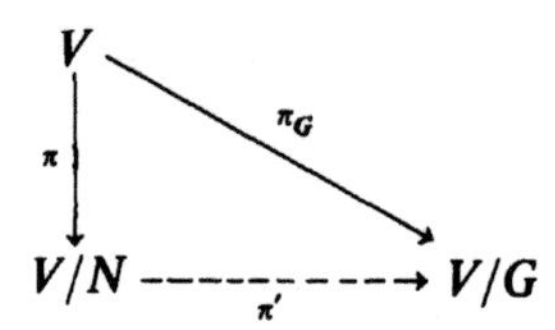

commutative. Clearly π' is an orbit map for the action of G/N on V/N. Since $\pi_G = \pi' \circ \pi$ is separable so also is π'. Hence 6.6 implies that π' is a quotient, because V/G is normal.

6.11 Corollary. *Let G be an affine k-group and let $N \subset M$ be closed subgroups of G defined over k such that N is normal in M. Then M/N acts k-morphically on G/N, the quotient exists and is isomorphic to G/M. For each point $x \in G/N$ the orbit map o_x is an isomorphism of M/N onto $x \cdot (M/N)$. If M and N are normal subgroups of G, these varieties are isomorphic as k-groups.*

Proof. To prove the first assertion we replace (V, G, N) in 6.10 by (G, M, N) with M acting via *right* translations. Then N acts trivially on G/N and we get by 6.10 a k-morphic action of M/N on G/N. Its fibres are the images of the left cosets $g \cdot M$ in G/N, hence the fibres of $G/N \to G/M$ and the first assertion follows. Clearly this action is free (1.8). But, since $m \mapsto g \cdot m$ is an isomorphism of M onto $g \cdot M$ in G, it follows that the differential of o_x is surjective hence o_x is an isomorphism. The last assertion is clear.

6.12 Proposition. *Let G be an affine k-group, and let M and N be closed subgroups defined over k. Let $\pi: G \to G/N$ be the quotient morphism. Then $L(M) \cap L(N) = L(M \cap N)$ if and only if π induces a separable morphism $\pi': M \to \pi(M)$. In this case $M \cap N$ is defined over k.*

As an immediate consequence we have:

Corollary. *If* $\mathrm{char}(k) = 0$ *then $L(M) \cap L(N) = L(M \cap N)$ and $M \cap N$ is defined over k.*

Proof. π' is just the map of M onto the M-orbit of $\pi(e) \in G/N$, and the stability group of $\pi(e)$ in M is $M \cap N$. Moreover, $\ker(d\pi')_e = L(M) \cap \ker(d\pi)_e = L(M) \cap L(N)$, so the proposition follows from 6.7.

6.13 Proposition. *Let G, G' be k-groups and $f: G \to G'$ a surjective k-morphism. Let H' be a k-subgroup of G' such that $\mathfrak{g}' = df(\mathfrak{g}) + \mathfrak{h}'$. Then $H = f^{-1}(H')$ is defined over k and f induces a k-isomorphism of G/H onto G'/H'.*

Let $\pi': G' \to G'/H'$ be the canonical projection. $\ker d\pi' = \mathfrak{h}'$, therefore $d(\pi' \circ f)$ is surjective and $\pi' \circ f: G \to G'/H'$ ise separable. Then $H = f^{-1}$ $(\pi'(H'))$ is defined over k. The map f induces a bijective k-morphism $f': G/H \to G'/H'$. The assumption implies that df' is surjective, hence f' is an k-isomorphism.

6.14 Remarks on Local Cross-Sections. Let G act k-morphically and freely (see 1.8) on the k-variety V and assume there is an orbit map $\pi: V \to W$ defined over k on a k-variety W. A (morphic) local cross-secton over k for π is a k-morphism $\sigma: U \to V$ where U is k-open in W, such that $\pi \circ \sigma = Id$. In that case, let $\tau: G \times U \to V$ be the map $(g, u) \mapsto g \cdot \sigma(u)$. It is bijective k-morphism of $G \times U$ onto $\pi^{-1}(U)$.

Lemma. *Let U, σ, τ be as above.*

(i) *The action of G on $\pi^{-1}(U)$ is principal if and only if τ is an isomorphism.*
(ii) *If U and $\pi^{-1}(U)$ are normal, then τ is an isomorphism.*

Proof. (i) We have, in the notation of 1.8, $\tau^{-1}(x) = (g(\sigma(\pi(x)), x), \pi(x))$ hence τ^{-1} is k-morphic if and only if $g(x, y)$ is k-morphic on the graph of the action of G on $\pi^{-1}(U)$, i.e. if and only if the action is principal.

(ii) The simple points $x \in \pi^{-1}(U)$ such that $\pi(x)$ is simple form an non-empty open set. If x is one, then the unique intersection point of $\sigma(U)$ and $G \cdot x$ is also one. Assume then that $y \in U$ is simple on W and $x = \sigma(y)$ is simple on V. The relation $\pi \circ \sigma = Id.$ shows that $d\sigma_y$ is injective and $d\pi_x$ is surjective. This will then also be true if we replace x by $g \cdot x$ for any $g \in G$. Since $d\pi$ annihilates the tangent space to $G \cdot x$ at $g \cdot x$ it follows that $d\tau_{(g,y)}$ is an isomorphism. That τ is an isomorphism then follows from AG.17.3, 18.2.

If τ is an isomorphism then the fibration π is said to be *trivial over U*. It is *locally trivial* if W is covered by such U's. In this case, the action of G is not only free, but also principal, by the lemma. The latter also shows that if V and W are normal, and W is covered by the domains of local cross-sections, then the action is principal.

It is clear that if W is covered by k-open subsets over each of which the map π admits a k-morphic cross-section, then $\pi(k): V(k) \to W(k)$ is surjective. This is in particular so if the fibration π is locally trivial over k, hence principal. Note however that a principal action is not always locally trivial, even in the case of a closed subgroup acting by left or right translations on a group.

As a simple example, let $V = \mathbf{GL}_n$, $W = \mathbf{GL}_1$ and π be defined by $\pi(v) = (\det v)^2$. Let $k = \mathbb{R}$. Then $\pi(V(k))$ consists of strictly positive numbers, hence $\pi(k)$ is not surjective. A fortiori, the fibration is not locally trivial over $\mathbb{R}$. In fact, it is neither over $\mathbb{C}$, because $G = \ker \pi$ consists of two connected components (elements of determinant ± 1) and if there were a cross section over an open set $U \subset W$, then $\pi^{-1}(U)$ would not be irreducible. This is to be contrasted with the fact that the fibration of a real Lie group by a closed subgroup is always locally trivial; but there, the neighborhoods are with respect to the ordinary manifold topology, whereas we deal here with Zariski-open subsets (and algebraic maps).

However, if G is a k-group and H a closed subgroup, then every $y \in G/H$ has a neighborhood U admiting an étale covering over which the induced fibration becomes trivial, a fact which has led Serre to introduce the notion of locally isotrivial fibrations [32].

6.15 Proposition. *Let G be a finite k-group and V an affine k-variety on which G operates k-morphically. Then V/G exists over k.*

Proof. Let $\varphi \in K[V]$. Then φ annihilates the polynomial $\prod_g (T - g(\varphi))$, the coefficients of which are in $K[V]^G$, therefore φ is integral over $K[V]^G$. Since $K[V]$ is finitely generated, so is $K[V]^G$ (cf. [11], 5, 1.9, Thm. 2). Let Y be the affine variety with coordinate ring $K[V]^G$ and π the morphism associated to the inclusion $\pi^o: K[V]^G \hookrightarrow K[G]$. It is surjective (AG.3.6) and constant on the orbits of G. Let now $x, y \in V$ be on different orbits. There exists $\varphi \in K[V]$ which is zero on x and 1 on $G \cdot y$. Then $\prod_g g(\varphi)$ is zero on $G \cdot x$ and one on $G \cdot y$. As consequence, $K[V]^G$ separates the orbits of G and the fibres of π are the orbits of G, i.e. π is an orbit map. The map π is closed because $K[V]$ is integral over $K[Y]$ (AG.3.6). It is therefore also open (6.4).

Let now $f \in K(V)^G$. We want to show that f is in the quotient ring of $K[Y] = K[V]^G$. More precisely, we claim:

(∗) *Assume that f is defined at $v \in V$. Then there exist $a, b \in K[V]^G$ such that b is a non-divisor of zero, $b(v) \neq 0$ and $f = a/b$.*

The function f is defined at all points of $G \cdot v$. For every g, there exists a non-divisor of zero $b_g \in K[V]$ such that $f \cdot b_g \in K[V]$ and $b_g(gv) \neq 0$. We can then write $G \cdot v$ as a disjoint union of subsets A_j $(j=1,\ldots,m)$ and find a regular function b_j which does not vanish on A_j, but is zero on A_k $(k>j)$. Then a suitable linear combination b_o of the b_j will be non-divisor of zero and not vanish at any point of $G \cdot v$. Consequently, the product b of the $g(b_o)$ is non-divisor of zero, which does not vanish at any point of $G \cdot v$ and belongs to $K[V]^G$. This proves $(*)$.

Therefore $K(Y)$, which is by definition the full quotient ring of $K[Y]$, is equal to $K(V)^G$. This implies that π is separable (AG.2.5). In order to prove that π is a quotient morphism, there remains to check condition (2) of 6.1. Let $U \subset V$ be open (non-empty) and $f \in K[U]$, which is constant on the intersections of U with the orbits of G. We have to show that $f \in \pi^o(K[\pi(U)])$. The condition on f implies that the function $g(f)$ defined on $g(U)$ by $gf(x)=f(g^{-1} \cdot x)$ coincides with f on $U \cap gU$ and it follows easily that we may use the $g(f)$ to extend f to a regular function on the union of the $g(U)$. We may assume therefore that U is G-stable. The function f belongs to $K(\bar{U})^G$, which is $K(\pi(\bar{U}))$. It suffices to show that, viewed as an element of the latter, it is defined on $\pi(U)$. But this follows from $(*)$, applied to the restriction of π to $\bar{U}$.

The space $K[V]$ is the union of G-invariant finite dimensional subspaces defined over k (1.9). Let E be one. Then E^G is defined over k_s, since all points of G are rational over k_s. The space $E^G(k_s)$ is also invariant under $\mathrm{Gal}(k_s/k)$, therefore it is defined over k. It follows that $K[V]$ has a k-structure. Therefore V/G and π are defined over k.

6.16 *Categorical Quotients.* Let V be a k-variety on which G acts k-morphically. A *categorical quotient* (over k) of V by G is a pair (π, W) consisting of a k-variety W and a surjective k-morphism $\pi: V \to W$, constant on the orbits of G and having the following universal property:

(CQ) *If $\sigma: V \to Z$ is a k-morphism constant on the orbits of G, then there exists a k-morphism $\tau: W \to Z$ such that $\sigma = \tau \circ \pi$.*

It is obvious that, if it exists, a categorical quotient is unique up to a unique isomorphism, which is defined over k. In [22] a quotient in the sense used so far here, is called *geometric*. We shall use this terminology in this subsection, but then drop again the adjective geometric. It is clear that a geometric quotient is categorical, but it may happen that a categorical quotient exists when a geometric one does not, e.g. when orbits are not all closed (an example will be given in 8.21).

There is an obvious transitivity for this notion, analogous to 6.10. Let N be a closed normal k-subgroup of G. Assume that the categorical quotient V' of V by N exists and is normal. Then the natural action of G/N on V' is k-morphic (6.10). Assume that the categorical quotient V'' of V' by G/N exists. Then the categorical quotient of V by G exists and is equal to V''.

This follows immediately from the definition. The assumption of normality was only used to insure that G/N operate k-morphically on V', but it can be dispensed with (see the Bibliographical Note).

Proposition. *Let X be a k-variety on which G acts k-morphically and $\pi: X \to Y$ a k-morphism of X into a k-variety Y, which is constant along the G-orbits. Assume*

(*a*) *π is surjective.*
(*b*) *Given U open in Y, $\pi^o K[U]$ is the ring of G-invariant regular functions on $\pi^{-1}U$.*
(*c*) *If F is closed G-invariant subspace of X, then $\pi(F)$ is closed. If (F_i) $(i \in I)$ is a collection of closed G-invariant subspaces of X, then $\pi(\cap F_i) = \cap \pi(F_i)$.*

Then Y is a categorical quotient over k of X by G.

Proof. Note first that (*b*) implies that the G-invariant regular functions on $\pi^{-1}(U)$ are constant on the fibres of π and separate them.

Let $\sigma: X \to Z$ be a k-morphism of X into a k-variety Z, which is constant on the G-orbits. Choose a finite open affine cover (V_i) $(i \in I)$ of Z. We prove first that there is an open cover (U_i) $(i \in I)$ of Y such that $\pi^{-1}(U_i) \subset \sigma^{-1}(V_i)$ for all $i \in I$. Let $F_i = X - \pi^{-1}(V_i)$. It is closed and G-invariant, hence, by (*c*), $\pi(F_i)$ is closed and $U_i = Y - \pi(F_i)$ is open. Obviously, $\pi^{-1}(U_i) \subset \sigma^{-1}(V_i)$. Moreover, $\{\sigma^{-1}(V_i)\}$ is a cover of X, therefore $\cap F_i = \phi$. By (*c*), $\cap \pi(F_i) = \phi$, and $\{U_i\}$ is indeed an open cover of Y. Fix i, for $f \in K[V_i]$, we have by (*b*)

$$\sigma^o f \,|\, \pi^{-1} U_i \subset \pi^o K[U_i] \tag{1}$$

therefore f is constant along the fibres of $\pi | U_i$. Since the elements of $K[V_i]$ separate the points of V_i and i is arbitrary, we see that σ is constant along the fibres of π. There exists therefore a map $\sigma': Y \to Z$ such that $\sigma = \sigma' \circ \pi$. Then, for f as before, (1) shows that

$$\pi^o(\sigma'^o f) | \pi^{-1} V_i = \pi^o \varphi \quad (\varphi \in K[U_i]),$$

whence $\sigma'^o f \,|\, U_i = \varphi$, which shows that σ' is a morphism. Similarly, if $f \in k[V_i]$, then $\sigma' f \,|\, U_i \in k[U_i]$, hence σ' is defined over k, and the proposition is proved.

Remark. It follows from (*a*) and (*c*) that a subset $V \subset Y$ is closed (resp. open) if and only if $\pi^{-1}(V)$ is closed (resp. open). In view of (*a*), it suffices to show this for V closed. If V is closed, then $\pi^{-1}(V)$ is closed. Suppose the latter. Since $\pi^{-1}(V)$ is G-invariant, $\pi(\pi^{-1} V)$ is closed by (*c*). But $V = \pi(\pi^{-1} V)$ by (*a*).

Bibliographical Note

6.10 is valid without assuming V/N and V/G to be normal (see [27, Prop. 2]). Similarly, it follows from Lemma 3 and Prop. 2 of [27] that the Cor. of 6.7 is true also if V_i/G_i is not normal $(i = 1, 2)$. However, these facts will not be needed in this book.

Theorem 6.8 is proved here for affine groups. It remains true however if affine k-groups is replaced by algebraic k-groups: the existence of a quotient

k-structure on G/H was proved by M. Rosenlicht [25] for k algebraically closed and by Weil [34] in general; the fact that G/H is quasi-projective is due to W.L. Chow (Algebraic Geometry and Topology, a Symposium in honor of S. Lefschetz, Princeton Univ. Press 1957, 122–128).

6.15 is proved in [22:III, §12]. See also [32:III, §12]. In both references it is pointed out that the proof extends to a general variety V provided that every orbit of G is contained in an open affine subset, a condition which is fulfilled if V is quasi-projective, but not always otherwise.

For a discussion of geometric and categorical quotients of schemes by group schemes, see [22]. Proposition 6.16 is Remark (6), p. 8 there.

§7. Algebraic Groups In Characteristic Zero

In this section it is assumed that $\mathrm{char}(k) = 0$ *and* G *is an affine* k*-group* G. Our aim is to obtain some basic results of Chevalley [12a] on the *algebraic Lie algebras* $\mathfrak{h}$ in $\mathfrak{g} = L(G)$, i.e. on those of the form $\mathfrak{h} = L(H)$ for some closed subgroup H of G.

7.1 *The operators* $\mathscr{A}$ *and* $\mathfrak{a}$. Recall (6.12) that, *if* H *and* N *are closed subgroups of* G *then*

$$L(H \cap N) = L(H) \cap L(N). \tag{1}$$

It follows that *if* H *is connected then*

$$H \subset N \Leftrightarrow L(H) \subset L(N). \tag{2}$$

In analogy with the notion of group closure in §2, we associate to a subset M of $L(G)$ the intersection of all closed subgroups H of G such that $M \subset L(H)$, to be denoted $\mathscr{A}(M)$. It is connected, and by 7.1, is the smallest closed subgroup of G whose Lie algebra contains M; its Lie algebra,

$$\mathfrak{a}(M) = L(\mathscr{A}(M)),$$

is therefore the smallest algebraic Lie algebra in $\mathfrak{g}$ containing M.

We can of course define $\mathscr{A}(M)$ in non-zero characteristic but then $L(\mathscr{A}(M))$ does not necessarily contain M, in fact may be zero even if M is not, and this notion seems uninteresting in that case.

7.2 Proposition. *Let* $\pi: G \to G'$ *be a surjective morphism of algebraic groups, and let* $M \subset L(G)$. *Then*

$$\pi(\mathscr{A}(M)) = \mathscr{A}(d\pi(M)) \quad \textit{and} \quad d\pi(\mathfrak{a}(M)) = \mathfrak{a}(d\pi(M)).$$

Proof. If H is a closed subgroup of G, then, since we are in characteristic zero, $\pi: H \to \pi(H)$ is separable, and therefore $d\pi(L(H)) = L(\pi(H))$.

It follows from 6.7 that, under $d\pi$, both the images and inverse images of algebraic Lie algebras are algebraic. In particular $d\pi(\mathfrak{a}(M))$ is an algebraic

Lie algebra containing $d\pi(M)$, and hence $\mathfrak{a}(d\pi(M))$. The reverse inclusion follows since $d\pi^{-1}(\mathfrak{a}(d\pi(M)))$ is an algebraic Lie algebra containing M, and hence $\mathfrak{a}(M)$.

Now $\pi(\mathscr{A}(M))$ and $\mathscr{A}(d\pi(M))$ are connected subgroups of G' with the same Lie algebra, and hence they are equal (see 7.1).

7.3 *The structure of $\mathscr{A}(X)$ for $X \in \mathfrak{gl}(V)$.*

(1) *X is nilpotent,* $\neq 0$. Define $\alpha: \mathbf{G}_a \to G = \mathrm{GL}(V)$ by

$$\alpha(t) = \exp(tX) = \sum_{n \geq 0} (n!)(tX)^n.$$

Since X is nilpotent, α is a polynomial map, and it is clearly a homomorphism. The minimal polynomial of X is a monomial, so it follows that the non-zero powers of X are linearly independent. Since $X \neq 0$ this implies α is injective. Hence α induces an isomorphism $\mathbf{G}_a \cong \alpha(\mathbf{G}_a) = H$ of algebraic groups, because we are in characteristic zero. Since $d\alpha: L(\mathbf{G}_a) \to \mathfrak{gl}(V)$ is the map $t \mapsto t \cdot X$, we conclude that $H = \mathscr{A}(X)$, for dimension reasons. Therefore

$$\mathscr{A}(X) \cong \mathbf{G}_a.$$

(2) $X = \mathrm{diag}(x_1, \ldots, x_n) \in L(\mathbf{D}_n)$.

This case is included for completeness, though it will not be needed elsewhere.

We shall invoke here some elementary results on tori to be proved below in §8. In particular we shall see in 8.2 that $H = \mathscr{A}(X)$ is the intersection of $\ker \chi$ for all characters $\chi \in X(\mathbf{D}_n)$ such that $\chi(H) = 1$, and the latter condition is equivalent to: $d\chi(L(H)) = 0$. If $\chi(\mathrm{diag}(t_1, \ldots, t_n)) = t_1^{m_1} \cdots y_n^{m_n}$ then $d\chi(\mathrm{diag}(s_1, \ldots, s_n)) = \sum m_i s_i$, where we identify $L(\mathbf{D}_n)$ with the diagonal matrices in $\mathfrak{gl}_n$. Thus, if

$$L = \{(m_i) \in \mathbb{Z}^n \mid \sum m_i x_i = 0\}$$

then

$$\mathscr{A}(X) = \{\mathrm{diag}(t_1, \ldots, t_n) \mid \prod t_{t_i}^{m_i} = 1 \text{ for all } (m_i) \in L\}$$

and

$$\mathfrak{a}(X) = \{\mathrm{diag}(s_1, \ldots, s_n) \mid \sum m_i s_i = 0 \text{ for all } (m_i) \in L\}.$$

(3) *If $X = X_s + X_n$ is the Jordan decomposition of X, then*

$$\mathscr{A}(X) = \mathscr{A}(X_s) \cdot \mathscr{A}(X_n) \text{ and } \mathfrak{a}(X) = \mathfrak{a}(X_s) + \mathfrak{a}(X_n).$$

This is clear, and, in fact, these products and sums are direct. In view of (1) and (2), we now have the structure of $\mathscr{A}(X)$ in general.

Remark. As a group analogue of (1), we have: *if $u \in \mathrm{GL}(V)$ is unipotent and $\neq I$, then $\mathscr{A}(u) \cong \mathbf{G}_a$.* To see this, write $x = I - u$. This is a nilpotent transformation; hence $X = \log u = \sum_{i>0} i^{-1}(-x)^i$ is a polynomial in x, with 0 as constant term, and is therefore nilpotent, too. By (1), $\mathscr{A}(X)$ is isomorphic to $\mathbf{G}_a$ and contains $u = \exp X$, hence $\mathscr{A}(u) \subset \mathscr{A}(X)$. But u has infinite order, and therefore $\dim \mathscr{A}(u) \geq 1$, whence $\mathscr{A}(u) = \mathscr{A}(X)$.

7.4 Lemma. *Let $\pi: G \to GL(E)$ be a rational representation, and let $N \subset M$ be vector subspaces of E. Put*

$$H = \{g \in G \mid \pi(g)N = N, \pi(g)M = M, \pi(g)_{M/N} = e\}.$$

Then

$$L(H) = \mathrm{tr}(M, N) = \{X \in \mathfrak{g} \mid d\pi(X)M \subset N\}.$$

In particular, $\mathrm{tr}(M, N)$ *is an algebraic Lie algebra.*

Proof. Clearly $\mathfrak{b} = \mathrm{tr}(M, N)$ is a Lie algebra containing $L(H)$. Conversely, supposing $X \in \mathfrak{b}$, it suffices to show that $\mathscr{A}(X) \subset H$. Since $\pi(\mathscr{A}(X)) = \mathscr{A}(d\pi(X))$ (by 7.2) we may replace G by $\pi(G)$. Since X_s and X_n are polynomials in X without constant term, it follows that $X_s, X_n \in \mathfrak{b}$ along with X, so we are reduced to the cases $X = X_s$ and $X = X_n$.

If $X = X_n$ then (see 7.3 (1)) $\mathscr{A}(X) = \{\exp(tX)\}$, which clearly lies in H. If $X = X_s$ we may assume X is diagonalized with respect to a basis $e_1, e_2, \ldots$ such that $e_1, \ldots, e_n$ span N, $e_1, \ldots, e_m$ span M, and $e_{n+1}, \ldots, e_m$ span $M' \subset \ker(X)$. It is then clear that $\mathscr{A}(X)$ lies in the group of diagonal matrices $\mathrm{diag}(d_1, \ldots, d_m, \ldots)$ for which $d_{n+1} = \cdots = d_m = 1$. Evidently these matrices belong to H. Q.E.D.

7.5 Proposition. *Let $(H_i)_{i \in I}$ be a family of closed smooth irreducible subvarieties of G such that, for each $i \in I$, $e \in H_i$, and $H_i^{-1} = H_j$ for some j. Let H be the subgroup generated by the H_i's. Then H is closed and $\mathfrak{h} = L(H)$ is spanned by the vector spaces*

$$\mathrm{Ad}(h)T(x^{-1}H_i)_e \quad (h \in H; x \in H_i; i \in I).$$

Proof. According to 2.2, H is closed, and there is a finite sequence $i_1, \ldots, i_s$ in I such that the product map $p: W = H_{i_1} \times \cdots \times H_{i_s} \to H$ is surjective. We shall change notation now and write H_j in place of H_{i_j} $(1 \leqq j \leqq s)$. Since p is separable (char 0) it follows that, for some $w = (w_1, \ldots, w_s) \in W$, $(df)_w: T(W)_w \to T(H)_v$ is surjective, where $v = p(w) = w_1 \ldots w_s$.

We now introduce $v_j = w_1 \cdots w_j$, $H'_j = w_j^{-1}H_j$, and $H''_j = v_j H'_j v_j^{-1}$, $(1 \leqq j \leqq s)$. Define $\alpha: W' = H'_1 \times \cdots \times H'_s \to W$ by $\alpha(x_1, \ldots, x_s) = (w_1x_1, \ldots, w_sx_s)$, and put $\beta = \mathrm{Int}(v_1) \times \cdots \times \mathrm{Int}(v_s): W' \to W'' = H''_1 \times \cdots \times H''_s$. We claim that the rectangle

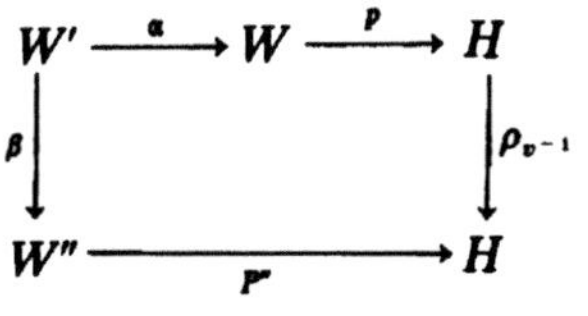

$$\begin{array}{ccccc} W' & \xrightarrow{\alpha} & W & \xrightarrow{p} & H \\ \Big\downarrow{\scriptstyle\beta} & & & & \Big\downarrow{\scriptstyle\rho_{v^{-1}}} \\ W'' & & \xrightarrow{\quad p'' \quad} & & H \end{array}$$

is commutative, where p'' is the product map. In fact,

$$\begin{aligned} p''\beta(x_1, \ldots, x_s) &= v_1x_1v_1^{-1}v_2x_2v_2^{-1} \cdots v_{s-1}^{-1}v_sx_sv_s^{-1} \\ &= w_1x_1w_2x_2 \cdots w_sx_sv^{-1}, \end{aligned}$$

(because $v_{j-1}^{-1}v_j = w_j (1 \leqq j \leqq s)$ and $v_s = v$); hence

$$p''\beta(x_1,\ldots,x_s) = (p\alpha(x_1,\ldots,x_s))v^{-1}.$$

Writing e also for $(e,\ldots,e) \in W'$, we have $\alpha(e) = w$. Since α and $\rho_{v^{-1}}$ are isomorphisms of varieties, and since $(dp)_w$ is surjective, it follows that the differential,

$$d(p''\circ\beta)_e : T(W')_e \to T(H)_e = \mathfrak{h},$$

of $p''\circ\beta = \rho_{v^{-1}}\circ p\circ\alpha$ at e is also surjective. If $X = (X_1,\ldots,X_s) \in T(W')_e$, then $d(p''\circ\beta)_e(X) = \sum d(\mathrm{Int}(v_j))_e(X_j) = \sum \mathrm{Ad}(v_j)(X_j)$. Thus $\mathfrak{h} = \sum \mathrm{Ad}(v_j)T(H'_j)_e$. Since $v_j \in H$ and since $H'_j = w_j^{-1}\cdot H_j$, with $w_j \in H_j$, the proposition is proved.

7.6 Theorem. *In 7.5, suppose that each H_i is a closed subgroup of G, with Lie algebra $\mathfrak{h}_i$. Then $\mathfrak{h}$ is spanned, as a vector space, by the spaces $Ad(h)(\mathfrak{h}_i)(h \in H; i \in I)$, and it is generated, as a Lie algebra, by the $\mathfrak{h}_i$ $(i \in I)$.*

Proof. If $x \in H_i$ then, since H_i is a group, $x^{-1}H_i = H_i$, so $T(x^{-1}H_i)_e = \mathfrak{h}_i$. Hence the first assertion is just 7.5.

Let M be the Lie subalgebra generated by the $\mathfrak{h}_i (i \in I)$. We must show that the inclusion $M \subset \mathfrak{h}$ is an equality. Thanks to the conclusion above it will suffice to show that M is stable under $\mathrm{Ad}(H)$, i.e. that $H \subset \mathcal{N}_G(M) = \mathrm{Tr}(M, M)$. Since the latter is a group, it suffices to show that it contains each H_i. But H_i is connected, so the latter follows if we show that $\mathfrak{h}_i \subset L(\mathrm{Tr}(M, M))$. According to 7.4 (applied to Ad, with $M = N$) we have $L(\mathrm{Tr}(M, M)) = \mathrm{tr}(M, M)$. Since M is a Lie algebra containing $\mathfrak{h}_i$ we have $[\mathfrak{h}_i, M] \subset M$. Q.E.D.

7.7 Corollary. *The following conditions on a Lie subalgebra $\mathfrak{h}$ of $\mathfrak{g}$ are equivalent:*

(1) $\mathfrak{h}$ *is algebraic.*
(2) *If $X \in \mathfrak{h}$ then $\mathfrak{a}(X) \subset \mathfrak{h}$.*
(3) $\mathfrak{h}$ *is spanned by algebraic Lie algebras.*
(4) $\mathfrak{h}$ *is generated as a Lie algebra by algebraic Lie algebras.*

The implications (1)⇒(2)⇒(3)⇒(4) are clear, and (4)⇒(1) follows immediately from 7.6.

7.8 Proposition. *Let $H = (M, N)$, where M and N are closed connected normal subgroups of G. Then $\mathfrak{h} = [\mathfrak{m}, \mathfrak{n}]$, where $\mathfrak{h}, \mathfrak{m}, \mathfrak{n}$ are the Lie algebras of H, M, N.*

Proof. For $x, y \in G$ write

$$c_x(y) = c'_y(x) = (x, y) = xyx^{-1}y^{-1}.$$

Then H is generated by the sets $H_a = c_a(N)$ and $H'_a = c'_a(N) = H_a^{-1}$ where a varies over M. These sets satisfy the hypotheses of 7.5 because N is a connected closed subgroup. Hence $\mathfrak{h}$ is spanned by subspaces of the form

$$\mathrm{Ad}(h)T(c_a(b)^{-1}H_a)_e \text{ and } \mathrm{Ad}(h)T(c'_a(b)^{-1}H'_a)_e,$$

for $h \in H$, $a \in M$, $b \in N$.

The inclusion $[\mathfrak{m}, \mathfrak{n}] \subset \mathfrak{h}$ follows from 3.12, and the invariance of $[\mathfrak{m}, \mathfrak{n}]$ under $\mathrm{Ad}(G)$ follows from the normality of M and N. Thus, it remains to show that, for $a \in M$ and $b \in N$, we have

$$T(c_a(b)^{-1}H_a)_e, T(c'_a(b)^{-1}H'_a)_e \subset [\mathfrak{m}, \mathfrak{n}].$$

Consider $f: N \to G$ defined by $f(x) = c_a(b)^{-1}c_a(bx)$ Then $f(e) = e$ and $(df)_e(\mathfrak{n}) = T(c_a(b)^{-1}c_a(N))_e$, clearly. Explicitly,

$$f(x) = bab^{-1}a^{-1}abxa^{-1}(bx)^{-1} = baxa^{-1}b^{-1}bx^{-1}b^{-1}$$
$$f(x) = (ba)\cdot x \cdot (ba)^{-1} \cdot bx^{-1}b^{-1}.$$

Thus $(df)_e = \mathrm{Ad}(ba) - \mathrm{Ad}(b) = \mathrm{Ad}(b)(\mathrm{Ad}(a) - 1)$. Since $\mathrm{Ad}(b)$ leaves $[\mathfrak{m}, \mathfrak{n}]$ stable it will suffice to show that

$$(\mathrm{Ad}(a) - 1)(\mathfrak{n}) \subset [\mathfrak{m}, \mathfrak{n}] \text{ for } a \in M.$$

(The case of $c'_a(b)^{-1}c'_a(N)$ follows from similar arguments which we omit.)

Let $\pi: \mathfrak{g} \to \mathfrak{g}' = \mathfrak{g}/[\mathfrak{m}, \mathfrak{n}]$ be the natural projection. Fix $X \in \mathfrak{n}$ and define $\alpha: M \to \mathfrak{g}'$ by $\alpha(a) = \pi((\mathrm{Ad}(a) - 1)(X))$. We must show that $\alpha = 0$.

Since $[\mathfrak{m}, \mathfrak{n}]$ is stable under $\mathrm{Ad}\, G$, the quotient $\mathfrak{g}'$ is also a G-module. For $a, a' \in M$ we have $\mathrm{Ad}(aa') - 1 = \mathrm{Ad}(a)(\mathrm{Ad}(a') - 1) + (\mathrm{Ad}(a) - 1)$, from which it follows that $\alpha(aa') = \mathrm{Ad}(a)\alpha(a') + \alpha(a)$. This implies that $P = \{a \in M \mid \alpha(a) = 0\}$ is a closed subgroup of M, and that $\alpha(aP) = \alpha(a)$ for all $a \in M$. Hence α can be factored through the quotient:

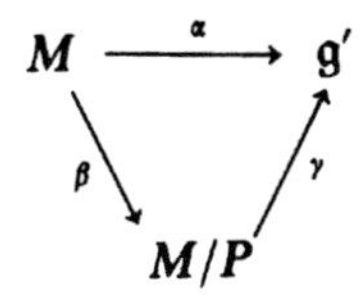

where β is the quotient morphism. Since γ is injective, it is an isomorphism of M/P onto its image (char 0). Since $\alpha(M) = M/P$ is connected we can show it is a single point by proving that $(d\alpha)_e = 0$.

We have $\alpha = \pi \circ \delta$ where $\delta(a) = (\mathrm{Ad}(a) - 1)(X)$ so $(d\alpha)_e = (d\pi)_0 \circ (d\delta)_e = \pi \circ (d\delta)_e$. (Since π is linear $(d\pi)_0 = \pi$.) Now using 3.9(2), we get $(d\delta)_e(Y) = \mathrm{ad}(Y)(X) = [Y, X]$, so $(d\delta)_e(\mathfrak{m}) = [\mathfrak{m}, X] \subset [\mathfrak{m}, \mathfrak{n}]$. Thus indeed $\pi \circ (d\delta)_e(\mathfrak{m}) = 0$. Q.E.D.

7.9 Corollary. *Let $\mathfrak{h}$ be a Lie subalgebra of $\mathfrak{g}$. Then $[\mathfrak{h}, \mathfrak{h}] = [\mathfrak{a}(\mathfrak{h}), \mathfrak{a}(\mathfrak{h})]$, and is an algebraic Lie algebra.*

Proof. $\mathfrak{h} \subset \mathrm{tr}(\mathfrak{h}, [\mathfrak{h}, \mathfrak{h}])$, clearly, and the latter is algebraic, by (7.4). Therefore $\mathfrak{a}(\mathfrak{h}) \subset \mathrm{tr}(\mathfrak{h}, [\mathfrak{h}, \mathfrak{h}])$, i.e. $[\mathfrak{a}(\mathfrak{h}), \mathfrak{h}] \subset [\mathfrak{h}, \mathfrak{h}]$. Therefore $\mathfrak{h} \subset \mathrm{tr}(\mathfrak{a}(\mathfrak{h}), [\mathfrak{h}, \mathfrak{h}])$, so again we see that $\mathfrak{a}(\mathfrak{h}) \subset \operatorname{tr}(\mathfrak{a}(\mathfrak{h}), [\mathfrak{h}, \mathfrak{h}])$, i.e. $[\mathfrak{a}(\mathfrak{h}), \mathfrak{a}(\mathfrak{h})] \subset [\mathfrak{h}, \mathfrak{h}]$. The opposite inclusion is obvious.

It follows from 7.8 that $[\mathfrak{a}(\mathfrak{h}), \mathfrak{a}(\mathfrak{h})]$ is the Lie algebra of $(\mathscr{A}(\mathfrak{h}), \mathscr{A}(\mathfrak{h}))$, and this shows that $[\mathfrak{h}, \mathfrak{h}]$ is algebraic.

Bibliographical Note

Linear algebraic groups over $\mathbb{C}$ were studied around the end of the XIX-th century by Maurer in a series of papers (see notably Sitz.-Ber. Bayer. Akad. 24 (1894)). One of his main results is the fact that such a group is a rational variety. Later, *E.* Cartan [C. R. Acad. Sci. Paris **120** (1895), 544–548] announced some further results on algebraic groups, notably Cor. 7.9 above, but never published the proofs. The topic then fell into oblivion. It was revived by Chevalley-Tuan and then by Chevalley [12]. The main results of this paragraph, in particular 7.6 to 7.9, are all proved in [12(a)]. The main tool of Chevalley is a formal exponential, which allows him to set up an analogue of the familiar correspondence between Lie algebras and Lie groups of Lie group theory. Because of this, he had to restrict himself to groundfields of characteristic zero. Here, we could dispense with this notion by using the structure of variety of the quotient space G/H of an algebraic group by a closed subgroup, and the separability of morphisms in characteristic zero. The latter fact was of course also used in [12], so that the main point is really the possibility of viewing G/H as an algebraic variety. To see this illustrated concretely, the reader may compare the proof of 7.1(2) given here with that of [12(a), p. 157].

Chapter III

Solvable Groups

In this chapter, all algebraic groups are affine, unless the contrary is explicitly allowed. G is a k-group.

§8. Diagonalizable Groups and Tori

8.1 Lemma. *Let H be an abstract group, and let X denote the set of homomorphisms $H \to K^*$. Then X is linearly independent as a set of functions from H to K.*

Proof. If not, let $n > 0$ be minimal such that there exist linearly dependent $\chi_1, \ldots, \chi_n \in X$; say

$$f = \left(\sum_{i<n} \alpha_i \chi_i \right) + \chi_n = 0.$$

Choose $h_0 \in H$ such that $\chi_n(h_0) \neq \chi_1(h_0)$ (clearly $n > 1$). Then, for all $h \in H$,

$$0 = f(h_o h) - \chi_n(h_o) f(h) = \sum_{i<n} \alpha_i(\chi_i(h_o) - \chi_n(h_o))\chi_i(h).$$

This is a non-trivial dependence relation with strictly less than n terms, contradicting the minimality of n.

8.2 *Diagonalizable groups.* Let $A = K[G]$. Then the character group $X(G)$ is a subset of A. We call G *diagonalizable* if $X(G)$ spans A (as K-module). If, further, $X(G)_k$ spans A, then we shall say G is *split over k*. Since $A = K \bigotimes_k A_k$ the latter condition is equivalent to $X(G)_k$ spanning $A_k = k[G]$, as a k-module.

Proposition. *Assume $Y \subset X(G)_k$ spans A_k. Then:*

(a) $Y = X(G)$. *In particular all characters of G are rational over k.*
(b) $A_k = k[X(G)]$, *the group algebra of the finitely generated abelian group*

$X(G)$. Moreover the Hopf algebra structure on A_k is induced by the diagonal map $X(G) \to X(G) \times X(G)$ and the inverse map $X(G) \to X(G)$.

(c) *If H is a closed subgroup of G, then H is a diagonalizable group defined and split over k, and H is defined by character equations (i.e. as the intersection of kernels of characters) in G. Moreover, every character on H extends to a character on G.*

(d) *If $\pi: G \to \mathbf{GL}_n$ is a k-rational representation, then $\pi(G)$ is conjugate over k to a subgroup of $\mathbf{D}_n$. In particular G is k-isomorphic to a closed subgroup of $\mathbf{D}_n$.*

Proof. (a) follows immediately from 8.1 applied to $X(G) \subset A$. Moreover 8.1 implies that $X(G)$ is linearly independent over k so that A_k is, indeed, the group algebra of $X(G)$. The description of the Hopf algebra structure follows directly from the definitions. For example, the diagonal is the comorphism of $G \times G \to G$, and the restriction of this comorphism to characters $X(G) \to X(G \times G) = X(G) \times X(G)$, is easily seen to be the diagonal map of $X(G)$. Since A_k is a finitely generated k-algebra it follows easily that $X(G)$ must be a finitely generated abelian group. This proves (b).

To prove (c) we first note that $B = K[H]$ is a residue class ring of A, and hence B is spanned by the image of $X(G)$. The latter elements are the restrictions to H of characters on G so it follows that H is diagonalizable, and therefore $B = K[X(H)]$. Since $p: A \to B$ is surjective and sends $X(G)$ to $X(H)$ it follows that p is just the group algebra homomorphism induced by a surjection $X(G) \to X(H)$. Since $X(G) = X(G)_k$ it follows that H is defined by character equations over k, and that $X(H) = X(H)_k$. This proves (c).

(d) A generating set of $X(G)$ gives an injective morphism of G into $(\mathbf{GL}_1)^d$ for some d, so G is a commutative group of semi-simple elements. It follows therefore, from 4.6, that for any rational linear representation $\pi: G \to GL(V)$, $\pi(G)$ is diagonalizable. The diagonal entries are then characters of G. Now suppose π is defined over k. Then, since each $\chi \in X(G)$ is defined over k, the eigenspace $V_\chi = \{x \in V \mid \pi(g)x = \chi(g)x, \forall g \in G\}$ is also defined over k (see 5.2). Thus $\pi(G)$ is diagonalizable in $GL(V)$ over k. In case π is immersive this yields a k-isomorphism of G with a closed subgroup of $\mathbf{D}_n$, and the existence of a k-rational immersive π is confirmed by 1.10. Q.E.D.

Corollary. *Let G be diagonalizable. Then G splits over k if and only if $X(G) = X(G)_k$. For any $g \in G$,*

$$\mathscr{A}(g) = \{h \in G \mid \chi(g) = 1 \Rightarrow \chi(h) = 1 \text{ for all } \chi \in X(G)\}.$$

The Lie algebra of G consists of semi-simple elements.

Example. Assume $k = \mathbb{Q}$, and let $m > 2$ be an integer. Let μ_m denote the kernel of $x \mapsto x^m$ in $\mathbf{GL}_1$. Thus $\mu_m(k')$ is the group of mth roots of unity in k' for any k-algebra k'. The definition makes it clear that μ_m is a diagonalizable

k-group split over k. Explicitly, $k[\mu_m] = k[t] = k[T]/(T^m - 1)$. Moreover $X(\mu_m) = \{1, t, \ldots, t^{m-1}\}$.

Let $\pi: \mu_m \to \mathbf{GL}_n$ be a faithful rational representation defined over k. Then the Proposition above guarantees that $\pi(\mu_m)$ is conjugate, *over* k, to a diagonal group. At first sight this appears unreasonable, since k does not contain the eigenvalues of a generator, x, of $\pi(\mu_m)$. The point is that $x \in \mathbf{GL}_n$ will *not* be in $\mathbf{GL}_n(k)$, even though π is defined over k, but x can nevertheless be diagonalized by conjugation by an element of $\mathbf{GL}_n(k)$.

8.3 Corollary. *The contravariant functor $G \mapsto X(G)$ is fully faithful (see proof for definition) from the category of diagonalizable groups split over k and their morphisms as algebraic groups to the category of finitely generated $\mathbb{Z}$-modules. In particular, all morphisms between such groups are defined over k.*

Proof. Let G, G' be two k-split diagonalizable groups, with affine rings A, A' respectively. Consider the commutative triangle

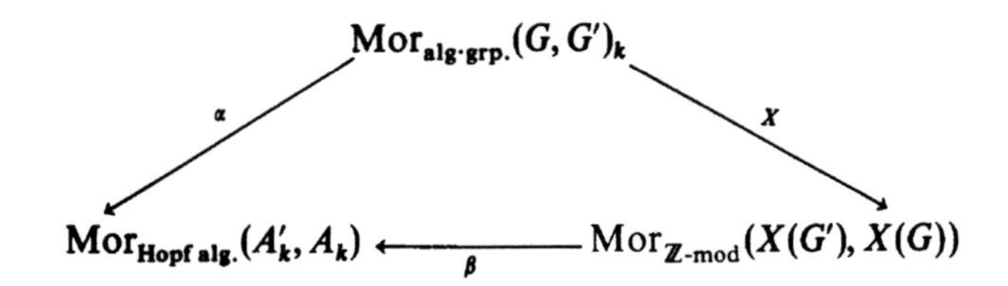

The bijectivity of α is essentially by definition (cf. 1.5). The existence and injectivity of β follows from part (b) of Proposition 8.2. It follows therefore that all three arrows are bijective. The first assertion of the corollary is the bijectivity of X. The last assertion follows because the target of X is independent of k, and therefore so is its source.

Remark. If X is a finitely generated abelian group, then $A = K[X]$ is a Hopf algebra, and therefore it defines a diagonalizable group with character group X provided A has no nilpotent elements. A has nilpotent elements if and only if $\mathrm{char}(k) = p > 0$ and X has elements of order p.

8.4 Proposition. *The following conditions are equivalent:*

(1) *G is diagonalizable.*
(2) *G is isomorphic to a subgroup of $\mathbf{D}_n$ for some $n > 0$.*
(3) *For any rational representation $\pi: G \to \mathbf{GL}_n$, $\pi(G)$ is conjugate to a subgroup of $\mathbf{D}_n$.*
(4) *G contains a dense commutative subgroup consisting of semi-simple elements.*

The corresponding assertion in the split case is:

Proposition′. *The following conditions are equivalent:*

(1′) *G is diagonalizable and split over k.*
(2′) *G is k-isomorphic to a subgroup of* $\mathbf{D}_n$ *for some* $n > 0$.
(3′) *If* $\pi: G \to \mathbf{GL}_n$ *is a rational representation defined over k then* $\pi(G)$ *is conjugate over k to a subgroup of* $\mathbf{D}_n$.

Proof. (1′)⇒(2′) follows from 8.2 (d), (2′)⇒(1′) from 8.2 (c), (1′)⇒(3′) from 8.2 (d), and (3′)⇒(2′) from the existence of an immersive k-rational representation $\pi: G \to \mathbf{GL}_n$ (see 1.10).

Proof. The equivalence of (1), (2) and (3) follows by taking $k = K$ above. (2)⇒(4) is obvious, and (4)⇒(3) follows from 4.6.

Corollary. *Suppose G is diagonalizable (and split over k). Then the same is true of each subgroup of G, and of the image of G under any morphism (defined over k).*

Proof. For subgroups use condition 8.2 (c). If $\pi: G \to G'$ is a morphism (defined over k) then embed G' in $\mathbf{GL}_n$ (over k) and apply condition (3) (resp. (3′)) to get $\pi(G)$ conjugate (over k) to a subgroup of $\mathbf{D}_n$. Then apply (2) (resp. (2′)).

8.5 *Tori.* The diagonal group $\mathbf{D}_n$ is a closed subgroup of $\mathbf{GL}_n$ which is evidently isomorphic, over the prime field, to $(\mathbf{GL}_1)^n$. An algebraic group isomorphic to $\mathbf{D}_n$ is called an *n-dimensional torus.*

The terminology stems from the fact that these groups play here a role analogous to that of topological tori (i.e. products of circle groups) in the theory of compact Lie groups. Note however that, if $K = \mathbb{C}$, the tori considered here are not compact. If they are defined over $\mathbb{R}$ their groups of real points may or may not be compact (see 8.16).

Proposition. *The following conditions on an algebraic group T are equivalent:*

(1) *T is an n-dimensional torus.*
(2) *T is a connected diagonalizable group of dimension n.*
(3) *T is a diagonalizable group and* $X(T) = \mathbb{Z}^n$.

Proof. (1)⇒(2) follows from 8.4(2). (2)⇒(3): Since $\mathbf{GL}_1$ is connected and of dimension 1, its only connected subgroups are $\{e\}$ and $\mathbf{GL}_1$. Applying this to images of characters we see that the character group of a connected group T is torsion free. If further T is diagonalizable then $K[T] = K[X(T)]$, and clearly $\dim T$ (i.e. $\operatorname{tr{\cdot}deg}_K K(T)$) is the rank of the free abelian group $X(T)$. (3)⇒(1): Let $\alpha_1, \ldots, \alpha_n$ be a basis for $X(T)$. Then $K[T] = K[\alpha_1, \alpha_1^{-1}, \ldots, \alpha_n, \alpha_n^{-1}]$ and evidently $\alpha: t \mapsto \operatorname{diag}(\alpha_1(t), \ldots, \alpha_n(t))$ gives the required isomorphism $T \to \mathbf{D}_n$. (For the comorphism $\alpha^0: K[\mathbf{D}_n] \to K[T]$ is visibly surjective, and both groups are connected, of dimension n.)

Corollary. *A closed connected subgroup S of a torus T is a torus and a direct factor.*

By the proposition, S is a torus and $X(S)$ is free. The restriction homomorphism $X(T) \to X(S)$ is surjective (8.2(c)) hence split, and S is a direct factor by 8.3.

8.6 *The multiplicative one-parameter subgroups* in a k-group G are the elements of

$$X_*(G) = \mathrm{Mor}(\mathbf{GL}_1, G).$$

Since $X(G) = \mathrm{Mor}(G, \mathbf{GL}_1)$ we can compose to obtain a map

$$X(G) \times X_*(G) \to \mathbb{Z} = X(\mathbf{GL}_1).$$

given by

$$\langle \chi, \lambda \rangle = m \quad \text{if} \quad (\chi \circ \lambda)(x) = x^m.$$

If G is commutative this is a bilinear map of abelian groups. It follows easily from 8.3 and 8.4 (or even directly) that:

Proposition. *If T is a torus then*

$$X(T) \times X_*(T) \to \mathbb{Z}$$

is a dual pairing over $\mathbb{Z}$.

8.7 Proposition. *Let G be diagonalizable and split over k. Then G is a direct product $G = G^o \times F$, where F is a finite group, and G^o is a torus defined and split over k.*

Proof. Thanks to 8.2(d) we may assume that G is a closed subgroup of some $\mathbf{D}_n$. Moreover 8.4 and 8.2(c) imply that all closed subgroups of $\mathbf{D}_n$ are defined and split over k, and 8.5 implies that G^o is a torus.

According to 8.2(c), the restriction homomorphism $X(\mathbf{D}_n) \cong \mathbb{Z}^n \to X(G^o)$ is surjective. Since G^o is connected, 8.5 implies that $X(G^o)$ is free, so the epimorphism splits. In other words we can find a basis $\chi_1, \ldots, \chi_n$ for $X(\mathbf{D}_n)$ so that $\chi_1, \ldots, \chi_d$ generate the group of characters which annihilate G^o. Then the k-automorphism $x \mapsto \mathrm{diag}(\chi_1(x), \ldots, \chi_n(x))$ of $\mathbf{D}_n$ maps G^o onto $\{\mathrm{diag}(x_1, \ldots, x_n) | x_i = 1, 1 \leqq i \leqq d\}$. Thus $\mathbf{D}_n = \mathbf{D}_d \times G^o$.

It follows that, as a group, $G = F \times G^o$ where $F = G \cap \mathbf{D}_d$. Then clearly $F \cong G/G^o$ so F is a finite group, and the product map $\alpha: F \times G^o \to G$ is a group isomorphism. That it is also an isomorphism of varieties follows because it is so on pairs of corresponding connected components.

8.8 Proposition. *If k is not an algebraic extension of a finite field then $T = (\mathbf{GL}_1)^n$ contains an element t, rational over k, that generates a dense subgroup.*

Proof. If $t=(t_1,\ldots,t_n)$, then t generates a dense subgroup if and only if no non-trivial character of T kills t. Since the characters are all of the form $t\mapsto t_1^{m_1}\cdots t_n^{m_n}$ this requirement is just that $t_1,\ldots,t_n$ be multiplicatively independent. Hence we want k^* to contain free abelian groups of arbitrarily large finite rank. If $\mathrm{char}(k)=0$ this follows from the infinitude of primes in $\mathbb{Z}$. If $\mathrm{char}(k)=p>0$, and if $x\in k$ is transcendental over the prime field $\mathbf{F}_p$, then this follows from the infinitude of primes in the polynomial ring $\mathbf{F}_p[x]$.

Remark. The Proposition above is valid without assuming that T is split over k, but the proof of the general case is somewhat more delicate. (See Tits, Yale lectures, 1967).

8.9 *Torsion in tori.* Let p denote the characteristic exponent of k, and let T be a d-dimensional torus defined over k. For $m\in\mathbb{Z}$ define

$$\alpha_m: T\to T, \qquad \alpha_m(x)=x^m.$$

Proposition. *Assume $m>0$.*

(a) *α_m is surjective.*
(b) *If m is a power of p, then α_m is bijective.*
(c) *If $(m,p)=1$ then α_m is separable,* $\ker(\alpha_m)\cong(\mathbb{Z}/m\mathbb{Z})^d$ *(as a group) and* $\ker(\alpha_m)\subset T(k_s)$.
(d) *If m is not a power of p then the union of the groups* $\ker(\alpha_{m^n})$ $(n>0)$ *is a dense subgroup of T.*

Proof. K^* is a divisible group in which the Frobenius map, $x\mapsto x^p$, is bijective. This implies (a) and (b).

(c): Since $(d\alpha_m): X\mapsto mX$ we see that α_m is separable because $(m,p)=1$. It follows that $\ker(\alpha_m)$ is defined over k (see 6.7, Remark), so its points rational over k_s are dense. Once we prove that $\ker(\alpha_m)$ is finite, it will follow therefore that all of its points are rational over k_s. Finally, the fact that $\ker(\alpha_m)\cong(\mathbb{Z}/m\mathbb{Z})^d$ follows from the fact that the mth roots of unity in K^* are a cyclic group of order m.

Part (d) follows from the case $d=1$. Then $T=\mathbf{GL}_1$ is irreducible of dimension one, so $\bigcup_{n>0}\ker(\alpha_{m^n})$ is dense as soon as it is infinite, and part (c) implies that this is the case if m is not a power of p.

Corollary. *Let G be diagonalizable. For each $m>0$ the elements of order dividing m in G form a finite group. The torsion subgroup of G is dense in G.*

This follows from 8.7 and the proposition above, for 8.7 says G is the direct product of a torus with a finite group.

8.10 *Rigidity of diagonalizable groups.* This refers to the fact that they do not admit a non trivial connected family of automorphisms. This property is shared by abelian varieties, and for that reason we do not require the algebraic groups in the following proposition to be affine.

Proposition. *Let $\alpha: V \times H \to H'$ be a morphism of varieties such that:*

(i) *H' is an algebraic group containing, for each $m > 0$, only finitely many elements of order m;*
(ii) *H is an algebraic group in which the elements of finite order are dense; and*
(iii) *V is a connected variety and, for each $x \in V$, $\alpha_x: h \mapsto \alpha(x, h)$ is a homomorphism.*

Then the map $x \mapsto \alpha_x$ is constant.

Proof. For $h \in H$ write $\beta_h(x) = \alpha(x, h)$. Then $\beta_h: V \to H'$ is a morphism from a connected variety. Its image, when h has finite order, is finite, by (i) and (iii). Hence β_h is constant when h has finite order. Therefore, if $x, y \in V$, the morphism $\gamma: H \to H'$, $\gamma(h) = \alpha_x(h)\alpha_y(h)^{-1}$, sends every element of finite order to e. Condition (ii) then implies that $\gamma(h) = e$ for all h.

Corollary 1. *Let $H \subset H'$ be closed subgroups of G, and let V be the connected component of e in* $\operatorname{Tran}(H, H') = \{g \in G \mid gHg^{-1} \subset H'\}$. *Suppose H' and H satisfy conditions* (i) *and* (ii) *above. Then $V = \mathscr{Z}_G(H)^o$.*

Proof. Apply the proposition to $\alpha(x, h) = xhx^{-1}$, to conclude that $\alpha(x, h) = \alpha(e, h)$ for all $x \in V$. This shows that $V \subset \mathscr{Z}_G(H)^o$; the reverse inclusion is obvious.

In case $H = H'$ is diagonalizable 8.9 permits us to conclude:

Corollary 2. *Let H be a diagonalizable subgroup of an algebraic group G. Then $\mathscr{N}_G(H)^o = \mathscr{Z}_G(H)^o$.*

8.11 Proposition. *Let G be diagonalizable. Then G splits over a finite separable extension of k.*

Proof. Choose a k-embedding $G \subset \mathbf{GL}_n$. Then it suffices to diagonalize $G(k_s)$ by conjugation by an element of $\mathbf{GL}_n(k_s)$, because $G(k_s)$ is dense in G. But the possibility of doing this follows directly from 4.6.

Remark. One can also argue as above using the elements of finite order in G. It follows from 8.9 that the latter are dense in G and lie in $G(k_s)$.

Corollary 1. *Let T be a torus defined over k and let $\Gamma = \operatorname{Gal}(k_s/k)$.*

(a) *$X(T) = X(T)_{k_s}$ and $X_*(T) = X_*(T)_{k_s}$. Hence $X(T)_k = X(T)^\Gamma$ and $X_*(T)_k = X_*(T)^\Gamma$.*

(b) *The natural pairing*

$$X(T) \times X_*(T) \to \mathbb{Z}$$

makes $X(T)$ and $X_*(T)$ a dual pair of Γ-modules.

Proof. Since T is split over k_s we have $X(T) = X(T)_{k_s}$. This implies (see 8.3) that $\mathrm{Mor}(G, T) = \mathrm{Mor}(G, T)_{k_s}$ for any diagonalizable group G split over k_s. With $G = \mathbf{GL}_1$ this gives $X_*(T) = X_*(T)_k$, thus proving (a).

The pairing in (b) is a separating duality over $\mathbb{Z}$ (see 8.6) so we need only to check its compatibility with the action of each $s \in \Gamma$. We must show that $\langle {}^s\alpha, {}^s\lambda \rangle = \langle \alpha, \lambda \rangle$ for $\alpha \in X(T)$ and $\lambda \in X_*(T)$. We have, for $x \in k_s^*$,

$$\begin{aligned} x^{\langle s_\alpha, s_\lambda \rangle} &= ({}^s\alpha \circ {}^s\lambda)(x) = ({}^s\alpha)(s\lambda(s^{-1}x)) = s\alpha(s^{-1}s\lambda(s^{-1}x)) = s(\alpha \circ \lambda)(s^{-1}x) \\ &= s(s^{-1}x)^{\langle \alpha, \lambda \rangle} = x^{\langle \alpha, \lambda \rangle}. \end{aligned}$$

Corollary 2. *With T as above, T is split over k if and only if $X_*(T) = X_*(T)_k$.*

Proof. Γ operates trivially on the free abelian group $X(T)$ if and only if it operates trivially on its dual, $X_*(T)$.

8.12 *The category of diagonalizable k-groups.* We have seen in 8.11 that such a group, G, is split over k_s. Thus, if $A = K[G]$, it follows from 8.2 that $A_{k_s} = k_s[X(G)]$, the group algebra of $X(G)$. If $s \in \Gamma = \mathrm{Gal}(k_s/k)$ then the action of s on this group algebra is given by

$${}^s(\Sigma\, a_\alpha \alpha) = \Sigma\, {}^s a_\alpha {}^s\alpha.$$

This action determines $A_k = A_{k_s}^\Gamma$, and thus we see that G, as a k-group, is completely determined by $X(G)$ with its structure as a Γ-module. For knowledge of the latter permits us to construct A_{k_s} and the action of Γ on A_{k_s}, and hence A_k. As a Γ-module, $X(G)$ is finitely generated as a $\mathbb{Z}$-module, and the action of Γ is continuous, i.e. some open subgroup of finite index in Γ acts trivially on $X(G)$ (because G is split by a finite extension of k). If $p = \mathrm{char}(k) > 0$, moreover, $X(G)$ has no p-torsion because $K[X(G)]$ is reduced.

Let $\alpha: G \to G'$ be a morphism of diagonalizable k-groups. It follows from 8.3 that α is defined over k_s. Moreover the following conditions are equivalent:

(1) α is defined over k.
(2) $\alpha^o: A'_{k_s} \to A_{k_s}$ (where $A' = K[G']$) is Γ-equivariant.
(3) $X(\alpha): X(G') \to X(G)$ is Γ-equivariant.

The equivalence of (1) and (2) follows from AG.14.3, and the equivalence of (2) and (3) follows from the description above of the action of Γ on the affine algebras, via its action on the character groups.

Now we can consider X to be a (contravariant) functor,

$$X: \mathscr{A} \to \mathscr{B}$$

where the two categories are defined as follows:

obj $\mathscr{A}$: diagonalizable k-groups.
mor $\mathscr{A}$: k-morphisms.
obj $\mathscr{B}$: finitely generated $\mathbb{Z}$-modules, without p-torsion if $\operatorname{char}(k) = p > 0$, on which Γ acts continuously.
mor $\mathscr{B}$: Γ-equivariant homomorphisms.

It follows from 8.3 and the remarks above that the functor X is fully faithful. In fact:

Proposition. *$X: \mathscr{A} \to \mathscr{B}$ is an equivalence of categories.*

There remains only to be shown that every $M \in \operatorname{obj} \mathscr{B}$ is the character module of some $G \in \operatorname{obj} \mathscr{A}$. The group algebra $A = K[M]$ is a Hopf algebra, and a reduced affine K-algebra, because M is finitely generated and without p-torsion. Hence $G = \operatorname{spec}_K(A)$ is an affine group. Moreover M is naturally a group of characters on G, so G is diagonalizable, with character group M (see 8.2).

We must now give G a k-structure inducing the given action of Γ on M. First we give A the k_s-structure $k_s[M]$. Let $s \in \Gamma$ operate on $k_s[M]$ by ${}^s(a\alpha) = {}^s a {}^s \alpha$ $(a \in k_s; \alpha \in M)$. This defines a continuous action of Γ on $k_s[M]$, i.e. one for which each element of $k_s[M]$ has an open isotropy subgroup. It follows therefore from AG.14.2 that $A_k = k_s[M]^\Gamma$ is a k-structure on A. This k-structure clearly meets our requirements.

8.13 Examples

(1) Suppose $M = \mathbb{Z}[\Gamma/U]$ where U is an open subgroup of Γ. Then, for any Γ-module N we have

$$\operatorname{Hom}_{\Gamma\text{-mod}}(M, N) = \operatorname{Hom}_{\mathbb{Z}\text{-mod}}(M, N)^\Gamma = N^U.$$

Let k' be any k-algebra and let Γ act on $k_s \bigotimes_k k'$ via its action on k_s. Then clearly $\left(k_s \bigotimes_k k'\right)^U = L \bigotimes_k k'$, where $L = k_s^U$, and therefore also $\left(k_s \bigotimes_k k'\right)^{*U} = \left(L \bigotimes_k k'\right)^*$, the notation referring to the groups of invertible elements in these algebras.

Now put $A = K[M]$ with k-structure $A_k = k_s[M]^\Gamma$. Then $G = \operatorname{spec}_K(A)$ is a diagonalizable k-group with character module M. The functor of points of G is described as follows, where k' is a variable k-algebra:

$$\begin{aligned}
G(k') &= \operatorname{Hom}_{k\text{-alg}}(S_k, k') \\
&= \operatorname{Hom}_{k_s\text{-alg}}\left(A_{k_s}, \left(k_s \bigotimes_k k'\right)\right)^\Gamma \\
&= \operatorname{Hom}_{\mathbb{Z}\text{-mod}}\left(M, \left(k_s \bigotimes_k k'\right)^*\right)^\Gamma \\
&= \left(k_s \bigotimes_k k'\right)^{*U} = \left(L \bigotimes_k k'\right)^*.
\end{aligned}$$

Thus we see that G is the multiplicative group $\mathbf{GL}_1(L)$ of the k-algebra L (see 1.6, Example (9)). It follows, in particular, that G is k-rational, i.e. that the function field $k(G)$ is purely transcendental (loc. cit.).

(2) Let T be a k-torus and put $N = X(T)$. Some open normal subgroup U of Γ operates trivially, so N is a $\mathbb{Z}$-free representation of finite rank of the finite group $\Gamma' = \Gamma/U$. Hence there is a monomorphism $\alpha^o: N \to M$ where M is a free $\mathbb{Z}[\Gamma']$-module. (For example one can take $M = N''$, where, for a Γ'-module H, we write $H' = \mathrm{Hom}_{\mathbb{Z}\text{-mod}}(H, \mathbb{Z}[\Gamma'])$.) The monomorphism α^o corresponds to an epimorphism $\alpha: S \to T$, where S is the torus with character module M. Thus we have an embedding of function fields $\alpha^o: k(T) \to k(S)$. It follows from example (1) above that $k(S)$ is purely transcendental. This shows that:

A *k-torus T is unirational over k*, i.e. $k(T)$ is contained in a purely transcendental extension of k. In particular, if k is infinite, $T(k)$ is dense in T (AG. 13.7).

8.14 *Anisotropic tori.* Write $\mathcal{B}_{\mathbb{Q}}$ for the category of finite dimensional $\mathbb{Q}$-modules on which Γ operates continuously, i.e. via finite quotient groups. Then (cf. Curtis–Reiner, for example) $\mathcal{B}_{\mathbb{Q}}$ is a semi-simple category, i.e. all short exact sequences split.

We have the exact functor $\mathcal{B} \to \mathcal{B}_{\mathbb{Q}}$ (see 8.12 for notation) which sends M to $M_{\mathbb{Q}} = \mathbb{Q} \bigotimes_{\mathbb{Z}} M$. If M is torsion free we can view M as embedded in $M_{\mathbb{Q}}$ as a lattice. Thus:

$$M^{\Gamma} = M \cap M_{\mathbb{Q}}^{\Gamma},$$
$$M^{\Gamma} = 0 \Leftrightarrow M_{\mathbb{Q}}^{\Gamma} = 0,$$

and

$$M^{\Gamma} = M \Leftrightarrow M_{\mathbb{Q}}^{\Gamma} = M_{\mathbb{Q}}.$$

A k-torus T is said to be *anisotropic over* k if $X(T)_k = \{1\}$, i.e. if $X(T)^{\Gamma} = \{1\}$. This is equivalent to the existence of no non-trivial Γ-fixed points in $\mathbb{Q} \bigotimes_{\mathbb{Z}} X(T)$, by the remarks above. The semi-simplicity of the category $\mathcal{B}_{\mathbb{Q}}$ implies that the functor "fixed points" is exact on $\mathcal{B}_{\mathbb{Q}}$. Thus:

Corollary. *Let $e \to T' \to T \to T'' \to e$ be an exact sequence over k of k-tori. Then T is split (resp., anisotropic) over k if and only if T' and T'' are split (resp., anisotropic) over k.*

8.15 *T_a and T_d.* Let T be a torus defined over k. The subtori of T correspond, in view of 8.5 and 8.12, to the Γ-module quotients of $X(T)$ which are torsion free.

One such quotient clearly is

$$X(T_a) = X(T)/X(T)_k = X(T)/X(T)^{\Gamma},$$

where (see 8.2 (c))

(a) $$T_a = \bigcap_{\alpha \in X(T)_k} \ker(\alpha).$$

By construction it is clear that $X(T_a)^\Gamma = \{1\}$, i.e. that T_a is anisotropic, and that it is the largest anisotropic subtorus of T.

There is also a largest split subtorus, T_d, of T. To obtain T_d as above we would first take the largest quotient of $X(T)$ on which Γ acts trivially, and then reduce this quotient modulo its torsion submodule. However, it is more natural here to work in the dual module $X_*(T)$ (see 8.11, Cor. 1). Then we can describe T_d as follows:

(d) $$X_*(T_d) = X_*(T)_k = X_*(T)^\Gamma;$$
T_d is the subgroup generated by
$$\{\mathrm{im}(\lambda) \mid \lambda \in X(T)_k\}.$$

The last description shows that T_d is a k-split torus, and that it contains all other such subtori of T.

A subtorus of $T_a \cap T_d$ must be both split and anisotropic (see 8.14) and hence trivial. Thus $(T_a \cap T_d)^o$ is trivial, and so $T_a \cap T_d$ is finite.

Let $r = \operatorname{rank} X(T)^\Gamma$ and let $r_* = \operatorname{rank} X_*(T)^\Gamma$. Then $\dim T_a = n - r$ (where $n = \dim T$) and $\dim T_d = r_*$. These ranks can be computed as $\mathbb{Q}$-dimensions after tensoring the modules with $\mathbb{Q}$. Moreover $X(T)_\mathbb{Q}$ and $X_*(T)_\mathbb{Q}$ remain a dual pair of $\mathbb{Q}$-Γ-modules. Since the trivial representations of Γ are self dual it follows that $X(T)_\mathbb{Q}^\Gamma$ and $X_*(T)_\mathbb{Q}^\Gamma$ have the same dimension (thanks to the fact that these are semi-simple Γ-modules). Thus $r = r_*$ and so $\dim T_a + \dim T_d = \dim T$. This implies, in view of the last paragraph, that the product morphism $T_a \times T_d \to T$ is surjective.

We summarize:

Proposition. *Let T be a torus defined over k, and let T_a and T_d be defined by (a) and (d) above.*

(1) (a) *T_a is the largest anisotropic subtorus of T defined over k.* (b) *T_d is the largest split subtorus of T defined over k.*
(2) *$T_a \cap T_d$ is finite and $T = T_a \cdot T_d$.*
(3) *If $\alpha: T \to T'$ is a k-morphism of k-tori then $\alpha T_a \subset T'_a$ and $\alpha T_d \subset T'_d$. In other words $T \mapsto T_a$ and $T \mapsto T_d$ are functorial.*

The last assertion is clear from the definitions, and all others were proved above.

8.16 *Examples over $k = \mathbb{R}$.* The group $\Gamma = \mathrm{Gal}(\mathbb{C}/\mathbb{R})$ has order two.

(1) if $\dim T = 1$ there are two possibilities: (a) *T is split.* $T(\mathbb{R}) = \mathbb{R}^*$ and $X(T) = \mathbb{Z}$ with trivial Γ-action. (b) *T is anisotropic.* $X(T) = \mathbb{Z}$ with the generator of Γ acting by $\chi \mapsto -\chi$. The group T is $\mathbf{SO}(2)$, the special

orthogonal group in two variables, and $T(\mathbb{R})$ is the *compact* group of orientation preserving rotations of the plane.

(2) In the general case T is anisotropic if and only if $T(\mathbb{R})$ is compact (in the real topology). This can be deduced easily from example (1), with the aid of 8.15, and using the fact that there are only two irreducible $\mathbb{Q}$-Γ-modules.

8.17 *Weights and roots of diagonalizable groups.* Let T be a diagonalizable group. We shall sometimes write the groups $X(T)$ of characters, and $X_*(T)$ of one parameter subgroups, *additively*. When doing this we shall employ an exponential notation, as follows:

$$t^\alpha = \alpha(t) \qquad (t \in T, \alpha \in X(T))$$
$$x^\lambda = \lambda(x) \qquad (x \in \mathbf{GL}_1, \lambda \in X_*(T))$$
$$(x^\lambda)^\alpha = x^{\langle \alpha, \lambda \rangle} \qquad \text{(see 8.5)}.$$

Let $T \to GL(V)$ be a rational representation of T. If $\alpha \in X(T)$ we write

$$V_\alpha = \{v \in V \mid t \cdot v = \alpha(t)v \text{ for all } t \in T\}.$$

Since T is diagonalizable V is the direct sum of the V_α's. Those α for which $V_\alpha \neq 0$ are called the *weights* of T in V. They are evidently finite in number.

Suppose T acts on G. Then T acts on $\mathfrak{g} = L(G)$, and the set $\Phi(T, G)$ of *non zero* weights of T in $\mathfrak{g}$ is called the set of *roots* of G relative to T. Thus

$$\mathfrak{g} = \mathfrak{g}^T \oplus \coprod_{\alpha \in \Phi(T,G)} \mathfrak{g}_\alpha.$$

In case T and G are given as subgroups of some larger group in which T normalizes G then $\Phi(T, G)$ will always refer to the action of T on G by conjugation. Of course, by taking the semi-direct product $T \cdot G$, the general case reduces to one of this type.

Suppose T acts on G as above and that H is a T-invariant closed subgroup of G. Then $\mathfrak{h} = L(H)$ is also T-invariant. For each $\alpha \in \Phi(T, G)$ we can write $\mathfrak{g}_\alpha = \mathfrak{h}_\alpha \oplus \mathfrak{g}'_\alpha$ for some complement $\mathfrak{g}'_\alpha$ of $\mathfrak{h}_\alpha = \mathfrak{h} \cap \mathfrak{g}_\alpha$. We shall write

$$\Phi(T, G/H) = \{\alpha \in \Phi(T, G) \mid \mathfrak{g}'_\alpha \neq 0\}$$
$$= \{\alpha \in \Phi(T, G) \mid \mathfrak{h}_\alpha \neq \mathfrak{g}_\alpha\}.$$

Then we have

$$\mathfrak{g} = (\mathfrak{g}^T + \mathfrak{h}) \oplus \coprod_{\alpha \in \Phi(T,G/H)} \mathfrak{g}'_\alpha.$$

In case $H \subset G^T$ we have $\mathfrak{h} \subset \mathfrak{g}^T$ and hence $\Phi(T, G/H) = \Phi(T, G)$. We shall sometimes refer to $\Phi(T, G/H)$ as the set of "roots of G outside of H (relative to T)," or of complementary roots of G, with respect to H.

8.18 Proposition. *We keep the notation of* 8.17. *Assume T to be connected and k infinite. Then there is $t \in T(k)$ such that*

$$\mathscr{Z}_H(t) = \mathscr{Z}_H(T), \quad \mathscr{Z}_\mathfrak{h}(t) = \mathscr{Z}_\mathfrak{h}(T).$$

Proof. We can assume $T \subset G = GL(V)$ for some vector space V defined over k. Write $V = V_1 \oplus \cdots \oplus V_n$, where the V_i are the eigenspaces for the distinct weights $\chi_1, \ldots, \chi_n$ of T on V. Since T is unirational over k (8.13 (2)) and k is infinite, $T(k)$ is dense in T, and we can choose $t \in T(k)$ such that $\chi_i(t) \neq \chi_j(t)$ for $i \neq j$. An obvious computation shows then that

$$\mathscr{Z}_H(t) = \mathscr{Z}_H(T) = M \cap H,$$
$$\mathscr{Z}_{\mathfrak{h}}(t) = \mathscr{Z}_{\mathfrak{h}}(T) = L(M) \cap \mathfrak{h},$$

where $M = GL(V_1) \times \cdots \times GL(V_n)$.

We end up §8 with some results on the existence of quotients. For the main part of this book, they are needed when G^o is a torus (8.21), but the proof under somewhat more general assumptions is the same, and the result is of independent interest. Therefore we adopt a more general framework in 8.19, 8.20.

8.19 Lemma. *Let G operate k-morphically on the affine k-variety V.*

(i) *$I = K[V]^G$ is defined over k.*
Assume that the representation of G in $K[V]$ is completely reducible.

(ii) *There exists an I-linear projection operator $\natural : K[V] \to I$, defined over k, which leaves every G-stable subspace invariant. The algebra I separates the G-stable disjoint closed subsets of V and is finitely generated.*

(iii) *If the orbits are closed, $K(V)^G$ is the full ring of fractions of I. More precisely, given $r \in K(V)^G$ and $v \in V$ at which r is defined, there exists a, $b \in I$, where b is a non-zero-divisor in I, which does not vanish at v and such that $r = a/b$.*

["completely reducible" means that every G-invariant finite dimensional subspace of $K[V]$ is a direct sum of G-invariant irreducible subspaces.]

Proof. (i) By 1.9, $K[V]$ is the union of G-invariant finite dimensional subspaces E defined over k. The group $G(k_s)$ is Zariski-dense in G (AG.13.3) therefore E^G is the solution space of a system of linear equations with coefficients in k_s and is defined over k_s. On the other hand $E^G(k_s)$ is invariant under the Galois group Γ of k_s over k. Therefore it is defined over k (AG, 14.4). Thus, I is the union of finite dimensional subspaces defined over k, hence has a k-structure.

(ii) The previous space E is the direct sum of E^G and of a unique G-invariant complementary subspace E', namely the sum of the isotypic subspaces of E corresponding to the non-trivial representations of G. It is defined over k because, over k_s, Γ permutes the non-trivial irreducible representations of $G(k_s)$. We let F be the inductive limit of these subspaces. It is the unique subspace of $K[V]$ stable under G and supplementary to I and it admits a k-structure, too. We let then $\natural$ be the projection of $K[V]$ onto I with kernel F. It is defined over k since I and F are. If Q is any invariant subspace, then, by looking at finite dimensional invariant subspaces, we see again that

$Q = Q \cap I \oplus Q \cap F$. Therefore $\natural$ leaves Q stable. If $a \in I$, then $a \cdot Q \cap I \subset I$ and $a \cdot (Q \cap F) \subset F$. Therefore for $b \in Q$, we have $(a \cdot b)^\natural = a \cdot b^\natural$, which shows that $\natural$ is I-linear and leaves Q-stable.

Let A and B be G-stable disjoint closed subvarieties of V and C, D their ideals in $K[V]$. Since $A \cap B$ is empty, $C + D = K[V]$ by the Hilbert Nullstellensatz (AG.3.8). Let $c \in C$ and $d \in D$ be such that $c + d = 1$. Then $c^\natural + d^\natural = 1$. Both C and D are stable under G, hence $c^\natural \in C$, $d^\natural \in D$ by (i). It follows that $c^\natural$ is zero on A and equal to one on B. It separates A and B.

To prove the last assertion of (ii) we may, by 1.12, assume V to be embedded in a finite dimensional vector space X defined over k on which G acts linearly, with action defined over k, leaving V invariant and inducing the given operation on V. The injection $V \to X$ induces a surjective comorphism $K[X] \to K[V]$. It is G-invariant and the uniqueness of the decomposition in (i) implies that I is a quotient of $K[X]^G$. We may therefore assume $V = X$. Then $K[X]$ is a polynomial algebra in $d = \dim X$ independent generators. Let J be the ideal generated by the invariant polynomials without constant term. By the Hilbert basis theorem, it has a finite generating set, say $\{f_1, \ldots, f_m\}$, and we may assume the $f_i \in I$ to be homogeneous. We claim that I is generated by the f_i. Let $a \in I$ be homogeneous of some degree m. There exist $b_i \in K[X]$ such that

$$a = b_1 \cdot f_1 + \cdots + b_m \cdot f_m \tag{1}$$

and we may assume the b_i to be homogeneous, so that $\deg b_i + \deg f_i = m$ for all i's. We get from (1)

$$a = a^\natural = \sum b_i^\natural \cdot f_i. \tag{2}$$

The space of homogeneous polynomials of a given degree is G-stable (since G acts linearly), hence degree $b_i^\natural = \deg b_i$ is $< m$. The last assertion of (ii) now follows by induction on the degree.

(iii) We now assume that the orbits are closed. Recall first that $K(V)$ is the direct sum of the fields $K(V_i)$, where V_i runs through the irreducible components of V. We have to find

$$s \in K[V] \text{ such that } t = f \cdot s \in K[V], s^\natural(v) \neq 0 \text{ and } s^\natural \text{ is not a divisor of zero,} \tag{3}$$

because we have then $t^\natural = f \cdot s^\natural$, with $s^\natural$ non-divisor of zero, which shows that f is in the quotient ring of I.

For the proof, we assume first that $G \cdot v$ is closed. Choose a non-divisor of zero $q \in K[V]$ such that $r = f \cdot q \in K[V]$ and $q(v) \neq 0$. Let E be the subspace of $K[V]$ generated by the $g(q)$ ($g \in G$). It is finite dimensional (1.9). Let J be the ideal of $K[V]$ generated by E and the functions $f_i - f_i(v)$ ($1 \leqq i \leqq m$), where the f_i's are as in (ii). Since E and the f_i are G-stable, so is J. We claim first that $1 \in J$. Assume this is not the case. Then the variety Y of zeroes of J is not empty. Since J is invariant under G, so is Y. For any $y \in Y$ we have $f_i(y) = f_i(v) (1 \leqq i \leqq m)$, hence also $s(y) = s(v)$ for any $s \in I$. Then (ii) implies that $v \in Y$, but this is absurd since $q(v) \neq 0$ by construction. Therefore $1 \in J$. There exists then

$$c_i, d_j \in K[V] \text{ and } g_j \in G \ (1 \leqq i \leqq m,\ 1 \leqq j \leqq n) \text{ such that}$$

$$\sum_i c_i(f_i - f_i(v)) + \sum_J d_j \cdot g_j(q) = 1 \tag{4}$$

Let $s = \sum_j d_j \cdot g_j(q)$. Then $\sum d_j g_j(r) = f \cdot s \in K[Y]$; applying $\natural$ to both sides of (4) and evaluating at v, we get $s^\natural(v) = 1 \neq 0$. We have now found s satisfying (3) except maybe for the last condition. In order to take care of this last point we proceed by induction on the number of irreducible components of V in the support of s. Assume the latter consists of b irreducible components of V and is $\neq V$. Let Z be an irreducible component of V not wholly contained in it and z a point of Z at which f is defined. The previous argument shows that we can find $s_1 \in K[V]$ such that $f \cdot s_1 \in K[V]$ and $s_1^\natural(z) \neq 0$. There exists $d \in K^*$ such that $s_1^\natural + d.s^\natural$ is not zero at v and not identically zero on any irreducible component of the support of $s^\natural$. Also

$$s_1^\natural(z) + d.s^\natural(z) = s_1^\natural(z) \neq 0.$$

As a consequence, $s_1 + d.s$ satisfies the first three conditions of (3) and the support of $(s_1 + ds)^\natural$ contains at least $b + 1$ irreducible components of V. This provides the induction step.

8.20 Proposition. *Let G act k-morphically on an affine k-variety V. Assume that the representation of G^o in $K[V]$ is completely reducible. Then there exists a categorical quotient* (6.15) *(π, Y) of V by G over k. It is affine, with coordinate ring isomorphic to $K[V]^G$. If all orbits of G are closed, then Y is the (geometric) quotient of V by G.*

Proof. We shall check that the three conditions of 6.16 are fulfilled. We first assume that G is connected. Let again $I = K[V]^G$. By 8.19, I is finitely generated and defined over k. Let Y be the affine k-variety with coordinate ring I and π the k-morphism associated to the inclusion $I \subset K[V]$. The morphism π is constant along the orbits of G. Let us show that π is surjective. Let $y \in Y$ and A the ideal of elements in I which are zero on y. We claim that $B = A \cdot K[V]$ is a proper ideal of $K[V]$. If not, there would exist $a_i \in A$, $c_i \in K[V]$ such that $\sum a_i \cdot c_i = 1$, hence such that

$$\sum_i (a_i \cdot c_i)^\natural = \sum_i a_i \cdot c_i^\natural = 1$$

(we have used 8.19(ii)), whence $1 \in A$, which is absurd. There exists then $v \in V$ which annihilates B, and therefore such that $\pi(v) = y$. This proves (*a*) of 6.16.

We now check (c). Let $F \subset V$ be a G-invariant closed subset and J the ideal of F in $K[V]$. It is G-stable. Of course, $K[F]$ may be identified to $K[V]/J$. By complete reducibility, $K[F]^G$ is the image of I, i.e. $K[F]^G = I/(I \cap J)$. Let Z be the subvariety of Y defined by $J \cap I$, hence with coordinate ring $K[F]^G$. Clearly $\pi(F) \subset Z$. But the first part of the proof, applied to F and G, shows that $\pi: F \to Z$ is surjective. Hence $\pi(F)$ is closed. This is the first condition in (c).

Let $y \in Y$. Its inverse image is G-stable, hence contains at least one closed orbit by 1.8, but at most one since I separates the G-invariant closed disjoint subsets of V (8.19(ii)). Hence each fibre of π contains exactly one closed orbit. Let F be a G-invariant closed subset of V whose image contains y. Then $F \cap \pi^{-1}(y)$ in non empty, G-invariant, closed, hence contains a closed orbit, and therefore F contains the unique closed orbit belonging to $\pi^{-1}(y)$. From this the second part of (c) follows.

We still have to see that 6.16(b) holds. Let $U \subset Y$ be open, $U' = \pi^{-1}U$ and $f \in K[U']^G$. We have to show that $f \in \pi^o K[U]$. It suffices to do this for a basis for the open sets in Y. We may therefore assume that U is a principal open set $Y_f (f \in K[Y], f \neq 0)$(AG.3.4). Viewing f as an element of $K[X]^G$, i.e. identifying it with $\pi^o f$, we have clearly $U' = X_f$, hence

$$K[U] = K[Y][f^{-1}], \quad K[U'] = K[X][f^{-1}],$$

and the claim follows from the obvious relation $(K[X][f^{-1}])^G = K[X]^G[f^{-1}]$.

We have proved that (π, Y) is a categorical quotient. Assume now that all orbits are closed. Then 8.19(ii) shows that π is an orbit map. 8.19(iii) implies $K(Y) = K(V)^G$, therefore π is separable (AG.2.5). The remark in 6.4 and the validity of 6.16(b) imply that π is open. Then the conditions (1), (2) of 6.1 are fulfilled and $Y = V/G$. This proves the proposition for G connected.

In the general case, the finite group G/G^o acts k-morphically in the obvious way on Y (6.10) and the quotient $Y/(G/G^o)$ exists over k by 6.14. It is the affine k-variety Z with coordinate ring, the ring of invariants of G/G^o in I, i.e. of G in $K[V]$. Assume the orbits of G are closed. Then $Y = V/G^o$, hence $Z = V/G$ by 6.10. Now drop that assumption. We can also view Z as the categorical quotient of Y by G/G^o. Then the obvious remark about the transitivity of the notion of categorical quotient (6.15) shows that Z is a categorical quotient.

8.21 Corollary. *Let G act k-morphically on an affine k-variety. Assume that G^o is a torus. Then the categorical quotient of V by G exists over k and is the affine variety with coordinate ring $I = K[V]^G$. If all orbits are closed, it is the quotient of V by G.*

Since all finite dimensional morphic representations of a torus are completely reducible, this follows from 8.20.

Remarks. This shows in particular that if $T \to \mathrm{GL}(E)$ is a finite dimensional rational representation of a torus T, then the categorical quotient of E by T exists. On the other hand, if the representation is not trivial, the orbits are not all of the same dimension and E/G does not exist.

Bibliographical Note

Tori are introduced in [1], the groups of their characters and of their one parameter subgroups in [13, Exp. 4]. That a k-torus T splits over a finite separable extension of k is pointed out while showing that T is unirational over k in [26, Prop. 10]. Another proof, due to J. Tate, is given in [4, §1]. The equivalence of categories 8.12, at least for tori, is proved in [24]. In fact, most of the results proved in this paragraph up to 8.18 may be found in one of these references.

8.19 for a torus and 8.21 are due to M. Rosenlicht [27]. The complete reducibility assumption holds for reductive groups (see §14) in characteristic zero. Therefore 8.20 is true for such groups. See Chap. I in [22], the first edition of which is in fact the original source for the theorem. The conclusion of 8.20 is also valid in arbitrary characteristic if G^o is reductive, but the proof requires other tools (see the discussion in Appendix 1A of [22]).

§9. Conjugacy Classes and Centralizers of Semi-Simple Elements

In this section it is shown that conjugacy classes of semi-simple elements are closed (9.2), and that their global and infinitesimal centralizers correspond (9.1.) The action of a semi-simple element s on a connected unipotent group U is studied and it is shown (9.3) that $\mathscr{Z}_U(s)$ is connected. Applications are then made to group actions of diagonalizable groups 9.4.

9.1 *The conjugacy class morphisms.* In this section we fix a closed subgroup H of G defined over k.

H acts on G by conjugation, and we write $C_H(s)$ for the orbit of an element $s \in G$; this is the *H-conjugacy class* of s.

$$\alpha: H \to C_H(s) \quad \alpha(h) = hsh^{-1},$$

is then the orbit map, and the isotropy group of s is the *centralizer* $\mathscr{Z}_H(s)$. We can now apply 6.7 to this to determine when α is a quotient morphism. In order to make the statement more explicit we shall first determine $(d\alpha)_e$. Since $\alpha(h)s^{-1} = (h, s)$ is the commutator, it follows from 3.16(a) that the latter has differential $(\mathrm{Id} - \mathrm{Ad}(s))|\mathfrak{h}$, which maps $\mathfrak{h}$ to $T(C_H(s)s^{-1})_e$. Thus its kernel is $\mathfrak{h} \cap \mathfrak{z}_{\mathfrak{g}}(s)$, where $\mathfrak{z}_{\mathfrak{g}}(s) = \ker(\mathrm{Id} - \mathrm{Ad}(s))$. Since translation is an isomorphism, we conclude also that $(d\alpha)_e: \mathfrak{h} \to T(C_H(s))_s$ has kernel $\mathfrak{h} \cap \mathfrak{z}_{\mathfrak{g}}(s)$. We shall denote the latter by $\mathfrak{z}_{\mathfrak{h}}(s)$, so that

$$\mathfrak{z}_{\mathfrak{h}}(s) = \{X \in \mathfrak{h} \mid \mathrm{Ad}(s)X = X\},$$

even though we have not assumed that $\mathrm{Ad}(s)$ leaves $\mathfrak{h}$ invariant. In any case we certainly have

$$L(\mathscr{Z}_H(s)) \subset \mathfrak{z}_{\mathfrak{h}}(s),$$

and we can now apply (6.7) to conclude:

(*) *Assume* $s \in G(k)$. *Then* $C_H(s)$ *is a smooth variety defined over* k, *and* α *is a* k*-morphism which induces a bijective* k*-morphism*

$$\alpha' : H/\mathscr{Z}_H(s) \to C_H(s).$$

The following conditions are equivalent: (a) α' *is an isomorphism*; (b) α *is separable*; (c) $L(\mathscr{Z}_H(s)) = \mathfrak{z}_{\mathfrak{h}}(s)$. *When these conditions hold,* $\mathscr{Z}_H(s)$ *is defined over* k.

Next we discuss the infinitesimal analogue of the above situation. Namely, H acts on $\mathfrak{g}$ via Ad_G, and we denote the H-orbit of an $X \in \mathfrak{g}$ by $\mathfrak{c}_H(X)$. Let

$$\beta : H \to \mathfrak{c}_H(X), \quad \beta(h) = \mathrm{Ad}(h)X,$$

be the orbit map. The stability group of X is denoted $\mathscr{Z}_H(X)$, and it is called the *centralizer of* X in H. Before applying 6.7 we again compute first the differential of β. Following β by translation by $-X$, and using 3.16(b), we see that the differential at e is $-\mathrm{ad}(X)$. Thus

$$\ker(d\beta)_e = \mathfrak{z}_{\mathfrak{h}}(X) = \{Y \in \mathfrak{h} \mid [X, Y] = 0\},$$

and this clearly contains $L(\mathscr{Z}_H(X))$. Now we apply 6.7 again to conclude:

(**) *Assume* $X \in \mathfrak{g}(k)$. *Then* $\mathfrak{c}_H(X)$ *is a smooth variety defined over* k, *and* β *is a* k*-morphism which induces a bijective* k*-morphism*

$$\beta' : H/\mathscr{Z}_H(X) \to \mathfrak{c}_H(X).$$

The following conditions are equivalent: (a) β' *is an isomorphism*; (b) β *is separable*; (c) $L(\mathscr{Z}_H(X)) = \mathfrak{z}_{\mathfrak{h}}(X)$. *When these conditions hold,* $\mathscr{Z}_H(X)$ *is defined over* k.

Note that, if $\mathrm{char}(k) = 0$, conditions (b) of (*) and of (**) are automatic. More generally, they hold if s and X are semi-simple and normalize H.

Proposition

(1) *If, in* (*), s *is semi-simple and normalizes* H, *then conditions* (a), (b), *and* (c) *hold.*
(2) *If, in* (**) X *is semi-simple and normalizes* H *then conditions* (a), (b), *and* (c) *hold.*

We say X *normalizes* H if $\mathrm{Ad}(h)X - X \in \mathfrak{h}$ for $h \in H$. This implies that X normalizes $\mathfrak{h}$, i.e. that $[X, \mathfrak{h}] \subset \mathfrak{h}$ (see 3.16(b)).

Proof. After choosing a k-rational immersive representation we can enlarge G and assume $G = GL(V)$ for some vector space V with k-structure.

Case 1. $H = G$.

(1) Write $V = V_1 \oplus \cdots \oplus V_t$ where the V_i are eigenspaces for the distinct eigenvalues of s (which is semi-simple). Then a simple direct calculation shows that $\mathscr{Z}_G(s) = GL(V_1) \times \cdots \times GL(V_t)$. If $Y \in \mathfrak{g} = \mathfrak{gl}(V)$ then $\operatorname{Ad}(s)Y = sYs^{-1}$ so we conclude similarly that $\mathfrak{z}_{\mathfrak{g}}(s) = \mathfrak{gl}(V_1) \oplus \cdots \oplus \mathfrak{gl}(V_t)$. The latter is just $L(\mathscr{Z}_G(s))$, thus establishing condition (c).

(2) The proof is completely parallel, using a decomposition of V for the semisimple endomorphism $X \in \mathfrak{gl}(V)$.

General case. Write $c: G \to M$, where $M = C_G(s) \cdot s^{-1}$ and $c(g) = gsg^{-1}s^{-1}$, and write $c': H \to M'$, where $c' = c|H$ and $M' = C_H(s)s^{-1}$. Then c' is just α followed by right translation by s^{-1}, so we have only to show that $(dc')_e: \mathfrak{h} \to T(M')_e$ is surjective (condition (b) of $(*)$). We know from case 1 that $(dc)_e: \mathfrak{g} \to T(M)_e$ is surjective. Since $(dc)_e = \operatorname{Id} - \operatorname{Ad}(s)$ we see therefore that $T(M)_e = \mathfrak{m}$, where $\mathfrak{g} = \mathfrak{z}_{\mathfrak{g}}(s) \oplus \mathfrak{m}$ and $\mathfrak{m}$ is the sum of the eigenspaces of $\operatorname{Ad}(s)$ corresponding to eigenvalues different from 1. Since s normalizes H, it follows that $\mathfrak{h} = \mathfrak{z}_{\mathfrak{h}}(s) \oplus \mathfrak{m}'$, where $\mathfrak{m}' = \mathfrak{m} \cap \mathfrak{h}$ is similarly defined. Since $(dc')_e = (dc)_e|\mathfrak{h}$ it follows that $(dc')_e(\mathfrak{h}) = \mathfrak{m}'$. Hence the proof of surjectivity of $(dc')_e$ will be finished once we show that $T(M')_e \subset \mathfrak{m}' = \mathfrak{m} \cap \mathfrak{h}$. Evidently $T(M')_e \subset \mathfrak{m} = T(M)_e$. On the other hand, since s normalizes H, we have $M' = C_H(s)s^{-1} \subset H$, and so $T(M')_e \subset \mathfrak{h}$ also.

The proof of (2) is similar to the proof of (1) above. We introduce $a: G \to \mathfrak{c}$, where $\mathfrak{c} = \mathfrak{c}_G(X) - X$ and $a(g) = \operatorname{Ad}(g)X - X$, and the morphism $a': H \to \mathfrak{c}' = \mathfrak{c}_H(X) - X$ where $a' = a|H$. We want to show that $(da')_e$ is surjective, and we know from case 1 that $(da)_e = -\operatorname{ad}(X): \mathfrak{g} \to T(\mathfrak{c})_0$ is surjective. It follows that $\mathfrak{g} = \mathfrak{z}_{\mathfrak{g}}(X) \oplus \mathfrak{m}$ where $\mathfrak{m} = T(\mathfrak{c})_0$ is the sum of the eigenspaces of $\operatorname{ad}(X)$ corresponding to eigenvalues different from 0. Since X normalizes $\mathfrak{h}$ we can similarly write $\mathfrak{h} = \mathfrak{z}_{\mathfrak{h}}(X) \oplus \mathfrak{m}'$ where $\mathfrak{m}' = \mathfrak{m} \cap \mathfrak{h}$. Since $(da')_e = (da)_e|\mathfrak{h}$ it follows that $(da')_e(\mathfrak{h}) = \operatorname{ad}(X)(\mathfrak{h}) = \mathfrak{m}'$. Hence the surjectivity of $(da')_e$ will follow once we show that $T(\mathfrak{c}')_0 \subset \mathfrak{m}' = \mathfrak{m} \cap \mathfrak{h}$. Evidently $T(\mathfrak{c}')_0 \subset \mathfrak{m} = T(\mathfrak{c})_0$. On the other hand, since X normalizes H, we have $\mathfrak{c}' = \mathfrak{c}_H(X) - X \subset \mathfrak{h}$, and so $T(\mathfrak{c}')_0 \subset \mathfrak{h}$ also.

Remark. Let $p: M' \times \mathscr{Z}_H(s) \to H$ be the product morphism, with differential $(dp)_{(e,e)}: \mathfrak{m}' \oplus L(\mathscr{Z}_H(s)) \to \mathfrak{h}$. The proof above shows that $L(\mathscr{Z}_H(s)) = \mathfrak{z}_h(s)$, and hence that $(dp)_{(e,e)}$ is an isomorphism. Moreover the differential of $c|M': M' \to M'$ at e is $\operatorname{Id} - \operatorname{Ad}(s)|\mathfrak{m}': \mathfrak{m}' \to \mathfrak{m}'$, which is clearly also an isomorphism.

9.2 Theorem. *We keep the notation of* 9.1.

(1) *If $s \in G$ is semi-simple and normalizes H, then $C_H(s)$ is closed.*
(*l*) *If $X \in g$ is semi-simple and normalizes H then $c_H(X)$ is closed.*

Recall that X normalizes H if $\operatorname{Ad}(h)X - X \in \mathfrak{h}$ for all $h \in H$.

Proof. After choosing a faithful representation, we may assume $G = GL(V)$. If $A \in \mathrm{End}(V)$ write $C(A, T)$ for the characteristic polynomial of A; and $M(A, T)$ for the minimal polynomial of A. With this notation we define

$$W = \{x \in \mathcal{N}_G(H) \mid M(s, x) = 0 \text{ and } C(\mathrm{Ad}\,x \mid \mathfrak{h}, T) = C(\mathrm{Ad}\,s \mid \mathfrak{h}, T)\}.$$

Clearly $s \in W$, and W is stable under H operating by conjugation. If $x \in W$ then $M(x, T)$ divides $M(s, T)$, and the latter is a product of distinct linear factors because s is semi-simple; hence x is likewise semi-simple. We can therefore apply (9.1) to obtain

$$\dim C_H(x) = \dim H - \dim \mathcal{Z}_H(x) = \dim H - \dim \mathfrak{z}_{\mathfrak{h}}(x) = \dim H - m_1(x),$$

where $m_1(x)$ is the multiplicity of 1 as an eigenvalue of $\mathrm{Ad}\,x \mid \mathfrak{h}$. But the second condition defining W implies that $m_1(x) = m_1(s)$. Therefore, under the action of H by conjugation on W, the orbits $C_H(x)$ have constant dimension. It therefore follows from the closed orbit lemma 1.8 that the orbits are closed in W. But evidently W is closed in $\mathcal{N}_G(H)$, and the latter is closed in G (see 1.7). This proves that $C_H(s)$ is closed.

The proof that $C_H(X)$ is closed is similar. It uses

$$W = \{Y \in \mathfrak{n}_{\mathfrak{g}}(H) \mid M(X, Y) = 0 \text{ and } C(\mathrm{ad}\,Y \mid \mathfrak{h}, T) = C(\mathrm{ad}\,X \mid \mathfrak{h}, T)\}.$$

Here $\mathfrak{n}_{\mathfrak{g}}(H)$ is the set of Y in $\mathfrak{g}$ that "normalize H" in the sense of 9.1. This set W is closed in $\mathfrak{g}$, it contains X, and it is stable (via Ad) under H. Using 9.1 one can argue as above to show that the H-orbits in W have constant dimension, and hence are closed.

Corollary. *Let L be a (not necessarily closed) subgroup of G, which is commutative, consists of semi-simple elements, and normalizes H. Then*

$$L(\mathcal{Z}_G(L) \cap H) = L(\mathcal{Z}_H(L)) = \mathfrak{z}_{\mathfrak{h}}(L) = L(\mathcal{Z}_h(L)) \cap L(H).$$

If either $L \subset G(k)$, or L is closed, defined over k, then $\mathcal{Z}_H(L)$ is defined over k.

Proof. By definition, $\mathcal{Z}_H(L) = \mathcal{Z}_G(L) \cap H$, whence the first equality. The third also follows by definition. We prove the second one. Clearly the right side contains the left one, so, in case $H^o \subset \mathcal{Z}_H(L)$, the left side equals $\mathfrak{h}$ and equality is forced. We complete the proof by induction on $\dim H$. Choose $s \in L$ so that $H' = \mathcal{Z}_H(s)$ does not contain H^o, and hence $\dim H' < \dim H$. Part (1) of 9.1 Proposition tells us that $\mathfrak{h}' = L(\mathcal{Z}_H(s)) = \mathfrak{z}_{\mathfrak{h}}(s)$, from which it follows that $\mathfrak{z}_{\mathfrak{h}}(L) = \mathfrak{z}_{\mathfrak{h}'}(L)$. Moreover it is clear that $\mathcal{Z}_H(L) = \mathcal{Z}_{H'}(L)$, and $L \subset H'$ because L is commutative. By induction we have $L(\mathcal{Z}_{H'}(L)) = \mathfrak{z}_{\mathfrak{h}'}(L)$, so this proves the first assertion. If $L \subset H(k)$, the same induction, and 9.1, show that $\mathcal{Z}_H(L)$ is defined over k. Let now L be closed, defined over k. Then, by the above $\mathcal{Z}_H(L(k_s))$ is defined over k_s. But $L(k_s)$ is Zariski dense in L (AG.13.3), hence $\mathcal{Z}_H(L(k_s)) = \mathcal{Z}_H(L)$. On the other hand, $\mathcal{Z}_H(L)$ is clearly k-closed. Therefore it is defined over k.

9.3 Proposition. *Let G be a k-group and let U be a connected unipotent subgroup defined over k. Let $s \in G(k)$ be a semi-simple element that normalizes U. Put $M = C_U(s)s^{-1}$ and write $c_s(g) = gsg^{-1}s^{-1}$ for $g \in G$, so that $M = c_s(U)$.*

(1) *$\mathscr{Z}_U(s)$ and M are closed subvarieties of U defined over k.*
(2) *The product morphism $\alpha: M \times \mathscr{Z}_U(s) \to U$ is an isomorphism of varieties. Hence $\mathscr{Z}_U(s)$ is connected.*
(3) *c_s induces an isomorphism of the variety M onto itself.*

Proof. It follows from 9.1 (1) that $\mathscr{Z}_U(s)$ and M are smooth varieties defined over k, and clearly $\mathscr{Z}_U(s)$ is closed. The fact that M is closed is just 9.2 (1). This proves (1).

It further follows from the Remark to the proof of 9.1 (1) that α and $c_s: M \to M$ are separable, once we know they are dominant. Therefore it suffices, to conclude the proof, to show that α and $c_s: M \to M$ are bijective. We shall do this in several steps. Write $Z = \mathscr{Z}_U(s)$.

(a) *$c_s(u) = c_s(v) \Leftrightarrow uZ = vZ$ for $u, v \in U$.* This is because c_s is conjugation of s followed by right translation by s^{-1}, and Z is the stability group of s under conjugation.

(b) *If $u, v \in U$ then $c_s(uv) = uc_s(v)u^{-1}c_s(u)$. Hence, if $u \in \mathscr{C}(U)$ we have $c_s(uv) = c_s(v)c_s(u) = c_s(vu)$, and $c_s(u^{-1}) = c_s(u)^{-1}$.* For $c_s(uv) = uvs(uv)^{-1}s^{-1} = u(vsv^{-1}s^{-1})u^{-1}(usu^{-1}s^{-1})$. If $u \in \mathscr{C}(U)$ then $uc_s(v)u^{-1} = c_s(v)$ and $uv = vu$, so the second assertion follows from the first. The third one clearly follows from the second one.

(c) *$M \cap Z = \{e\}$.* Suppose $z = c_s(u) \in Z$ with $u \in U$. Then $zs = usu^{-1}$ is the Jordan decomposition of the semi-simple element usu^{-1}, so the unipotent part, z, equals e.

(d) *$\alpha: M \times Z \to U$ is bijective if U is abelian.* Part (b) implies that $c_s: U \to U$ is a homomorphism. It has kernel Z, by (a), and image M, so $\dim U = \dim M + \dim Z$. Moreover part (c) implies that α is injective. (Note that α is a group morphism now by (b).) Since U is connected, the dimension formula above implies that α is also surjective.

(e) *α is bijective.* Since U is nilpotent we can find a connected central subgroup $N \neq \{e\}$ of U normalized by s (e.g. the last non trivial term of the descending central series of U). If $N = U$ we can apply (d). If not, we can assume, by induction on dimension, that the analogue of our assertion is valid for the pairs (s, N) and (s', U'), where $U' = U/N$ and s' is the image of s in the quotient modulo N of the normalizer of N. Let π denote this quotient morphism.

Put $Z' = \mathscr{Z}_{U'}(s')$ and $M' = c_{s'}(U') = \pi(M)$. The induction hypothesis says that the product morphism $\alpha': M' \times Z' \to U'$ is bijective. Similarly, $c_s(N) \times \mathscr{Z}_N(s) \to N$ is bijective.

To show that α is injective suppose we have $xa = yb$ with $a, b \in Z$ and $x, y \in M$. Replacing a by ab^{-1} we can assume $b = e$. Applying π we conclude from the injectivity of α' that $\pi(a) = e$, so that $a \in N$. If $x = c_s(u)$ and $y = c_s(v)$

we have $usu^{-1}s^{-1}a = vsv^{-1}s^{-1}$ and $a(\varepsilon N \cap Z)$ centralizes U and s. Therefore we have $(usu^{-1})a = (vsv^{-1})$, which is the Jordan decomposition of the semi-simple element vsv^{-1}, so the unipotent part, a, equals e. This shows that α *is injective.*

We next claim that the inclusion $c_s(N) \subset M \cap N$ is an equality. For any $n \in N$ can be written as ma with $m \in c_s(N)$ and $a \in \mathscr{Z}_N(s)$, by induction. If also $n \in M$, the injectivity of α implies $a = e$.

Now we will show that $\pi: Z \to Z'$ is surjective. Suppose $x \in U$ and $\pi(x) \in Z'$. Then $c_{s'}(\pi(x)) = e$ so $c_s(x) \in \ker(\pi) \cap M = N \cap M = c_s(N)$, by the last paragraph. Say $c_s(x) = c_s(n)$ with $n \in N$. Since $N \subset \mathscr{C}(U)$ it follows from (b) that $c_s(n^{-1}x) = c_s(x)c_s(n^{-1}) = c_s(x)c_s(n)^{-1} = e$. Thus $n^{-1}x \in Z$ and $\pi(n^{-1}x) = \pi(x)$, so we have lifted $\pi(x)$ to an element of Z, as required.

Now we can show that α is surjective:

$$\begin{aligned} U &= MZN, && (\text{because } U' = M'Z', \pi M = M', \\ & && \text{and } \pi Z = Z'), \\ &= MNZ, && (N \subset \mathscr{C}(U)), \\ &= Mc_s(N)\mathscr{Z}_N(s)Z, && (\text{by induction}), \\ &= MZ, && (\text{because } \mathscr{Z}_N(s) \subset Z) \text{ and} \\ & && Mc_s(\mathscr{C}(U)) = M, \text{ by (b)}). \end{aligned}$$

(f) $c_s: M \to M$ *is bijective.* Using part (a) and the surjectivity of α we have $M = C_s(U) = c_s(MZ) = c_s(M)$. If $u, v \in M$ and $c_s(u) = c_s(v)$ then (a) implies $u = vz$ for some $z \in Z$. Thus $\alpha(u, e) = \alpha(v, z)$ so the injectivity of α implies that $z = e$. Q.E.D.

9.4 *Group actions of diagonalizable groups.* We fix a diagonalizable group T, a morphic action of T on G, and a T-invariant closed subgroup H of G containing $G^T = \mathscr{Z}_G(T)$. With respect to the action of T on the Lie algebras $\mathfrak{g} = L(G)$ and $\mathfrak{h} = L(H)$ we have (see 8.17 for the notation)

$$\mathfrak{g} = \mathfrak{g}^T \oplus \coprod_{\alpha \in \Phi(T,G)} \mathfrak{g}_\alpha$$

$$\mathfrak{g} = (\mathfrak{g}^T + \mathfrak{h}) \oplus \coprod_{\alpha \in \Phi(T,G/H)} \mathfrak{a}_\alpha,$$

where $\mathfrak{a}_\alpha$ is a complement for $\mathfrak{h}_\alpha$ in $\mathfrak{g}_\alpha$. Finally, we write $T_\alpha = \ker(\alpha), (\alpha \in X(T))$.

Proposition

(1) *We have* $L(G^T) = \mathfrak{g}^T$, *and hence* $\mathfrak{g}^T \subset \mathfrak{h}$. *If G is connected and unipotent, then G^T is connected.*
(2) *The following conditions on a subgroup S of T are equivalent*: (a) $(G^S)^o \subset H$; (b) $\mathfrak{g}^S \subset \mathfrak{h}$; (c) *$S$ is contained in no T_α for* $\alpha \in \Phi(T, G/H)$.
(3) *If G^S is connected, then $G^S = G^T \Leftrightarrow S$ is contained in no T_α* $(\alpha \in \Phi(T, G))$.
(4) *If G is connected and if $G \neq G^T$, then G is generated by the subgroups* $\mathscr{Z}_G(T_\alpha)^o$, $(\alpha \in \Phi(T, G))$.

Proof (1). The first assertion is a special case of the Corollary to 9.2 (applied to the semi-direct product of T and G).

The second assertion of (1) is proved by induction on $\dim G$: If $G = G^T$, then G^T is connected, by hypothesis. Otherwise choose s such that $G \not\subset \mathscr{Z}_G(s)$. By 9.3 (2), G^s is connected. One argues then as in the proof of the Corollary to 9.2.

(2): (a)$\Rightarrow$(b). If $(G^S)^o \subset H$ then $L(G^S) \subset \mathfrak{h}$, and (1) implies that $L(G^S) = \mathfrak{g}^S$.

(b)$\Rightarrow$(a). Since $\mathfrak{h} \subset \mathfrak{g}$ we obtain $\mathfrak{h}^S \subset \mathfrak{g}^S$, clearly, and $\mathfrak{g}^S \subset \mathfrak{h}$ implies $\mathfrak{h}^S = \mathfrak{g}^S$. Thus the dimension equality implies $(H^S)^o = (G^S)^o \subset H$, using (1).

(b)$\Rightarrow$(c). Writing

$$\mathfrak{g} = \mathfrak{h} \oplus \coprod_{\alpha \in \Phi} \mathfrak{a}_\alpha \qquad (\Phi = \Phi(T, G/H))$$

we have

$$\mathfrak{g}^S = \mathfrak{h}^S \oplus \coprod_{\alpha \in \Phi} \mathfrak{a}_\alpha^S = \mathfrak{h}^S \oplus \coprod_{\alpha \in \Phi, \alpha(S) = \{1\}} \mathfrak{a}_\alpha.$$

Thus $\mathfrak{g}^S \subset \mathfrak{h} \Leftrightarrow \alpha(S) \neq \{1\}$ for all $\alpha \in \Phi$, as claimed.

(3) Since $G^T \subset G^S$ this follows by applying (1) and (2) ((a)$\Leftrightarrow$(c)) with $H = G^T$.

(4) Let G' denote the subgroup generated by all $G^{T_\alpha}(\alpha \in \Phi(T, G))$. The condition $G^T \neq G$ implies that $\Phi(T, G)$ is not empty. Since $L(H^{T_\alpha})$ equals $\mathfrak{g}^{T_\alpha}$ by (1), and hence contains $\mathfrak{g}^T + \mathfrak{g}_\alpha$, it follows that $L(G')$ contains $\mathfrak{g}^T + \sum_{\alpha \in \Phi_T(G)} \mathfrak{g}_\alpha = \mathfrak{g}$. Hence $G' \supset G^o = G$.

This completes the proof.

9.5 Corollary. *Keep the notation of* 9.4.

(1) *If $\lambda \in X_*(T)$ and if $S = \operatorname{im}(\lambda)$, then $(G^S)^o \subset H$ if and only if $\langle \alpha, \lambda \rangle \neq 0$ for all $\alpha \in \Phi(T, G/H)$. In particular $(G^S)^o = (G^T)^o$ if and only if $\langle \alpha, \lambda \rangle \neq 0$ for all $\alpha \in \Phi(T, G)$.*

(2) *Suppose T is a torus and $G \neq G^T$. Then G^o is generated by the $(G^{T_\alpha^o})^o$. Moreover, if k is infinite there is a $t \in T(k)$ such that $t^\alpha \neq 1$ for all $\alpha \in \Phi(T, G)$, and for such a t* we have $\mathscr{Z}_G(t)^o = (G^T)^o$.

Part (1) follows directly from 9.4 (2). Since the centralizer of T^o contains that of T, the first assertion of (2) is a consequence of 9.4 (4). The existence of t follows from the fact that $T(k)$ is dense in T (8.13 (2)). The last equality of (2) then follows from 9.4 (2), applied to $H = G^T$ and to the subgroup S generated by t.

9.6 Proposition. *Let $\pi: G \to G'$ be a surjective and T-equivariant morphism of k-groups on which the diagonalizable group T acts. Then the induced homomorphism $(G^T)^o \to (G'^T)^o$ is surjective.*

Proof. Since $N = \ker(\pi)$ is T-invariant there is an action of T on G/N, and π factors through a T-equivariant and bijective morphism $G/N \to G'$. Hence we may assume that $G' = G/N$. In this case $(d\pi)_e : \mathfrak{g} \to \mathfrak{g}'$ is surjective. Since T is diagonalizable $\mathfrak{g}^T \to \mathfrak{g}'^T$ is also surjective. According to 9.4 (1), however, the latter is the differential of $G^T \to G'^T$, so the proposition follows.

Remark. The proof shows even that $(G^T)^o \to (G'^T)^o$ is a quotient morphism if π is a quotient morphism.

Bibliographical Note

The Proposition in 9.1 and Theorem 9.2 are proved in [4, §10] for groups, and in [2, 3] for Lie algebras. Proposition 9.3 is proved in [1, lemme 9.6] when U is commutative, and in [4, §11.1] in the general case. In [1], there is a counterpart where s is unipotent, and U is a torus, but it will not be needed in this book. 9.5 generalizes a result proved in [13, Exp. 9, No. 1] for actions of tori on unipotent groups.

§10. Connected Solvable Groups

The analysis of a general affine group proceeds via a study of its connected solvable subgroups. This is because the latter have a number of special properties which make them easier to work with. The main ones are the fixed point theorem 10.4 and the structure theorem 10.6.

10.1 *Complete varieties.* We shall collect here some properties of complete varieties to be used below. Recall (AG.7.4) that V is complete if, for all varieties X, the projection $V \times X \to X$ is a closed map. Properties (1), (2), and (3) which follow are taken from (AG.7.4).

(1) *A closed subvariety of a complete variety is complete. The image of a complete variety under a morphism is closed and complete. Products of complete varieties are complete.*
(2) *A morphism from a connected complete variety into an affine variety is constant.*
(3) *Projective varieties are complete.*
(4) *Let $\alpha: V \to W$ be a bijective morphism. If W is normal and complete then V is also complete.* (See (AG.18.3).)

Finally, from (AG.18.5(d)) we have:

(5) *Let $\alpha: V \to W$ be a morphism from an irreducible smooth curve V into a complete variety W. Then α extends to a morphism $\bar{\alpha}: \bar{V} \to W$ from the complete smooth curve $\bar{V}$ containing V.*

10.2 *A composition series for* $\mathbf{T}_n$. Recall the following subgroups of $\mathbf{GL}_n$:

$$\mathbf{T}_n = \{g = (g_{ij}) | g_{ij} = 0 \text{ for } j < i\}$$
$$= \left\{ \begin{pmatrix} * & & * \\ & \ddots & \\ 0 & & * \end{pmatrix} \in \mathbf{GL}_n \right\}$$
$$\mathbf{U}_n = \{g \in \mathbf{T}_n | g_{ii} = 1, 1 \leqq i \leqq n\}$$
$$= \left\{ \begin{pmatrix} 1 & & * \\ & \ddots & \\ 0 & & 1 \end{pmatrix} \right\}$$
$$\mathbf{D}_n = \{g \in \mathbf{GL}_n | g_{ij} = 0 \text{ for } i \neq j\}$$
$$= \{\operatorname{diag}(t_1, \ldots, t_n) | t_i \in K^*\}.$$

The following facts are readily checked:

$$\mathbf{U}_n = (\mathbf{T}_n)_u = (\mathbf{T}_n, \mathbf{T}_n),$$
$$\mathbf{D}_n \cong (\mathbf{GL}_1)^n, \quad \mathbf{T}_n = \mathbf{D}_n \cdot \mathbf{U}_n.$$

$\mathbf{T}_n$ is the group of invertible elements in the algebra A of all upper triangular matrices. The set N of matrices in A with zeroes on the diagonal is an ideal (in fact the radical) of A. The two sided ideal N^h is spanned by the basic matrices e_{ij} for which $j \geqq i + h$. Moreover, the image of $e_{i,i+h}$ in N^h/N^{h+1} spans a one dimensional two sided ideal in A/N^{h+1} since it is killed by N and is an eigenvector for the diagonal matrices. Thus the vector space A_{hl} spanned by $\{e_{ij} | j > i + h, \text{ or } j = i + h \leqq l\}$ is a two sided ideal in A for $0 \leqq h < n$ and $1 \leqq l \leqq n - h$. If we order the pairs (h, l) lexicographically we obtain a descending chain of two sided ideals, starting with $A = A_{0,0}$, ending with $A_{n-1,1} = Ke_{nn}$, and such that each has codimension one in the next larger one. Writing $\mathbf{T}_{hl} = \{g \in \mathbf{T}_n | g \equiv I \bmod A_{hl}\}$ we therefore obtain a descending chain of normal subgroups of $\mathbf{T}_n$. Note that $N = A_{1,n-1}$, and hence $\mathbf{T}_{1,n-1} = \mathbf{U}_n$. Thus one sees that the first n quotients are isomorphic to $\mathbf{GL}_1$, and the remaining ones are isomorphic to $\mathbf{G}_a$. In summary:

$\mathbf{T}_n$ is filtered by a chain of normal subgroups with successive quotients isomorphic to $\mathbf{GL}_1$ or $\mathbf{G}_a$.

10.3 *Grassmannians and flag varieties.* Let V be an n-dimensional vector space. We propose to put on the set $G_d(V)$ of d-dimensional subspaces of V the structure of a projective variety. Define

$$f : G_d(V) \to \mathbf{P}(\Lambda^d V)$$

by sending W to the point in the projective space corresponding to the line $\Lambda^d W \subset \Lambda^d V$. It is easily checked (and well known) that f is injective (cf. 5.1 Lemma), so we need only show that its image is closed.

$\mathbf{P}(\Lambda^d V)$ is covered by (affine) open sets U of the following type relative to a suitable basis $e_1, \ldots, e_n$ for $V:U$ consists of all points whose homogeneous coordinate in the basis of $\Lambda^d V$ defined by $e_1, \ldots, e_n$, are such that the coefficient of $e = e_1 \wedge \cdots \wedge e_d$ is not zero. Thus U is the complement of a linear variety.

Write $V = W_o \oplus W'_o$ where W_o and W'_o are spanned by $e_1, \ldots, e_d$ and $e_{d+1}, \ldots, e_n$, respectively. Then, for $W \in G_d(V)$, we have $f(W) \in U$ if and only if the projection maps W is isomorphically onto W_o. In this case W has a unique basis of the form $e_1 + x_1(W), \ldots, e_d + x_d(W)$ with $x_i(W) \in W'_o$. Say $x_i(W) = \sum_{j>d} a_{ij} e_j$. Then $f(W)$ is the projection into $\mathbf{P}(\Lambda^d V)$ of the vector $e + \left(\sum_{1 \leq i \leq d} e_1 \wedge \cdots \wedge x_i(W) \wedge \cdots \wedge e_d \right) + (*)$, where $(*)$ involves basis vectors omitting two or more of $e_1, \ldots, e_d$. Now

$$e_1 \wedge \cdots \wedge x_i(W) \wedge \cdots \wedge e_d = \sum_{j>d} a_{ij} e_1 \wedge \cdots \wedge e_j \wedge \cdots \wedge e_d,$$

so we see that, in $(e_1 + x_1(W)) \wedge \cdots \wedge (e_d + x_d(W))$, we recover a_{ij} as the coefficient of the basis vector $e_1 \wedge \cdots \wedge e_j \wedge \cdots \wedge e_d$, $(1 \leq i \leq d;\ j > d,\ e_j$ at the *ith* place), and these coefficients, which determine W, may be prescribed arbitrarily. The coefficients of the remaining vectors in $\Lambda^d V$ are polynomial functions of the a_{ij}. Thus, $f(G_d(V))$ is essentially the graph of a morphism from the space of (a_{ij})'s to another linear space. In particular, it is closed.

Suppose $W \in G_d(V)$ and $W' \in G_{d'}(V)$, with $d \leq d'$. Then the condition that $W \subset W'$ can be expressed by algebraic equations on the coordinates in $\mathbf{P}(\Lambda^d V) \times \mathbf{P}(\Lambda^{d'} V)$. Thus $\{(W, W') \in G_d(V) \times G_{d'}(V) \mid W \subset W'\}$ is a closed subvariety. The *flag variety*, $\mathscr{F}(V)$, is

$$\{(V_1, \ldots, V_n) \in G_1(V) \times \cdots \times G_n(V) \mid V_i \subset V_{i+1}, 1 \leq i < n\}.$$

The remarks above show that *$\mathscr{F}(V)$ is a projective variety*. Hence, by 10.1 (3), *$\mathscr{F}(V)$ is complete.*

The following remarks on $\mathrm{GL}(V)$ illustrate certain theorems to be proved below for arbitrary connected groups.

If $e_1, \ldots, e_n$ is a basis for V we can define

$$\varphi: \mathrm{GL}(V) \to \mathscr{F}(V)$$

by $\varphi(g) = (V_1, \ldots, V_n)$, where V_i is the space spanned by $ge_1, \ldots, ge_i$ $(1 \leq i \leq n)$. It is clear that $\mathrm{GL}(V)$ operates transitively on the flags in V, the operation being such that φ is equivariant. Therefore φ induces a bijective morphism $\alpha: \mathrm{GL}(V)/B \to \mathscr{F}(V)$, where B is the isotropy group of the flag $\varphi(e)$. Under the isomorphism $\mathrm{GL}(V) \to \mathbf{GL}_n$ defined by the basis above one sees that B corresponds to $\mathbf{T}_n$. Write U^- for the unipotent subgroup corresponding to lower triangular matrices. It is easy to check that $U^- \cdot B$ contains an open set in $\mathrm{GL}(V)$. In fact, it corresponds to the set of $g = (g_{ij})_{1 \leq i,j \leq n}$ in $\mathbf{GL}_n$ such that, for each $d \leq n$, $\det(g_{ij})_{1 \leq i,j \leq d} \neq 0$, and this is clearly open.

In terms of the projective coordinates introduced on each $G_d(V)$, we see that the coordinates of $\varphi(g)$ are given by $(ge_1, ge_1 \wedge ge_2, \ldots, ge_1 \wedge \cdots \wedge ge_n)$.

If $g \in U^-$ then $ge_i = e_i + \sum_{j>i} a_{ij} e_j$, so

$$ge_1 \wedge \cdots \wedge ge_i = (ge_1 \wedge \cdots \wedge ge_{i-1} \wedge e_i) + \left(\sum_{j>i} a_{ij} ge_1 \wedge \cdots \wedge ge_{i-1} \wedge e_j \right).$$

Thus we see, by induction on i, that the ge_i can be determined algebraically from the projective coordinates of $\varphi(g)$, for $g \in U^-$. As a consequence, φ induces an isomorphism of U^- onto its image. Since, as we saw above, $\varphi(U^-)$ contains an open set in $\varphi(\mathrm{GL}(V))$, it follows that the differential of φ is surjective, i.e. φ is separable. This proves:

$\varphi: \mathrm{GL}(V) \to \mathscr{F}(V)$ *induces an isomorphism of varieties* $\alpha: \mathrm{GL}(V)/B \to \mathscr{F}(V)$. *In particular,* $\mathrm{GL}(V)/B$ *is a projective variety.*

The above proof is a little sketchy, but an independent, and much more general, one will be given in 11.1.

10.4 Theorem. *Let G be a connected solvable group operating morphically on a non-empty complete variety V. Then G has a fixed point in V.*

Proof. We argue by induction on $d = \dim G$. If $d = 0$ then $G = \{e\}$, so assume $d > 0$. Then $N = (G, G)$ is connected and of smaller dimension, so the set F of fixed points of N in V is a non-empty, closed, and hence complete, variety. Since N is normal in G, it follows that F is stable under G.

By the closed orbit lemma 1.8 there is an $x \in F$ such that $G(x)$ is closed. Since $N \subset G_x$, it follows that G_x is normal in G. Thus

$$G/G_x \to G(x)$$

is a bijective morphism from a connected affine variety to a complete one. Since $G(x)$ is smooth, and hence normal, it follows from 10.1 (4) that G/G_x is complete. Now 10.1 (2) implies that G/G_x is a point. Q.E.D.

10.5 Corollary (Lie–Kolchin Theorem). *If $\pi: G \to \mathrm{GL}(V)$ is a linear representation of a connected solvable group, then $\pi(G)$ leaves a flag in V invariant. i.e. $\pi(G)$ can be put in triangular form.*

Proof. G has a fixed point for the action induced by π on the variety $\mathscr{F}(V)$ because, by 10.3, $\mathscr{F}(V)$ is complete.

Here is a purely algebraic corollary:

Corollary. *Let m be a solvable, not necessarily closed, subgroup of* $\mathrm{GL}(V)$. *Then some subgroup of finite index in M can be put in triangular form.*

Proof. Let $H = \mathscr{A}(M)$. We know from 2.4 Cor. 2 that H is solvable. Now apply the last corollary to H^o. Then $H^o \cap M$ has finite index in M, and hence solves our problem.

10.6 Theorem. *Let G be connected, solvable.*

(1) *G_u is a connected normal k-closed subgroup of G containing $\mathscr{D}G = (G, G)$.*
(2) *G/G_u is a torus, and G_u contains a chain of closed connected subgroups, normal in G, such that the successive quotients have dimension one.*
(3) *G is nilpotent if and only if G_s is a subgroup of G. In this case, G_s is a closed subgroup defined over k, and G is the direct product $G_s \times G_u$.*
(4) *The maximal tori in G are conjugate by $\mathscr{C}^{\infty} G$. If T is a maximal torus, then $G = T \cdot G_u$ (semi-direct product). $L(G_u)$ is the union of the nilpotent elements of $L(G)$.*
(5) *Let S be a not necessarily closed subgroup of G, consisting of semi-simple elements. Then*
   (i) *S is contained in a torus, and*
   (ii) *$G^S = \mathscr{Z}_G(S)$ is connected and equal to $\mathscr{N}_G(S)$.*
(6) *Let T be a maximal torus of G. Then any semisimple element of G is conjugate to a unique element of T.*

Proof. (1) Using the Lie–Kolchin Theorem we can embed G in $\mathbf{T}_n$. Then $U_n = (\mathbf{T}_n)_u = \mathscr{D}\mathbf{T}_n$ is a closed normal subgroup of $\mathbf{T}_n$, so G_u is a closed normal subgroup of G containing $\mathscr{D}G$. It follows from 4.5 that G_u is k-closed. Let $\pi: G \to G' = G/\mathscr{D}G$ be the canonical projection. By 4.7, $G' = G'_s \times G'_u$, hence G'_u is connected. We claim that $G_u = \pi^{-1}(G'_u)$. If $x \in G_u$, then $\pi(x) \in G'_u$ by 4.4. Let now $x \in \pi^{-1}(G'_u)$ and $x = x_s \cdot x_u$ its Jordan decomposition. Then, by 4.4, x_s, $x_u \in \pi^{-1}(G'_u)$, and $x_s \in \mathscr{D}G$. But $\mathscr{D}G \subset G_u$, hence $x_s = e$ and $x \in G_u$, which shows that $G_u = \pi^{-1}(G'_u)$. Since $\mathscr{D}G$ and G'_u are connected, it follows that G_u is connected.

(2) G/G_u injects into $\mathbf{T}_n/\mathbf{U}_n \cong \mathbf{D}_n$ so G/G_u is a commutative connected group consisting of semi-simple elements, and hence is a torus (see 8.4 and 8.5). Starting with a chain of connected normal subgroups N_i of $\mathbf{T}_n$ contained in $\mathbf{U}_n$ and with successive quotients isomorphic to $\mathbf{G}_a$ (see 10.2) we obtain from the groups $(N_i \cap G)^o$ a chain of connected normal subgroups of G contained in G_u with successive quotients of dimension $\leqq 1$. Once repetitions are eliminated, the successive quotients will have dimension one.

(3) Suppose first that G_s is a subgroup of G. It projects injectively into G/G_u, so G_s is commutative. Hence we can use 4.6 to diagonalize G_s under some faithful rational representation of G in a $GL(V)$. It is then clear that the closure of G_s is a diagonalizable subgroup of G, necessarily equal to G_s, clearly. By rigidity 8.10 we have $\mathscr{Z}_G(G_s)^o = \mathscr{N}_G(G_s)^o$. But evidently G_s is normal in G, so, since G is connected, G_s is central in G. The quotient G/G_s is unipotent, and hence nilpotent 4.8, so it follows that G is nilpotent, as claimed.

Suppose, conversely, that G is nilpotent. We claim then that G_s lies in the center of G.

Let $a \in G_s$ and put $U = G_u$. Write $c_a(x) = xax^{-1}a^{-1}$ for $x \in G$ and put $M = c_a(U)$. According to 9.3 (3) c_a induces a bijection $M \to M$, so $M \subset \mathscr{C}^{\infty} G$. Since G is nilpotent we conclude that $M = \{e\}$, i.e. that a centralizes U. Hence

G^a contains $\mathscr{D}G$, so it is a normal subgroup. In order to prove that $G^a = G$, it suffices then, in view of 4.4, 4.7, to show that $G^a \supset G_s$.

Suppose then that $t \in G_s$. Then $c_a(t) \in U$, so a commutes with $c_a(t)$. Therefore $c_a(t)a = tat^{-1}a^{-1}a = tat^{-1}$ is the Jordan decomposition of the semi-simple element tat^{-1}, so the unipotent part, $c_a(t)$, is e.

Now that G_s is central it follows as above that G_s is a closed diagonalizable subgroup of G. The Jordan decomposition in G and in $L(G)$ shows that $G = G_s \times G_u$ (group direct product) and that $L(G_s) \cap L(G_u) = 0$. Thus G is the direct product of G_s and G_u as an algebraic group, because $G_s \times G_u \to G$ is bijective and separable.

It remains to be shown that G_s is defined over k.

(a) $p = \operatorname{char} k = 0$. The Jordan factors of a $g \in G(\bar{k})$ are in $G_s(\bar{k}) \times G_u(\bar{k})$, and the action of $\Gamma = \operatorname{Gal}(\bar{k}/k)$ evidently preserves Jordan decomposition. Therefore G_s and G_u each have dense sets of $\bar{k}$-points which are Γ-stable, so they are subgroups defined over k.

(b) $p > 0$. There is a $q = p^r (r > 0)$ such that $u^q = e$ for all $u \in G_u$. (If $G \subset \mathbf{GL}_n$ then $r = n - 1$ works.) Then the Jordan decomposition shows that $g \mapsto g^q$ provides a morphism $G \to G_s$, evidently defined over k. It follows from 8.9 (b) that its restriction to G_s is bijective. Thus G_s, being the image of a k-morphism, is defined over k.

(4) We first claim, by induction on $\dim G$, that *there is a torus T in G that projects onto G/G_u*. It will then follow that $G = T \cdot G_u$ (semi-direct product of algebraic groups) because the Jordan decompositions imply that $T \cap G_u = \{e\}$ and $L(T) \cap L(G_u) = 0$.

If G is nilpotent we take $T = G_s$ as in (3). If not then there is an $s \in G_s$ which is not central, so $\dim G^s < \dim G$, where we write $G^s = \mathscr{Z}_G(s)$. Moreover it follows from (9.6) that $(G^s)^o \to (G/G_u)^o = G/G_u$ is surjective. Hence we can find the required T in $(G^s)^o$, by induction.

Next we claim:

(*) *Suppose $G = T \cdot G_u$ as above. Then every $s \in G_s$ is conjugate by an element of $\mathscr{C}^\infty G$ to an element of T.*

We prove (*) by induction on $\dim G$. In case G is nilpotent it follows from part (3) above that G_s is the unique maximal torus, so we may assume G is not nilpotent, i.e. that $\mathscr{C}^\infty G \neq \{e\}$. Let N be the identity component of the center of $\mathscr{C}^\infty G$. Then $N \neq \{e\}$, for $\mathscr{C}^\infty G$ is connected and unipotent, and N contains the last non-trivial term of the descending central series of $\mathscr{C}^\infty G$, which is also connected (see 2.3).

Let $\pi: G \to G' = G/N$ be the natural projection. Then $G' = T' \cdot G'_u$ (semi-direct) where $T' = \pi(T)$. By induction, there is a $g' \in \mathscr{C}^\infty G'$ such that $g'\pi(s)g'^{-1} \in T'$. Choosing $g \in \mathscr{C}^\infty G$ such that $\pi(g) = g'$, and replacing s by gsg^{-1}, therefore, we may assume $s \in T \cdot N$. We want to conjugate s into T by an element of N.

Write $s = nt$ with $n \in N$ and $t \in T$. We apply 9.3 to t and N in order to write $n = c_t(u)z$ where $u \in N$, $c_t(u) = utu^{-1}t^{-1}$, and where $z \in \mathscr{Z}_N(t)$. Thus $s =$

$utu^{-1}t^{-1}zt = utu^{-1}z$. Since z is unipotent and commutes with t and u, the equation $s = (utu^{-1})z$ is the Jordan decomposition of the semi-simple element s, and hence $z = e$. Thus $u^{-1}su = t \in T$, thus proving (*).

To conclude the proof of the first assertion in (4) suppose T' is another maximal torus in G. Choose $s \in T'$ so that $s^\alpha \neq 1$ for all $\alpha \in \Phi(T', G)$. Then it follows from 9.4 that s and T' have centralizers in G with the same connected component of e. Using (*) above we can conjugate s into T with an element of $\mathscr{C}^\alpha G$. Conjugating T' likewise we are reduced to the case, therefore, where $T \subset (G^s)^o = (G^{T'})^o$. From (*) we conclude that each element of T' is conjugate in $(G^{T'})^o$ to an element of T. But T' is central in $(G^{T'})^o$, so we have $T' \subset T$, and hence $T' = T$, by maximality.

Let $\pi: G \to G/G_u$ be the canonical projection. Its restriction to T is an isomorphism of T onto G/G_u. In particular $L(G/G_u)$ consists of semi-simple elements, and 4.4 shows that if $X \in \mathfrak{g}$ is nilpotent, then $X \in \ker d\pi = L(G_u)$. Since $L(G_u)$ consists of nilpotent elements (4.8), this ends the proof of (4).

(5) Let S be a subgroup of G consisting of semi-simple elements, and let $\pi: G \to G/G_u$ be the canonical projection. Then the restriction of π to S is injective, hence S is commutative since G/G_u is. Moreover, if $n \in G$ normalizes S, then $\pi(n)$ centralizes $\pi(S)$, and therefore (since $\pi|_S$ is injective) n centralizes S. This proves that $\mathscr{Z}_G(S) = \mathscr{N}_G(S)$. The group $\bar{S} = \mathscr{A}(S)$ is a closed diagonalizable subgroup of G, and we have $\mathscr{Z}_G(\bar{S}) = \mathscr{Z}_G(S)$, which reduces us to the case where S is closed for the proof of the remaining assertions. Let T be a maximal torus of G.

Case 1. S is central. Then $G^S = G$ is connected. If $s \in S$, some conjugate of s lies in T by (*) in (4), so $s \in T$. Thus $S \subset T$.

Case 2. S is not central. Choose a non central $s \in S$. Replacing T by a conjugate we can assume $s \in T$. Then $T \subset G^s = T \cdot G_u^s$. By 9.3, G_u^s is connected. Therefore G^s is connected, has smaller dimension than G, and contains S. By induction S is conjugate in G^s to a subgroup of T, and $(G^s)^S = G^S$ is connected. This completes the proof of (5).

(6) By (5), any semisimple element s of G is conjugate to at least one element $t \in T$. Let $\pi: G \to G' = G/\mathscr{R}_u G$ be the canonical projection. It is an isomorphism of T onto G'. This implies that $\pi(s) = \pi(t)$ and the uniqueness of t.

10.7 *Curves with a connected group of automorphisms.* The structure theorem 10.6 has one glaring deficiency; it gives no accounting of groups of dimension one. These appear as the "composition factors" of G_u in part (2) of the theorem.

In fact the only one dimensional connected groups are $\mathbf{GL}_1$ and $\mathbf{G}_a$. This will be shown in 10.9. We shall deduce this fact from the following proposition, which, in turn, is a corollary of the classification of one-dimensional groups.

The proof we give of the proposition uses facts about Jacobians of curves, which are in the spirit of, but outside the framework of, these notes.

Proposition. *Let X be a complete, smooth, irreducible algebraic curve. Suppose* *that a connected group G of dimension $\geqq 1$ operates non trivially on X with a fixed point. Then X is isomorphic to the projective line,* $\mathbf{P}_1$.

Proof. We must show that the genus, $\operatorname{gen} X$ of X, is zero. Let $f: X \to J$ be the canonical morphism of X into its Jacobian J. (See Lang, Abelian varieties, Ch. II, §2.) J is an abelian variety (= complete connected algebraic group) whose dimension equals $\operatorname{gen} X$, and $f(X)$ generates J. Moreover (loc. cit., Theorem 9) any rational map $h: X \to A$, where A is an abelian variety, induces a unique homomorphism $\alpha: J \to A$ such that $h(x) = \alpha(f(x)) + a$ for some $a \in A$ independent of $x \in X$. (We are, of course, writing $+$ for the group operation in the abelian varieties here.) In fact this "universal mapping property" clearly determines f up to translation by an element of J. We shall normalize f so that $f(p) = 0$, where $p \in X$ is some fixed point of G (which is assumed to exist).

If $g \in G$ the universal mapping property implies that $f \circ g: X \to J$ is of the form $\alpha_g \circ f + a_g$ for some group morphism $\alpha_g: J \to J$ and some $a_g \in J$. Evaluating at $p = g(p)$ shows that $a_g = 0$, so $f \circ g = \alpha_g \circ f$. We are now tempted to assert that we have an action $G \times J \to J$, giving a connected family of automorphisms of J, and to invoke the rigidity of abelian varieties (cf. 8.10).

Rather than justify that assertion we argue directly: If $a \in J$ define $\beta_a: G \to J$ by $\beta_a(g) = \alpha_g(a)$. In case $a = f(x)$ for some $x \in X$ this is the composite map $G \xrightarrow{\beta_x} X \xrightarrow{f} J$, where $\beta_x(g) = g(x)$, and this is a morphism. In general we can write $a = \Sigma f(x_i)$ for suitable $x_i \in X$ so $\beta_a = \Sigma \beta_{f(x_i)}$ is again a morphism.

Let ${}_mJ = \ker(a \mapsto ma)$ in J, where m is a positive integer. Then ${}_mJ$ is finite (loc. cit.), and it is clearly stable under each α_g. Hence $\beta_a(G)$ is finite for each $a \in {}_mJ$. Since G is connected and $\beta_a(e) = a$, it follows that $\beta_a(G) = \{a\}$.

Thus, for $g \in G$, $\alpha_g: J \to J$ fixes all elements of finite order. But the latter are dense in J (loc. cit), so $\alpha_g = 1_J$.

We conclude that $f: X \to J$ is a G-equivariant map with G operating trivially on the right. Hence f collapses each G-orbit in X to a point. Since G acts non trivially on the irreducible curve X some G-orbit must contain an open dense set. The complement of the latter is finite, so X has only finitely many G-orbits. It follows that J is generated by a *finite* set $f(X)$. This is clearly impossible unless $J = \{0\}$, i.e. unless $\dim J (= \operatorname{gen} X) = 0$. Q.E.D.

10.8 *The automorphism group of* $\mathbf{P}_1$ *is* $\mathbf{PGL}_2$. We shall write

$$G = \mathbf{PGL}_2 = \mathbf{GL}_2/S,$$

where

$$S = \mathscr{C}(\mathbf{GL}_2) = \{aI \mid a \in K^*\}.$$

is the group of scalar 2×2 matrices. The projection $\mathbf{GL}_2 \to G$ will be denoted

$$\begin{pmatrix} a & b \\ c & d \end{pmatrix} \to \begin{bmatrix} a & b \\ c & d \end{bmatrix}.$$

In order to avoid confusion the projection

$$\mathfrak{gl}_2 \to \mathfrak{g} = \mathfrak{pgl}_2 = \mathfrak{gl}_2/K \cdot I$$

will be denoted

$$\begin{pmatrix} a & b \\ c & d \end{pmatrix} \to \begin{bmatrix} a & b \\ c & d \end{bmatrix}_L.$$

Consider the torus $T = \mathbf{D}_2/S$ in G. We have the isomorphism

$$\lambda: \mathbf{GL}_1 \to T, \quad a \mapsto a^\lambda = \begin{bmatrix} a & 0 \\ 0 & 1 \end{bmatrix}.$$

Let $\alpha \in X(T)$ be such that $\langle \alpha, \lambda \rangle = 1$, i.e. such that $(a^\lambda)^\alpha = a$ for $a \in \mathbf{GL}_1$. Next define

$$u_\alpha, u_{-\alpha}: \mathbf{G}_a \to G$$

by

$$u_\alpha(b) = \begin{bmatrix} 1 & b \\ 0 & 1 \end{bmatrix} \quad \text{and} \quad u_{-\alpha}(c) = \begin{bmatrix} 1 & 0 \\ c & 1 \end{bmatrix}.$$

The images of u_α and $u_{-\alpha}$ will be denoted U_α and $U_{-\alpha}$, respectively. A direct calculation shows that

$$tu_\alpha(b)t^{-1} = u_\alpha(t^\alpha b),$$

and

$$(1) \qquad tu_{-\alpha}(c)t^{-1} = u_{-\alpha}(t^{-\alpha}c)$$

for $t \in T$ and $b, c \in \mathbf{G}_a$. The resulting commutator formulas,

$$(t, u_\alpha(b)) = u_\alpha((t^\alpha - 1)b),$$

and

$$(t, u_{-\alpha}(c)) = u_{-\alpha}((t^{-\alpha} - 1)c),$$

show that the derived group $\mathscr{D}G$ contains U_α and $U_{-\alpha}$. The subgroup generated by U_α and $U_{-\alpha}$ clearly has dimension > 2. Since $\dim G = 3$ and G is connected we conclude:

(2) *$G = \mathscr{D}G$, and G is generated by U_α and $U_{-\alpha}$.* The Lie algebras $L(T)$, $L(U_\alpha)$, and $L(U_{-\alpha})$ are spanned by

$$(3) \qquad H = \begin{bmatrix} 1 & 0 \\ 0 & 0 \end{bmatrix}_L, \quad X_\alpha = \begin{bmatrix} 0 & 1 \\ 0 & 0 \end{bmatrix}_L, \quad \text{and} \quad X_{-\alpha} = \begin{bmatrix} 0 & 0 \\ 1 & 0 \end{bmatrix}_L,$$

respectively. Moreover it follows from (1) that X_α and $X_{-\alpha}$ are semi-invariants of weights α and $-\alpha$, respectively, for T under Ad_G. Therefore

$$\begin{aligned} \mathfrak{g} &= L(T) \oplus L(U_\alpha) \oplus L(U_{-\alpha}) \\ &= \mathfrak{g}^T \oplus \mathfrak{g}_\alpha \oplus \mathfrak{g}_{-\alpha} \end{aligned}$$

is the root space decomposition of $\mathfrak{g}$ relative to the torus T, and

$$\Phi(T, G) = \{\alpha, -\alpha\}.$$

Write the elements of K^2 as column vectors, and denote the projection

$$K^2 - \{0\} \to \mathbf{P}_1$$

by

$$\begin{pmatrix} a \\ b \end{pmatrix} \to \begin{bmatrix} a \\ b \end{bmatrix}.$$

The action (by left multiplication) of $\mathbf{GL}_2$ on K^2 induces an action of $\mathbf{GL}_2$ on $\mathbf{P}_1$ so that the above projection is equivariant. Since S operates trivially on $\mathbf{P}_1$ we deduce an action of G on $\mathbf{P}_1$ by

$$\begin{bmatrix} a & b \\ c & d \end{bmatrix} \begin{bmatrix} x \\ y \end{bmatrix} = \begin{bmatrix} ax + by \\ cx + dy \end{bmatrix}.$$

Embed K into $\mathbf{P}_1$ by the identification

$$x \mapsto \begin{bmatrix} 1 \\ x \end{bmatrix},$$

and write $\infty = \begin{bmatrix} 0 \\ 1 \end{bmatrix}$, so that

$$\mathbf{P}_1 = K \cup \{\infty\}.$$

We next introduce the open set

$$V = \{(x, y, z) \in (\mathbf{P}_1)^3 \mid x, y, \text{ and } z \text{ are distinct}\}$$

and define

$$\varphi : G \to V \quad \text{by} \quad \varphi(g) = (g(0), g(1), g(\infty)).$$

Thus φ is just the G-orbit map for $(0, 1, \infty) \in V$.

Contention. *φ is an isomorphism of varieties. In particular G operates simply transitively on triples of distinct points in $\mathbf{P}_1$.*

If

$$g = \begin{bmatrix} a & b \\ c & d \end{bmatrix}$$

then

$$\varphi(g) = \left(\begin{bmatrix} a \\ c \end{bmatrix}, \begin{bmatrix} a + b \\ c + d \end{bmatrix}, \begin{bmatrix} b \\ d \end{bmatrix} \right).$$

Thus

$$\begin{aligned} &g(\infty) = \infty \Leftrightarrow b = 0, \\ (4) \qquad &g(0) = 0 \Leftrightarrow c = 0, \quad \text{and} \\ &g(1) = 1 \Leftrightarrow a + b = c + d. \end{aligned}$$

Therefore $\varphi(g) = (0, 1, \infty)$ implies $g = \begin{bmatrix} a & 0 \\ 0 & a \end{bmatrix} = e$, so φ is injective.

To see that φ is surjective suppose we are given $(x, y, z) \in V$. Since $\mathbf{GL}_2$ is clearly doubly transitive on lines in K^2 we can first transform (x, y, z) into an element of the form $\left(0, \begin{bmatrix} a \\ d \end{bmatrix}, \infty\right)$. The fact that $\begin{bmatrix} a \\ d \end{bmatrix}$ is distinct from 0 and ∞ means that $a \neq 0 \neq d$. Therefore, we can transform $\left(0, \begin{bmatrix} a \\ d \end{bmatrix}, \infty\right)$ to $(0, 1, \infty)$ with $\begin{bmatrix} a & 0 \\ 0 & d \end{bmatrix}^{-1}$.

Finally we must show that $(d\varphi)_e : \mathfrak{g} \to T(V)_{(0,1,\infty)}$ is surjective. We have:

$$u_{-\alpha}(c)(0, 1, \infty) = (c, 1 + c, \infty).$$

Since $du_{-\alpha}(1) = X_{-\alpha}$ this yields $(d\varphi)_e(X_{-\alpha}) = (1, 1, 0)$. By symmetry, we have $(d\varphi)_e(X_\alpha) = (0, 1, 1)$. Next we have

$$a^\lambda(0, 1, \infty) = (0, a^{-1}, \infty).$$

Since $\mathrm{d}\lambda(1) = H$, it follows that $(d\varphi)_e(H) = (0, -1, 0)$. Q.E.D.

Remark. If we worked with $\mathbf{SL}_2$ in place of $\mathbf{GL}_2$ it would still be true that $\mathbf{SL}_2 \to \mathbf{PGL}_2$ is surjective. However, it is not separable in characteristic two. We would have to replace T by the image of the group T' of matrices of the form $\mathrm{diag}(a, a^{-1})$ in $\mathbf{SL}_2$. But $L(T')$ is spanned by $\begin{bmatrix} 1 & 0 \\ 0 & -1 \end{bmatrix}$, and $\begin{bmatrix} 1 & 0 \\ 0 & -1 \end{bmatrix}_L$ vanishes in characteristic two.

Proposition. *Let H be a k-group acting k-morphically on $\mathbf{P}_1$. Then this action is induced by a unique k-morphism $\alpha: H \to \mathbf{PGL}_2$.*

Proof. Define $\beta: H \to V$ by $\beta(h) = (h(0), h(1), h(\infty))$, and let $\alpha = \varphi^{-1} \circ \beta$. Then $\alpha(h)(i) = h(i)$, $i = 0, 1, \infty$, and α is clearly a k-morphism. To show, finally, that $\alpha(h)$ and h yield the same automorphism of $\mathbf{P}_1$ it suffices to show that an automorphism g of $\mathbf{P}_1$ fixing $0, 1$, and ∞ is the identity.

But $K(\mathbf{P}_1) = K(x)$ where x is the unique rational function on $\mathbf{P}_1$ with a zero of order one at 0, a pole of order one at ∞, and no other singularities, and $x(1) = 1$. Since $x \circ g$ must have the same properties we see that g induces the identity on $K(\mathbf{P}_1)$; hence g is the identity.

10.9 Theorem. *Let G be a connected affine group of dimension one. Then G is isomorphic to either $\mathbf{GL}_1$ or to $\mathbf{G}_a$.*

Proof. G is a dense open set in a unique complete smooth curve $\bar{G}$ (see AG.18.5(d)). It follows from (AG. 18.5(f)) that the action of G on itself by translation extends uniquely to an action of G on $\bar{G}$. Since $\bar{G} - G$ is a finite set stable under the connected group G it follows that G fixes the points of $\bar{G} - G$. Since G is affine the number, m, of such points is > 0. Hence it follows from 10.7 that $\bar{G} \cong \mathbf{P}_1$. Choose an identification of $\bar{G}$ with $\mathbf{P}_1$ so that $\infty \notin G$. Then we obtain from 10.8 an embedding of G into $\mathbf{PGL}_2$ so that G lies in the isotropy group $\left\{\begin{bmatrix} a & 0 \\ c & d \end{bmatrix}\right\}$ of ∞ (see 10.8 (4)). It further follows from 10.8 that G fixes at most two points of $\mathbf{P}_1$, i.e. that $m \leqq 2$.

Case 1. $m = 2$. Choose projective coordinates so that the fixed point other than ∞ is 0. Then G lies in the torus $T \cong \mathbf{GL}_1$ of elements $\left\{\begin{bmatrix} a & 0 \\ 0 & d \end{bmatrix}\right\}$. For dimension reasons $G = T$.

Case 2. $m = 1$. G acts on the affine line $K = \mathbf{P}_1 - \{\infty\}$ by transformations of the form $x \mapsto ax + c$. These form a solvable group, so G is solvable. Since $\mathscr{D}G$ is connected and $\dim \mathscr{D}G < \dim G = 1$ we conclude that G is abelian. Write $G = G_s \times G_u$ (see 4.7). For dimension reasons again, we must have $G = G_s$ or $G = G_u$. If $G = G_s$, then (see 8.4 and 8.5) G is a one dimensional torus, i.e. $G \cong \mathbf{GL}_1$. Let now $G = G_u$. If $g(x) = a_g x + c_g$ then $g \mapsto a_g$ is a morphism $G \to \mathbf{GL}_1$. It must be trivial because G is unipotent. Hence $g \mapsto c_g$ gives an embedding $G \to \mathbf{G}_a$, and dimension count again shows that this must be an isomorphism.

Remark. In case $G \cong \mathbf{GL}_1$, it follows from 8.11 that such an isomorphism exists over k_s. Suppose, on the other hand, that $G \cong \mathbf{G}_a$. Let $\bar{G}$ be the complete non-singular curve defined over k containing G. The argument above shows that $\bar{G} - G$ consists of a single point, P, so P must be rational over $L = k^{p^{-\infty}}$. It is known then (see Serre, Corps locaux, Ch. X, §6, Ex. 1) that $\bar{G}$ is isomorphic over L to $\mathbf{P}_1$, and we can choose this isomorphism to carry P to ∞ in $\mathbf{P}_1$. This done, the isomorphism of G with $\mathbf{G}_a$ obtained above can be seen to be rational over L.

10.10 *Group actions on* $\mathbf{G}_a$. The points of $\mathbf{G}_a$ and of its Lie algebra $\mathfrak{g}_a$ both coincide with K. An endomorphism of $\mathbf{G}_a$ as a curve is given by an endomorphism of its affine algebra, $K[T]$, and the latter is defined by a polynomial $f(T)$. This will be a group morphism if and only if f is additive: $f(T + H) = f(T) + f(H)$. Let $p = \operatorname{char}(K)$.

(i) If $p = 0$ then $f(T) = cT$ for some $c \in K$.
(ii) If $p > 0$ then $f(T) = \Sigma_i c_i T^{p^i}$.

These follow by applying d/dT to the addition formula to conclude that $f'(T)$ is a constant. Subtracting the linear term from $f(T)$ one obtains, in case (ii), a polynomial $g(T^p)$ and g is additive of lower degree, so induction applies to establish (ii).

In either case it is easily seen that an automorphism of $\mathbf{G}_a$ corresponds to a polynomial $f(T) = cT$ $(c \in K^*)$, and that we thus obtain an isomorphism of $\mathbf{GL}_1$ with the automorphism group of $\mathbf{G}_a$.

If we view $\mathbf{G}_a$ as $\mathbf{P}_1 - \{\infty\}$ and note that group automorphisms fix $0 \in \mathbf{G}_a$, then we can also obtain the automorphism group as the intersection in $\mathbf{PGL}_2$ of the isotropy groups of 0 and ∞. This intersection is the torus $T = \left\{\begin{bmatrix} a & 0 \\ 0 & 1 \end{bmatrix}\right\}$ introduced in 10.8.

Let G be any group acting as automorphisms on $\mathbf{G}_a$. Then it follows from the above remarks that there is a character $\alpha: G \to \mathbf{GL}_1$ through which G acts:

$$g(x) = g^\alpha x \quad (g \in G, x \in \mathbf{G}_a).$$

For such an action the induced action on $\mathfrak{g}_a$ is clearly given by the same character:

$$g(X) = g^\alpha \cdot X, \quad (g \in G, X \in \mathfrak{g}_a).$$

Bibliographical Note

10.4 and 10.6 are proved in [1]. The original Lie theorem states that a connected linear solvable Lie group over the complex numbers can be put in triangular form. The generalization to algebraic groups in 10.5 is due to Kolchin [19]. The proof given here is taken from [1].

It seems somewhat surprising that the proof $G \cong \mathbf{GL}_1, \mathbf{G}_a$ if G is connected, one-dimensional (10.9) is not more elementary, or at any rate more self-contained. The result has been known for quite a while, of course. However, the author would be hard put to refer to a complete proof antedating the one given by Grothendieck in [13, Exp. 7]. The latter proof is quite different from the one described above, and is much more algebraic. It makes use of some results of §§10, 11, and will be sketched in 11.6. More elementary proofs may be found in [32:2.6] or [17:20].

Chapter IV

Borel Subgroups; Reductive Groups

Throughout this chapter G denotes a connected affine group, and all algebraic groups are understood to be affine.

§11. Borel Subgroups

11.1 *A Borel subgroup* of G is one which is maximal among the connected solvable subgroups. They clearly exist, for dimension reasons.

Theorem. *Let B be a Borel subgroup of G. Then all Borel subgroups are conjugate to B, and G/B is a projective variety.*

Proof. Let R be a Borel subgroup of maximal dimension. Choose a faithful representation $\pi: G \to \mathrm{GL}(V)$ with a line $V_1 \subset V$ such that R is the stability group of V_1 in G and $L(R)$ is the stability Lie algebra of V_1 in $L(G)$. (See Theorem 5.1.) Applying 10.5 to the induced representation of R on V/V_1, we obtain a flag $F = (V_1, V_2, \ldots, V_n)$ in V stabilized by R. Let $\mathscr{F}(V)$ denote the flag variety of V, on which G operates via π. Then the canonical map from G/R to the orbit, $G(F)$, of F in $\mathscr{F}(V)$ is an isomorphism of varieties. This follows because the map from G/R to the orbit of V_1 in the projective space $\mathbf{P}(V)$ is already an isomorphism of varieties (See Theorem 6.8 and proof.)

Suppose $F' \in \mathscr{F}(V)$ has stability group R' in G. Since R' leaves a flag invariant, it is solvable. The maximality of $\dim R$ therefore implies $\dim R' \leqq \dim R$, and hence $\dim G/R \leqq \dim G/R'$. Thus $G(F)$ is a G-orbit in $\mathscr{F}(V)$ of minimal dimension, so the closed orbit lemma (1.8) implies that $G(F)$ is closed. This proves that G/R is a projective variety.

Letting B operate on G/R, in the natural way, we see, using 10.4, that B has a fixed point, i.e. that $BxR \subset xR$ for some $x \in G$. But this means $x^{-1}BxR \subset R$, hence $x^{-1}Bx \subset R$. Since B is maximal connected solvable, this implies $x^{-1}Bx = R$.

11.2 *A parabolic subgroup* P of G is a closed subgroup such that G/P is a complete variety. Since the homogeneous space G/P is always quasi-projective (see 6.8), it is complete if and only if it is a projective variety.

Corollary. *A closed subgroup P of G is parabolic if and only if it contains a Borel subgroup.*

Proof. If P contains a Borel subgroup B then $G/B \to G/P$ is a surjective morphism from a complete variety, so G/P is complete. Conversely, by 10.4, a Borel subgroup B has a fixed point in the complete variety G/P, so some conjugate of B lies in P.

11.3 Corollary. (1) *The maximal tori in G coincide with the maximal tori in the various Borel subgroups of G, and they are all conjugate.*

(2) *The maximal connected unipotent subgroups of G are each the unipotent part of a Borel subgroup, and they are all conjugate.*

Proof. (1) A maximal torus T is connected and solvable so it lies in some Borel subgroup B. Evidently it is a maximal torus in B, so 10.6(4) implies that $B = T \cdot B_u$ (semi-direct), and that all maximal tori of B are conjugate to T. Since any two Borel subgroups are conjugate, part (1) follows.

(2) If U is connected and unipotent then U is nilpotent (see 4.8), so U lies in a Borel subgroup B. According to 10.6(2), B_u is a connected subgroup of B, evidently containing U, and hence $U = B_u$ if U is maximal. The conjugacy of the B_u's follows immediately from that of the B's.

11.4 Corollary. *Let B be a Borel subgroup of G.* (1) *If an automorphism a of G fixes the elements of B, then a is the identity.*

(2) *If $x \in G$ centralizes B then $x \in \mathscr{C}(G)$.*

Proof. Part (2) follows from (1) with $a = \mathrm{Int}(x)$. To prove (1) consider the morphism $f: G \to G$, $f(g) = a(g)g^{-1}$. Then f factors through $G \to G/B$, so $f(G)$ is complete, and affine, hence a point (see 10.1).

11.5 Corollary. *Let B be a Borel subgroup of G.*

(1) *If $B = B_s$, then G is a torus.*
(2) *If B contains no torus $\neq \{e\}$, then G is unipotent. In either case $G = B$.*
(3) *The following conditions are equivalent:*
  (a) *G has a unique maximal torus.*
  (b) *Some maximal torus lies in $\mathscr{C}(G)$.*
  (c) *G is nilpotent.*
  (d) *B is nilpotent.*

Proof. (1) Using 10.6(4), we can write $B = T \cdot B_u$, with T a maximal torus. If $B = B_s$ then $B = T$ is commutative, and hence, by 11.4, central in G. But then

G/B is an affine connected group which is also a complete variety, hence $= \{e\}$.

(2) On the other hand, if $T = \{e\}$, so that $B = B_u$ is nilpotent, then $\mathscr{Z}_B(B)^o = H \neq \{e\}$. By 11.4, H is central in G. Since B/H is a unipotent Borel subgroup in G/H we conclude by induction on $\dim G$ that $G/H = B/H$, i.e. that $B = G$.

(3) (a)$\Rightarrow$(b). If T is the unique maximal torus, then T is normal in G, and the rigidity of tori (8.10) implies that T is central.

(b)$\Rightarrow$(c). If T is a central maximal torus, then let T' be the inverse image in G of a torus in G/T. Since T and T'/T both consist of semi-simple elements, so also does T'. Hence it follows from part (1) above that T' is a torus. (We have used the fact that T' is connected, which follows because T and T'/T are connected.) Now by maximality of T, we must have $T' = T$. In conclusion, this argument shows that G/T contains no non-trivial tori. Hence part (2) above implies that G/T is unipotent, and hence also nilpotent. Since T is central in G, the group G is also nilpotent.

(c)$\Rightarrow$(d) is obvious.

(d)$\Rightarrow$(a). If B is nilpotent then 10.6(3) implies that $B = T \times B_u$ with $T = B_s$, a maximal torus in G. Now $T \subset \mathscr{C}(B)$, and $\mathscr{C}(B) \subset \mathscr{C}(G)$ by 11.4. Hence T has a unique conjugate (itself), and (a) follows from 11.3(1).

Corollary. *Suppose G contains a normal torus T such that G/T is also a torus. Then G is a torus.*

Proof. The hypotheses clearly imply that $G = G_s$, so part (1) above implies G is a torus.

11.6 Corollary. *If $\dim G \leqq 2$, then G is solvable.*

Proof. Write $B = T \cdot B_u$ as above. If $B \neq G$ then $\dim B \leqq 1$, so we must have $B = T$ or $B = B_u$. But 11.5 then implies that $G = B$; contradiction.

Remark. We now sketch the proof of 10.9 given in [13, Exp. 7], alluded to at the end of §10. Let G be one-dimensional. The Corollary above implies that G is solvable. Then $\dim(G, G) < \dim G$, hence G is *commutative.* By 10.6(4), we have $G = T \cdot G_u$ where T is a maximal torus. Dimension considerations then show that either $G = T$, in which case $G \cong \mathbf{GL}_1$, or $G = G_u$. It follows from the proof of 10.6(2) (using an embedding of G in the unipotent part of some $\mathbf{T}_n$) that G admits a non-trivial morphism $\pi: G \to \mathbf{G}_a$. Since $\mathbf{G}_a$ is connected of dimension one it follows that π is an isogeny, i.e. that π is surjective and that $N = \ker(\pi)$ is finite.

Let p be the characteristic exponent of K. Then every element of the unipotent group G has order a power of p. It follows that π is an isomorphism if $p = 1$ (i.e. if $\operatorname{char}(K) = 0$). If $p > 1$ then N is a finite group of order p^n for some $n \geqq 0$. One proves that $G \cong \mathbf{G}_a$ by induction on n.

If $n=0$, i.e. if π is bijective, then, by (AG.18.2), $K(G)$ is a purely inseparable extension of $K(\mathbf{G}_a)=K(x)$. Taking a high pth power one concludes that $K(G)$ is isomorphic to a subfield of $K(x)$, so Lüroth's theorem (see e.g. van der Waerden, *Algebra*, vol. 1) implies that $K(G)$ is purely transcendental. Thus G is, as a variety, isomorphic to an open subset of the projective line, and one concludes the proof by embedding G into $\mathbf{PGL}_2$ as in the proof of Theorem 10.9.

If $n>0$ we can factor out a subgroup of index p in N and apply induction to reduce to the case $n=1$. We can further use the case $n=0$ to conclude that $G/N \cong \mathbf{G}_a$ and so arrange that π is separable. In this case $K(G)$ is a Galois extension of degree p of $K(x)$, to which one can apply Artin–Schreier theory. (See [13, Exp. 7], Lemma 3, for details.)

11.7 Corollary. *Let T be a maximal torus of G. Then $C=\mathscr{Z}_G(T)^o$ is nilpotent, and $C=\mathscr{N}_G(C)^o$.*

The conjugacy theorem 11.3 shows that T is the unique maximal torus of C, therefore by 11.5, C is nilpotent. Moreover, T is normal in $\mathscr{N}_G(C)$, and consequently by 8.10, T is centralized by $\mathscr{N}_G(C)^o$.

11.8 Proposition. *Let $X\in L(G)$. Then X is semi-simple if and only if it is tangent to a torus in G.*

A torus is isomorphic to a diagonal group, hence its Lie algebra consists of semi-simple elements, which proves the "if" part of the proposition.

Assume now X to be semi-simple. By 9.1, the Lie algebra $\mathfrak{h}$ of $H=\mathscr{Z}_G(X)$ is equal to $\mathfrak{z}_{\mathfrak{g}}(X)$; in particular, it contains X. Let T be a maximal torus of H and $C=\mathscr{Z}_H(T)^o$. Then $L(C)=\mathfrak{z}(T)$ by (9.2, Cor.), hence $X\in L(C)$. By 11.7, C is nilpotent, hence $C=T\times C_u$, in view of 10.6. Since $L(C_u)$ consists of nilpotent elements, it follows that $X\in L(T)$.

11.9 Lemma. *Let H be a closed subgroup of G, and put*

$$X={}^G H=\bigcup_{g\in G} gHg^{-1}.$$

(1) *If G/H is complete, then X is closed.*

(2) *Assume there is an $h\in H$ having only finitely many fixed points in G/H, i.e. such that $\{x\in G \mid h\in xHx^{-1}\}$ constitutes a finite number of cosets of H. Then X contains a dense open set in G.*

Proof. Consider the morphisms

$$G\times G \xrightarrow{\alpha} G\times G \xrightarrow{\beta} (G/H)\times G$$

where $\alpha(x,y)=(x,xyx^{-1})$ and $\beta=\pi\times 1_G$, with $\pi: G\to G/H$ the quotient morphism. Put $M=\beta(\alpha(G\times H))=\{(\pi(x),z) \mid x\in G, x^{-1}zx\in H\}$.

(i) *M is closed.* If $x^{-1}zx\in H$, then $(xh)^{-1}z(xh)\in H$ for all $h\in H$, so it follows that $\beta^{-1}(M)=\alpha(G\times H)$. Since α is an isomorphism of varieties, and since

$\beta: G \times G \to (G \times G)/(H \times \{e\})$ is a quotient morphism, and hence open, we conclude that M is closed because $\beta^{-1}(M)$ is closed.

(ii) $X = \mathrm{pr}_G(M)$, *so* X *is closed if* G/H *is complete.* For $\mathrm{pr}_G(M) = \{y | x^{-1}yx \in H \text{ for some } x \in G\} = X$.

(iii) $\dim M = \dim G$ *at each point of* M.

The fibre over $\pi(x)$ of the surjective morphism $\mathrm{pr}_{G/H}: M \to G/H$ is isomorphic to xHx^{-1}, so the dimension of each fibre is $\dim H$. Hence $\dim M = \dim G/H + \dim H = \dim G$ at each point.

The fibre of $\mathrm{pr}_G: M \to G$ over y is

$$\begin{aligned}\{\pi(x) | x^{-1}yx \in H\} &= \{\pi(x) | y \in xHx^{-1}\} \\ &= \{\pi(x) | y \cdot \pi(x) = \pi(x)\}.\end{aligned}$$

(In the latter, the dot refers to the natural action of G on G/H.) In view of this, the hypothesis of (2) says simply that the fibre of $\mathrm{pr}_G: M \to G$ over some $h \in H$ is finite (and $\neq \phi$). Therefore, if N is an irreducible component of M such that $h \in \mathrm{pr}_G(N)$, the fibres of pr_G in N are "generically finite," i.e. they are each finite over some dense open set in $\overline{\mathrm{pr}_G(N)}$ (see (AG.10.1)). Since $\dim N = \dim G$ and G is connected it follows that $\mathrm{pr}_G: N \to G$ is dominant. Thus X, which contains $\mathrm{pr}_G(N)$, contains a dense open set in G.

11.10 Theorem. *Let B be a Borel subgroup of G, T a maximal torus of G, and $C = \mathscr{Z}_G(T)^o$. Then the union of the conjugates of B (resp. B_u, resp. T, resp. C) is G (resp. G_u, resp. G_s, resp. contains a dense open set of G).*

By 11.7, C is nilpotent. Since T is a maximal torus, it follows then from 10.6 that $C = T \times C_u$. By 8.18, there exists $t \in T$ such that $\mathscr{Z}(t)^o = \mathscr{Z}(T)^o = C$. Let $g \in G$ be such that $gtg^{-1} \in C$. Then $gtg^{-1} \in T$, *and* $\mathscr{Z}(gtg^{-1}) \supset \mathscr{Z}(T)$. For dimension reasons, we have then $\mathscr{Z}(gtg^{-1})^o = C$, hence $g \in \mathscr{N}(C)$. Since $\mathscr{N}(C)^o = C$, by 11.7, it follows that the set of conjugates of t contained in C is finite. This is condition (2) of 11.9, taking C to be the subgroup H, hence ${}^G C$ contains a dense open subset of G. Since C is nilpotent, it is contained in some Borel subgroup B' of G. Then ${}^G B'$ contains a dense open set. But G/B' is complete (11.1), hence (11.9) ${}^G B'$ is closed. Consequently $G = {}^G B'$. By the conjugacy of Borel subgroups, we have also $G = {}^G B$. The remaining part of the theorem then follows from 10.6.

11.11 Corollary. *$\mathscr{C}(G)$ is the center of each Borel subgroup. $\mathscr{C}(G)_s$ is the intersection of all maximal tori in G.*

Proof. Let $g \in \mathscr{C}(G)$ and let B be a Borel subgroup. Some conjugate of g lies in B, so $g \in B$, i.e. $\mathscr{C}(G) \subset \mathscr{C}(B)$. The reverse inclusion follows from 11.4.

If $g \in \mathscr{C}(G)_s$ then $g \in B$, as we saw above, and 10.6(5) implies that g belongs to a maximal torus T in B. Now part (1) of 11.3 implies g belongs to every maximal torus. Thus $\mathscr{C}(G)_s \subset H$, the intersection of all maximal tori. Since H is a closed subgroup of a torus it is a diagonalizable group, and it is clearly normal in G. Hence by rigidity (8.10), H is central in G. With the inclusion proved above this shows that $H = \mathscr{C}(G)_s$, as claimed.

11.12 Corollary. *Let S be a subtorus of G and $a \in \mathscr{Z}_G(S)$. Then $\{a_s\} \cup S$ is contained in a torus of G. The group $\mathscr{Z}_G(S)$ is connected. For any $g \in G$, the element g belongs to $\mathscr{Z}_G(g_s)^o$.*

Proof. We show first that $\{a\} \cup S$ is contained in a Borel subgroup of G. Let B such a group and F, the fixed point set of a in G/B, under the natural action. By 11.10, a is contained in a conjugate of B, hence F is not empty. Since S centralizes a, it leaves F stable. By 10.4, S has a fixed point in F, say x. The stability group B' of x is then a Borel subgroup of G containing $\{a\} \cup S$.

This reduces the proof of the first assertion to the case where G is solvable, in which case it follows from 10.6(5). By 10.6(5), also, the group $\mathscr{Z}_{B'}(S)$ is connected, whence $a \in \mathscr{Z}_G(S)^o$, and the second assertion. Let now $g \in G$. By 11.10, g belongs to a Borel subgroup B of G. Then $g \in \mathscr{Z}_B(g_s)$. But the latter group is connected (10.6)(5), hence $g \in \mathscr{Z}_G(g_s)^o$.

11.13 Definition. *A Cartan subgroup of G is the centralizer of a maximal torus.*

The Cartan subgroups of G are connected by 11.12, nilpotent by 11.7, and conjugate to each other by 11.3. In view of 10.6, the map $T \mapsto \mathscr{Z}_G(T)$ is a bijection of the set of maximal tori onto the set of Cartan subgroups, and $\mathscr{Z}_G(T) = T \times \mathscr{Z}_G(T)_u$. Finally, by 11.10, the union ${}^G C$ of the conjugates of a Cartan subgroup contains a dense open set of G.

11.14 Proposition (1) *Let $\alpha: G \to G'$ be a surjective morphism of algebraic groups, and let $B = T \cdot B_u$ be a Borel subgroup of G, with T a maximal torus. Then $\alpha(B) = \alpha(T) \cdot \alpha(B_u)$ is a Borel subgroup of G', and every such subgroup is obtained in this way. Moreover $\alpha(T)$ is a maximal torus in G' and $\alpha(B_u) = \alpha(B)_u$.*

(2) *Let H be a connected subgroup of G and let B_o be a Borel subgroup of H. Then $B_o = (H \cap B)^o$ for some Borel subgroup B of G. If H is normal, the Borel subgroups of H are the groups $(B \cap H)^o$, where B ranges over all Borel subgroups of G.*

The analogous assertions hold for maximal tori and for maximal connected unipotent subgroups.

Proof. The composite $G \to G' \to G'/\alpha(B)$ induces a surjective morphism $G/B \to G'/\alpha(B)$, so the latter is complete, i.e. $\alpha(B)$ is parabolic. Therefore $\alpha(B)$ contains a Borel subgroup (11.2). But $\alpha(B)$ is connected and solvable, so $\alpha(B)$ is itself a Borel subgroup. The semi-direct product decomposition $\alpha(B) = \alpha(T) \cdot \alpha(B_u)$ and the fact that $\alpha(B_u) = \alpha(B)_u$ follow from the conservation of Jordan decomposition. In particular $\alpha(T)$ is a maximal torus in $\alpha(B)$, and hence also in G'. The conjugacy theorem in G' implies that all Borel subgroups, all maximal tori, and all maximal connected unipotent subgroups of G' are obtained in this way.

(2) Extend the connected solvable group B_o to a Borel subgroup B of G. Then $B_o \subset (H \cap B)^o$, and the latter is a connected solvable subgroup of H. Hence it coincides with B_o. The argument for tori and connected unipotent groups is similar.

Corollary 1. *Let S be a torus and $f: G \to S$ a surjective morphism. Then any maximal torus T of G contains a torus S' such that $f: S' \to S$ is an isogeny.*

By the proposition, $f: T \to S$ is surjective. By 8.5, Cor., the identity component of $\ker f|_T$ is a direct factor in T, whence the corollary.

Corollary 2. *Let G be connected, $f: G \to G'$ a surjective morphism and S a torus in G. Then $f(\mathscr{Z}_G(S)) = \mathscr{Z}_{G'}(f(S))$.*

Let $s \in S$ act on G by Int s and on G' by Int $f(s)$. Then f is S-equivariant and 9.6 shows that $f(\mathscr{Z}_G(S)^o) = \mathscr{Z}_{G'}(f(S))^o$. Since $\mathscr{Z}_G(S)$ and $\mathscr{Z}_{G'}(f(S))$ are connected by the theorem, the corollary follows.

11.15 Proposition. *Suppose G acts transitively on a variety D so that the isotropy groups of the points in D are Borel subgroups of G. Let T be a torus in G. Then $G^T = \mathscr{Z}_G(T)$ stabilizes and acts transitively on each irreducible component of D^T. If B is a Borel subgroup of G normalized by T, then B^T is a Borel subgroup of G^T and every Borel subgroup of G^T is of this form.*

Proof. Clearly G^T stabilizes D^T, and hence also each irreducible component of D^T since G^T is connected (11.12). Let X be an irreducible component of D^T and let $B_o \in \mathscr{B}$ be the stability group of some $x_o \in X$. The orbit map $\pi: G \to D$, $\pi(g) = gx_o$, induces a bijective morphism $G/B_o \to D$, so D is complete.

We must show that the inclusion $G^T x_o \subset X$ is an equality. Since X is connected and since π has connected fibres ($\cong B_o$) it follows that $Y = \pi^{-1}(X)$ is connected If $y \in Y$ then $\pi(y) \in D^T$ so $y^{-1}Ty \subset B_o$. Let $\alpha: Y \times T \to B_o/(B_o)_u$ be the composite of $(y, t) \mapsto y^{-1}ty$ with the projection $B_o \to B_o/(B_o)_u$. The rigidity of tori (8.10) implies now that $\alpha(y, t)$ is independent of y. Since $T \subset B_o$ we have $e \in Y$ and hence for $y \in Y$ we have $y^{-1}ty \equiv t \bmod (B_o)_u$ for all $t \in T$. Thus $y^{-1}Ty \subset T \cdot (B_o)_u$. The latter is connected so the conjugacy of its maximal tori implies that $y^{-1}Ty = g^{-1}Tg$ for some $g = tb \in T \cdot (B_o)_u$, and we can replace g by b. If $s \in T$ then, modulo $(B_o)_u$, we have $y^{-1}sy \equiv s \equiv b^{-1}sb$. But $y^{-1}Ty \to B_o/(B_o)_u$ is injective, so the congruence implies $y^{-1}sy = b^{-1}sb$, i.e. $yb^{-1} \in G^T$. Thus $\pi(y) = yx_o \sim yb^{-1}x_o \in G^T x_o$ (because $b \in B_o \sim G_{x_0}$), and $G^T x_o$ contains $\pi(Y) = X$, as claimed. Since X is complete (being a closed set in D), it follows from (AG.18.3) that G^T/B_o^T is complete, for we have a bijective morphism $G^T/B_o^T \to X$. But B_o^T is connected and solvable, so B_o^T is a Borel subgroup of G^T. It follows then from 11.14 that every Borel subgroup of G^T is so obtained.

Corollary. *If G^T is solvable, then T normalizes only finitely many Borel subgroups of G, and each of those contains G^T. This occurs, for instance, if T is a maximal torus.*

Proof. Let $Y = G/B$, for some Borel subgroup B of G. The orbits of G^T on the complete variety Y^T are the irreducible components of Y^T, by the proposition. But G^T is connected (11.12). If it is solvable, then the fixed point theorem (10.14) implies that G^T has a fixed point on each irreducible component of Y^T, so each of the latter reduces to a point, and Y^T is finite. Now $x \mapsto G_x$ is a bijection of Y^T on the set of Borel subgroups normalized by T, and each such G_x contains G^T. The last assertion follows from 11.7.

11.16. Theorem (C. Chevalley). *Every parabolic subgroup P of G is equal to its normalizer in G and is connected.*

Proof. We first reduce the proof of the case of a Borel subgroup. Assume then that every Borel subgroup is equal to its normalizer. Let $n \in \mathcal{N}_G(P)$ and B be a Borel subgroup of P. Then ${}^nB$ is also one, hence there exists by 11.1 an element p of P such that ${}^{pn}B = B$. But B is also a Borel subgroup of G, since P is parabolic, hence, by our assumption $pn \in B$, and therefore $n \in P$. The identity component P^o of P is also a parabolic subgroup, and P belongs to its normalizer, hence $P^o = P$ and P is connected.

There remains to show that every Borel subgroup B of G is equal to its normalizer. Note that if $\dim \leqq 2$, then G is solvable (11.6) and there is nothing to prove. Arguing by induction on $\dim G$ we may therefore assume our assertion to be true for connected groups of dimension $< \dim G$.

Let $N = \mathcal{N}_G(B)$. Let T be a maximal torus of B, hence of G. We claim first that $N \subset B \cdot (\mathcal{N}_G(T) \cap N)$. Let $n \in N$. Then ${}^nT$ is a maximal torus of B, hence there exists $b \in B$ such that $bn \in \mathcal{N}_G(T)$ and our assertion follows. It suffices therefore to show that if $n \in N \cap \mathcal{N}_G(T)$, then $n \in B$. Let S be the fixed point set of $\operatorname{Int} n$ in T. We distinguish three cases:

(i) $S^o \subset \mathscr{C}(G)$ and is of dimension $\geqq 1$. We note first that S^o, being central, belongs to B (11.11). Let $G' = G/S^o$ and $\pi: G \to G'$ be the natural projection. Then $\pi(n) \in \mathcal{N}_G(\pi(B^o))$. Since $\pi(B)$ is a Borel subgroup of G' (11.14), we have $\pi(n) \in \pi(B)$ by the induction assumption, hence $n \in \pi^{-1}(\pi(B))$. But $\pi^{-1}(\pi(B)) = B$ since $S^o \subset B$.

(ii) *S^o is not central, of dimension* $\geqq 1$. Then $\mathscr{Z}(S^o)$, is $\neq G$ and is connected by 11.12. By construction, $n \in \mathscr{Z}(S^o)$, hence n normalizes $\mathscr{Z}(S^o) \cap B$, which, by 11.15, is a Borel subgroup of $\mathscr{Z}(S)$. Our induction assumption then implies that $n \in \mathscr{Z}(S^o) \cap B$, whence $n \in B$.

(iii) *S is finite.* This is the essential case. Let ϕ be the map $t \mapsto n \cdot t \cdot n^{-1} \cdot t^{-1}$. It is a morphism of T into itself, whose kernel is S. Since S is finite, ϕ is surjective. Therefore every element of T is a commutator in N, and T belongs to the derived group $\mathscr{D}N$ of N. By 5.1, we can find a morphism $\sigma: G \to GL(E)$ such that E contains a line D whose full stability group in G is N. Then D

is the space of a one dimensional representation of N, given by a rational character χ. The image of χ is contained in $\mathbf{GL}_1$, which is commutative and consists of semisimple elements. Therefore $\ker\chi$ contains B_u and $\mathscr{D}N$, hence also T. As a consequence, $B \subset \ker\chi$. For $d \in D$, the orbit map $g \mapsto g\cdot d$ defines a morphism of the irreducible projective variety G/B into affine space, hence the image is reduced to a point and $G = B$. A fortiori $N = B$.

11.17 Corollary. *Let B be a Borel subgroup of G.*

(i) *Every parabolic subgroup is conjugate to one and only one parabolic subgroup containing B.*
(ii) *B is maximal among the solvable (not necessarily closed or connected) subgroups of G.*
(iii) *Let P, Q, R be parabolic subgroups and assume that ${}^gQ = R$ for some $g \in G$. If $Q \cap R$ contains P, then $Q = R$. If $P \supset Q, R$ then $g \in P$.*

Proof. (i) In view of the conjugacy of Borel subgroups 11.1 and of 11.2, every parabolic subgroup is conjugate to at least one containing B. For the uniqueness, we have to show that if $P, Q \supset B$ and $P = {}^gQ$ for some $g \in G$, then $P = Q$. By 11.14, there exists $p \in P$, such that ${}^{pg}B = B$, whence $pg \in B$ by the theorem and $g \in P, P = Q$.

(ii) Suppose $B \subset H \subset G$, with H a solvable subgroup. Then $\bar{H}$ is solvable (2.4), so we can assume that H is closed. It then follows that $B = H^o$, hence $H \subset \mathscr{N}_G(B) = B$ and $H = B$.

(iii) P contains a Borel subgroup, hence the first assertion follows from (i). Assume now that $P \supset Q, R$. Then ${}^gP$ and P contain R, hence are equal, and $g \in \mathscr{N}_G P$, hence $g \in P$ by the theorem.

Caution. A maximal solvable subgroup need not be a Borel subgroup. For instance, if $G = \mathbf{SO}(n)$, $(n \geqq 3, p \neq 2)$, the group of diagonal matrices in G is isomorphic to $(\mathbb{Z}/2\mathbb{Z})^{n-1}$, but not contained in any Borel subgroup.

11.18 *The "variety" $\mathscr{B} = \mathscr{B}(G)$ of all Borel subgroup of G.* It is first of all a set on which G operates by conjugation. The conjugacy Theorem 11.1 says G acts transitively. The stability group of $B \in \mathscr{B}$ is $\mathscr{N}_G(B)$ which, by the normalizer Theorem 11.16, is just B.

If H is a subgroup of G then its fixed point set in $\mathscr{B}$ is

$$\mathscr{B}^H = \{B \in \mathscr{B} \mid H \subset B\},$$

again because of the normalizer theorem.

Fix $B_o \in \mathscr{B}$ and let $\pi: G \to G/B_o$ be the quotient morphism. If $x = \pi(g)$ then the stability group of x is

$$G_x = \{h \mid hgB_o \sim gB_o\} = \{h \mid g^{-1}hg \in B_o\} = {}^gB_o \in \mathscr{B}.$$

Thus we can define

$$\varphi: G/B_o \to \mathscr{B}, \varphi(x) = G_x.$$

Since $\varphi(\pi(g)) = {}^gB_o$ it follows from the conjugacy theorem that *φ is surjective.* Moreover $\varphi(\pi(g)) = \varphi(\pi(h)) \Leftrightarrow {}^gB_o \sim {}^hB_o \Leftrightarrow g^{-1}h \in \mathcal{N}_G(B_o) = B_o$ (normalizer theorem) $\Leftrightarrow \pi(g) = \pi(h)$. Thus *$\varphi$ is also injective.* With the aid of the bijection φ one can therefore give $\mathcal{B}$ the variety structure of G/B_o. Moreover the conjugacy theorem implies that this structure does not depend on the choice of B_o. This follows from the fact that *φ is G-equivariant.*

For if $g, h \in G$ we have $\varphi(h \cdot \pi(g)) = \varphi(\pi(hg)) = {}^{hg}B_o \sim {}^h({}^gB_o) = {}^h\varphi(\pi(g))$. A further consequence of this is that: *If H is a subgroup of G then φ induces a bijection*

$$(G/B_o)^H \to \mathcal{B}^H.$$

Thus the fixed points of H in G/B_o correspond bijectively to the set of Borel subgroups containing H.

11.19 *Simple transitivity of the Weyl group.* If T is a torus in G then

$$W = W(T, G) = \mathcal{N}_G(T)/\mathcal{Z}_G(T),$$

is called the *Weyl group* of G relative to T. The Weyl groups of maximal tori are isomorphic, by virtue of the conjugacy of maximal tori, and they are called, simply, "Weyl groups of G."

We know from the rigidity of tori that $\mathcal{Z}_G(T) = \mathcal{N}_G(T)^o$, so *$W$ is a finite group.*

Proposition. *Assume T is a maximal torus in G.*

(a) *A Borel subgroup containing T also contains $\mathcal{Z}_G(T)$.*
(b) *Via conjugation by $\mathcal{N}_G(T)$, the Weyl group W acts simply transitively on the set $\mathcal{B}^T$ of Borel subgroups containing T. In particular* $\operatorname{card}\mathcal{B}^T = [W:1]$, *is finite.*
(c) *The group of inner automorphisms of G is transitive on the pairs (B, T) consisting of a Borel subgroup and a maximal torus T of B.*

Proof. (a) follows from (11.15) Corollary.

(b) $\mathcal{N}_G(T)$ operates by conjugation on $\mathcal{B}^T$, and part (a) implies that G^T operates trivially; therefore W operates. Suppose $B, B' \in \mathcal{B}^T$. We can write $B = {}^gB'$ for some $g \in G$. Then T and ${}^gT$ are maximal tori in B, so ${}^gT = {}^bT$ for some $b \in B$. Thus $g = bn^{-1}$ with $n = g^{-1}b \in \mathcal{N}_G(T)$. Now $B' = {}^{g^{-1}}B = {}^{nb^{-1}}B = {}^nB$. This proves that $\mathcal{N}_G(T)$ (and hence W) acts transitively on $\mathcal{B}^T$.

Suppose now that $n \in \mathcal{N}_G(T)$ and ${}^nB = B$, i.e. $n \in \mathcal{N}_B(T)$. Simple transitivity of W then requires that we show that $n \in G^T$. Since $n \in B$ this follows from 10.6(5).

(c) follows from (b) and the conjugacy of maximal tori.

Remark. We shall see in §13 that the Borel subgroups containing T generate G.

11.20 Proposition. *Let $\alpha: G \to G'$ be a surjective morphism of algebraic groups, and let T be a maximal torus in G. Then $T' = \alpha(T)$ is a maximal torus in G', and α induces surjective maps*

$$\mathscr{B}^T \to \mathscr{B}'^{T'}, \quad (\mathscr{B}' = \mathscr{B}(G')), \tag{1}$$

$$W(T, G) \to W(T', G'). \tag{2}$$

If the kernel of α lies in every Borel subgroup of G, then (1) and (2) are bijective.

Proof. It follows from 11.14 that T' is a maximal torus in G', and that $\mathscr{B} \to \mathscr{B}'$ is surjective. If $B' \in \mathscr{B}'^{T'}$ then every Borel subgroup of $\alpha^{-1}(B')^o$ is a Borel subgroup of G mapping onto B', and one of them contains the maximal torus $T \subset \alpha^{-1}(B')^o$. This shows that (1) is surjective.

Choose $B \in \mathscr{B}^T$ and put $B' = \alpha(B)$. Writing W and W' for the two Weyl groups we obtain a commutative square

$$\begin{array}{ccc} W & \xrightarrow{(2)} & W' \\ \downarrow & & \downarrow \\ \mathscr{B}^T & \xrightarrow[(1)]{} & \mathscr{B}'^{T'}, \end{array} \tag{3}$$

where the verticals are the orbit maps $w \mapsto {}^wB$ and $w' \mapsto {}^{w'}B'$, respectively. According to 11.19 the latter are bijective, so the surjectivity of (2) follows from that of (1).

Finally, if $\ker(\alpha)$ is contained in every Borel subgroup then $\mathscr{B} \to \mathscr{B}'$ is injective, and hence (1) is injective. The argument above with diagram (3) then shows that (2) is also injective.

11.21 *The radicals; reductive and semi-simple groups.* The group

$$\mathscr{R}G = \left(\bigcap_{B \in \mathscr{B}} B \right)^o$$

is called the *radical* of G. It is evidently a connected solvable normal subgroup of G, and it contains all other such subgroups. Its unipotent part

$$(\mathscr{R}G)_u \text{ (sometimes denoted } \mathscr{R}_u G),$$

is called the *unipotent radical* of G. It is a connected unipotent normal subgroup of G, and it contains all other such subgroups. This follows from the analogous property of $\mathscr{R}G$.

If $\pi: G \to G'$ is a surjective morphism with solvable kernel, then it follows immediately from the definition and 10.6 that $\pi(\mathscr{R}G) = \mathscr{R}G'$ and $\pi^{-1}(\mathscr{R}G')^o = \mathscr{R}G$. We shall see later (14.11) that the first equality is true without assumption on $\ker \pi$.

One says that G is *semi-simple* if $\mathscr{R}G = \{e\}$, and *reductive* if $\mathscr{R}_u G = \{e\}$. Evidently $G/\mathscr{R}G$ is semi-simple, and $G/\mathscr{R}_u G$ is reductive, and these are the largest quotient groups of G with these properties.

By considering the derived series in $\mathscr{R}G$ and the descending central series in $\mathscr{R}_u G$ we see that: G is semi-simple (resp., reductive) if and only if G has no connected abelian (resp., unipotent abelian) normal subgroup $\neq \{e\}$.

Proposition. *If G is reductive then $\mathscr{R}G = (\mathscr{C}G)^o$, and this group is a torus.*

Proof. Evidently $\mathscr{R}G \supset \mathscr{C}G$. Since G is reductive we have $\mathscr{R}G = (\mathscr{R}G)_s$, so 10.6 implies that $\mathscr{R}G$ is a torus. By rigidity of tori, a normal torus in a connected group is central, and so $\mathscr{R}G \subset (\mathscr{C}G)^o$.

11.22 Definition. A Levi subgroup of G is a connected subgroup L such that G is the semi-direct product of L and $\mathscr{R}_u G$.

A Levi subgroup maps isomorphically onto $G/\mathscr{R}_u G$, hence is reductive. It is maximal among reductive subgroups and provides a cross-section to the projection map $\pi: G \to G/\mathscr{R}_u G$. In characteristic zero, by a result of G.D. Mostow, Levi subgroups exist and are conjugate. In positive characteristic however, they need not exist, nor be conjugate. But they do in parabolic subgroups of reductive groups, as we shall see, and this is our chief reason to introduce this terminology.

11.23 Proposition. (i) *Let L be a Levi subgroup of G. Then the identity component S of the center of L is a maximal torus of the radical $\mathscr{R}G$ of G and is equal to $L \cap \mathscr{R}G$.*

(ii) *Assume that any maximal torus of $\mathscr{R}G$ is a Cartan subgroup of $\mathscr{R}G$. Then G has Levi subgroups. They are the centralizers of the maximal tori of $\mathscr{R}G$. Any two are conjugate under a unique element of $\mathscr{R}_u G$.*

(i) Let again $\pi: G \to G/\mathscr{R}_u G$ be the canonical projection. By definition, $G = L \cdot \mathscr{R}_u G$. On the other hand, S is the radical of L and $L = \mathscr{D}L \cdot S$ (with $\mathscr{D}L \cap S$ finite). Therefore $G = \mathscr{D}L \cdot S \cdot \mathscr{R}_u G$. Since S normalizes $\mathscr{R}_u G$, the semi-direct product $S \cdot \mathscr{R}_u G$ is a connected solvable subgroup, invariant under $\mathscr{D}L$, hence normal in G and consequently contained in $\mathscr{R}G$. Since $\mathscr{R}_u G$ is connected, solvable, $\mathscr{R}G = \pi^{-1}(\pi(S)) = S \cdot \mathscr{R}_u G$ and S is a maximal torus of $\mathscr{R}G$. Let $M = L \cap \mathscr{R}G$. Its identity component belongs to the radical of L, hence to S, and therefore is equal to S. But in $\mathscr{R}G$, the normalizer of a torus is connected (10.6), hence $M = S$.

(ii) Let S be a maximal torus of $\mathscr{R}G$. We have already pointed out that $\pi(S)$ is in the center of $G/\mathscr{R}_u G$. Therefore π is S-equivariant, S acting by inner automorphisms on G and trivially on $G/\mathscr{R}_u G$. By 9.6, $L = \mathscr{Z}_G(S)$ maps onto $G/\mathscr{R}_u G$ under π. Thus $G = L \cdot \mathscr{R}_u(G)$. Since $\mathscr{Z}_G(S) \cap \mathscr{R}_u G = \{1\}$, it follows also (see 9.2) that $L(\mathscr{Z}_G(S)) \cap L(\mathscr{R}_u G) = \{0\}$ hence $L(G)$ is direct sum of the Lie algebras of L and $\mathscr{R}_uG$. Therefore L is a Levi subgroup. The second assertion follows from (i) and 10.6(4)(5).

Bibliographical Note

Up to 11.14, and except for 11.8, the results of this paragraph are proved in [1]. The terminology, however, was introduced later in [13], and was kindly used by the notetaker of the first edition. Most of the other results of this paragraph are due to Chevalley [13]. The proof of the normalizer theorem 11.16, however, is quite different from that of Chevalley [13: Exp. 9] which was reproduced in the first edition of this book. I gave it first in a course in Buenos Aires, in 1973, and it was included in the books of Humphreys and Springer.

The variety $\mathscr{B}$ of 11.18 can be introduced in an intrinsic way and be given a structure of variety defined over k, even if G does not contain any Borel subgroup defined over k (see [3: §7] or [15: Exp. 12]).

The terminology Levi subgroup was introduced in [4]. It was suggested by a theorem of E.E. Levi on real Lie algebras, although it represents a minor deviation from it (besides the fact that it deals with groups rather than Lie algebras). More precisely, Levi showed that any real Lie algebra $\mathfrak{h}$ is a semi-direct sum of its radical and a semi-simple Lie subalgebra, and Malcev proved that the latter is determined up to conjugacy (see e.g. [8]). Globally, if the connected real Lie group H is simply connected, then it is the semi-direct product of its radical $\mathscr{R}H$ by a maximal semisimple closed subgroup S, determined up to conjugacy. If H is not simply connected however, $S \cap \mathscr{R}H$ may be non-trivial. In the algebraic group case, it has been found more convenient to look for complements to the unipotent radical rather than to the full radical.

§12. Cartan Subgroups; Regular Elements

12.1 *Properties of Cartan subgroups.* Recall from 11.13 that a *Cartan subgroup* of G is the centralizer of a maximal torus in G.

Theorem.

(a) *The Cartan subgroups are all conjugate.*
(b) *Their union contains a dense open set in G. Let C be a Cartan subgroup.*
(c) $C = \mathscr{N}_G(C)^o$.
(d) $C = C_s \times C_u$, *where* $T = C_s$ *is a maximal torus in G, the unique one contained in C, and* $C = G^T$.
(e) *C is a connected nilpotent group, and it is maximal among such subgroups of G.*

Proof. (a) to (d) have already been proved in 11.7, 11.13.

It remains to show that C is maximal connected nilpotent in G. By virtue of (*c*) this follows from the following lemma, reminiscent of its analogue for finite groups.

Lemma. *If G is nilpotent and H is a proper closed subgroup then* $\dim H < \dim \mathcal{N}_G(H)$.

For the application above take $H = C$ and, for G, a connected nilpotent subgroup properly containing C.

Proof. Let $Z = \mathscr{C}(G)^o$. If $Z \not\subset H$ then the conclusion follows because $ZH \subset \mathcal{N}_G(H)$. If not we apply induction on dimension to H/Z in G/Z, the inverse image of whose normalizer is $\mathcal{N}_G(H)$.

12.2 *Regular elements; rank.* The dimension of a Cartan subgroup of G is called the *rank* of G. If $g \in G$ then g_s belongs to a maximal torus T so $\dim \mathscr{Z}_G(g_s) \geqq \dim G^T = \operatorname{rank} G$, and we call g *regular* if the former is an equality. Thus g is regular if and only if g_s is regular.

The set of regular elements of G will be denoted G_{reg}. An element in $G - G_{\mathrm{reg}}$ is called *singular.*

Lemma. *Let T be a maximal torus in G. The following conditions on $t \in T$ are equivalent*:

(a) *t is regular*; (b) $\mathscr{Z}_G(t)^o = G^T$; (c) $t^\alpha \neq 1$ *for all roots* $\alpha \in \Phi(T, G)$.

Proof. Since G^T is a connected subgroup of $\mathscr{Z}_G(t)$ the equivalence of (a) and (b) follows by dimension count. The equivalence of (b) and (c) follows from 9.4.

It follows from (c) that *the regular elements in T form a dense open set in T.* In particular regular elements exist.

Proposition. *The following conditions on a semi-simple element $g \in G$ are equivalent*:

(1) *g is regular.*
(2) *$\mathscr{Z}(g)^o$ is a Cartan subgroup.*
(3) *$\mathscr{Z}(g)^o$ is nilpotent.*
(4) *g belongs to a unique maximal torus.*
(5) *g belongs to only finitely many maximal tori.*

Proof. Let T be a maximal torus containing g. Then (1)$\Leftrightarrow$(2) follows from (a)$\Leftrightarrow$(b) in the Lemma, and (2)$\Leftrightarrow$(3) is (12.1(e)).

Since a connected nilpotent group contains a unique maximal torus (10.6(3)), it follows that (3)$\Leftrightarrow$(4). Moreover (4)$\Rightarrow$(5) is obvious.

Let $H = \mathscr{Z}_G(g)^o$ Condition (5) implies, by virtue of the conjugacy of maximal tori in H, that $H/\mathcal{N}_H(T)$ is a *finite* connected variety, hence a single point. Thus T is normal in the connected group H. By rigidity (8.10), T is central in H, i.e. $H \subset G^T$. But $G^T \subset \mathscr{Z}_G(g)^o$ so $G^T = H$. This proves that (5) implies (2), and hence completes the proof.

12.3 Theorem. (1) *An element $g \in G$ is regular if and only if it belongs to a unique Cartan subgroup.*

(2) *G_{reg} contains a dense open set in G.*

Proof. (1) Suppose g is regular. Then g_s belongs to a unique Cartan subgroup $C = \mathscr{Z}_G(g_s)^o$ (see 12.2), and 11.12 implies $g \in C$. If C' is a Cartan subgroup containing g then $g_s \in C'_s$ so $C' = \mathscr{Z}_G(C'_s)$ (see 12.1(d)) is contained in $\mathscr{Z}_G(g_s)$, and hence equals C.

Suppose, conversely, that g belongs to a unique Cartan subgroup C. Since $g_s \in C_s \subset \mathscr{C}(C)$ it follows that $H = \mathscr{Z}_G(g_s)^o$ contains C, and C is clearly then a Cartan subgroup of H. The others are conjugate in H to C and hence contain $g_s \in \mathscr{C}(H)$. Since g lies in a unique one, the same is therefore true of $g_u = g_s^{-1}g$. Now the regularity of g_s, and hence of g, follows, in view of 12.2, from the

Lemma. *Suppose a connected group H has a unipotent element h belonging to a unique Cartan subgroup C. Then H is nilpotent.*

Proof. Write $C = H^T$ with T a maximal torus, and embed C in a Borel subgroup $B = T \cdot B_u$. It suffices, by 11.5(3), to show that B is nilpotent. This will follow by showing that $B \subset C$, which, in turn, results if $B_u \subset C$. Let $B_u = N_m \supset N_{m-1} \supset \cdots \supset N_0 = \{e\}$ be the descending central series of B_u. We will show, by induction on i, that $N_i \subset C$, and we may assume $i > 0$, clearly. If $x \in N_i$ then $h^{-1}xhx^{-1} \in N_{i-1}$ because $h \in B_u$, so $xhx^{-1} \in hN_{i-1} \subset C$, by induction. Thus $N_i \subset \mathscr{N}_H(C)$, so $N_i \subset \mathscr{N}_H(C)^o = C$ because N_i is connected. This completes the proof of the lemma.

(2) Let $C = G^T = T \times C_u$ be a Cartan subgroup, and let $T_0 = \{t \in T \mid t^\alpha \neq 1$ for all $\alpha \in \Phi(T, G)\}$. Then $C_0 = T_0 \times C_u$ is open dense in C, and the Lemma of 12.2 implies that $C_0 = C \cap G_{\text{reg}}$. Since every regular element belongs to a Cartan subgroup it follows that G_{reg} is the image of the morphism

$$f: G \times C_0 \to G, \quad f(g, c) = gcg^{-1}.$$

Since $C_0 \subset \text{im}(f) = G_{\text{reg}}$ it follows that $C = \bar{C}_0 \subset \bar{G}_{\text{reg}}$. Since $\bar{G}_{\text{reg}}$ is stable under conjugation it therefore contains ${}^G C$, and 12.1(b) implies the latter is dense in G. Since $G \times C_0$ is irreducible it follows that f is dominant. Thus $G_{\text{reg}} = \text{im}(f)$ contains a dense open set.

12.4 Proposition. *Let $\alpha: G \to G'$ be a surjective morphism of algebraic groups.*

(1) *The Cartan subgroups of G' are the images of those in G.*

(2) $\alpha(G_{\text{reg}}) \subset G'_{\text{reg}}$.

Proof. (1) Let $C = G^T$ be a Cartan subgroup of G. By conjugacy, it suffices to show that $\alpha(C)$ is a Cartan subgroup of G'. But $T' = \alpha(T)$ is a maximal torus (11.14) and it follows from 9.6 and 11.12 that $G^T \to G'^{T'}$ is surjective.

(2) If $g \in G_{\text{reg}}$ and $t = g_s$ then $\alpha(t) = \alpha(g)_s$, and 9.6 implies that $Z_G(t)^o \to Z_{G'}(\alpha(t))^o$ is surjective. Since $Z_G(t)^o$ is a Cartan subgroup, part (1) implies that $Z_{G'}(\alpha(t))^o$ is one also, so $\alpha(g)$ is regular.

12.5 Proposition. *Let H be a not necessarily connected nilpotent algebraic group, and let $T = (H^o)_s$ be the maximal torus in H^o* (cf. 10.6(3)). *Then T is central in H.*

Proof. Since $H^o = T \times (H^o)_u$ (see 10.6(3)) it follows that T is central in H^o and normal in H. Consider the isomorphism,

$$X_*: \text{End}_{\text{alg-grp}}(T) \to \text{End}_{\mathbb{Z}\text{-mod}}(X_*(T)),$$

of endomorphism rings (see 8.3 and 8.6). If $h \in H$ write $I(h)$ for $\text{Int}(h)$ on T, and $x(h)$ for $X_*(I(h))$. If we think of T additively, then commutating with h, i.e. $t \mapsto (h,t) = hth^{-1}t^{-1}$, is the endomorphism $I(h) - id$. Since H is nilpotent, it follows that $I(h) - id$, and hence $x(h) - id$, are nilpotent, i.e. $x(h)$ is unipotent. Thus $x(H)$, being an image of H/H^o, is a finite unipotent group in $\text{Aut}_{\mathbb{Z}\text{-mod}}(X_*(T)) \cong \mathbf{GL}_n(\mathbb{Z})$ for some $n \geqq 0$. But in characteristic zero, there are no non-trivial unipotent elements of finite order (see, e.g., 7.3). Thus $X_*(H) = \{id\}$, and this implies that H centralizes T.

12.6 *Chevalley's definition of a Cartan subgroup.* It is condition (2) of the following theorem. Its interest is that it makes sense for an abstract group.

Theorem. *The following conditions on a (not necessarily closed) subgroup C of G are equivalent:*

(1) *C is a Cartan subgroup.*
(2) (a) *C is a maximal nilpotent subgroup; and*
  (b) *every subgroup of finite index in C has finite index in its normalizer (in G).*
(3) *C is a closed connected nilpotent subgroup, and $C = \mathcal{N}_G(C)^o$.*

Proof. (1)$\Rightarrow$(2). If H is a nilpotent group containing $C = \mathcal{Z}_G(T)$, we can assume H to be closed. Since T is a maximal torus in G it is also one in H^o, so 12.5 implies that $T \subset \mathcal{C}(H)$, i.e. $H \subset \mathcal{Z}_G(T) = C$.

If H is a subgroup of finite index in C then H is dense in C, because C is connected. Hence $\mathcal{N}_G(H) \subset \mathcal{N}_G(C)$. But $\mathcal{N}_G(C)^o = C$, by (12.1), hence the chain $H \subset C \subset \mathcal{N}_G(C)$ shows that H has finite index in $\mathcal{N}_G(H)$.

(1)$\Rightarrow$(3) is contained in (12.1).

(3)$\Rightarrow$(1). Write $C = S \times C_u$ with $S = C_s$. Embed C in a Borel subgroup B, and let T be a maximal torus in B containing S; then $B = T \cdot B_u$. Put

$M = \mathscr{Z}_B(S)$. Then M is connected (11.12) and $M = T \cdot M_u$, clearly. Now S is central in M, so $S \cdot M_u$ is connected nilpotent, and contains C. Since, by hypothesis, $C = \mathscr{N}_G(C)^o$, it follows from the lemma in 12.1 that $S \cdot M_u = C$. Since M is connected and solvable we have $(M, M) \subset M_u \subset C$, so C is normal in M. But $C = \mathscr{N}_G(C)^o$ and M is connected, so $C = M$. Thus $C \supset T$, hence $S = T$, and $C = \mathscr{Z}_B(T)$ is a Cartan subgroup of B, therefore also of G.

(2)$\Rightarrow$(1). Suppose $C \subset G$ satisfies (a) and (b). Since $\bar{C}$ is nilpotent whenever C is, (a) implies that C is closed. Now (b) implies that C^o has finite index in $\mathscr{N}_G(C^o)$, so we have $C^o = \mathscr{N}_G(C^o)^o$. Since C^o is nilpotent it follows from (3)$\Rightarrow$(1) that C^o is a Cartan subgroup of G. But then (1)$\Rightarrow$(2) implies C^o to be maximal nilpotent, so $C^o = C$.

Bibliographical Note

Chevalley's "abstract group theoretic" definition of Cartan subgroups is given in [12b], where Cartan subgroups are studied in characteristic zero. For the results of this section, see [1].

§13. The Borel Subgroups Containing a Given Torus

If H is a closed connected subgroup of G then the set $\mathscr{B}^H$ of Borel subgroups containing H is empty unless H is solvable. When it is not empty we shall write

$$I(H) = I_G(H) = \left(\bigcap_{B \in \mathscr{B}^H} B \right)^o.$$

If T is a maximal torus then, since $I(T)$ is connected solvable, we can write $I(T) = T \cdot I(T)_u$. The main objective of this paragraph is to prove that $I(T)_u$ is the unipotent radical $\mathscr{R}_u(G)$ (see 13.16) of G.

This fact has several important consequences for reductive groups. Together with some information on groups of "semi-simple rank 1" in 13.14, it goes a long way toward showing (in 13.18) that $\Phi(T, G)$ is a root system when G is reductive. The final proof of this fact in 14.8 requires further information about actions of tori on unipotent groups.

A further consequence is the construction of the "big cell" associated with a pair of "opposite" Borel subgroups (see 14.1).

13.1 *Regular, semi-regular, and singular tori.* Let S be a torus in G.

S is *regular* if S contains a regular element. Thus *maximal tori are regular* (see proof of 12.3).

S is *semi-regular* if $\mathscr{B}^S$ is finite.

S is *singular* if $\mathscr{B}^S$ is infinite.

Let S be regular. Then, if $s \in S$ is regular, $\dim \mathscr{Z}_G(s) \leqq \dim \mathscr{Z}_G(t)$ for any $t \in S$. On the other hand (8.18), there exists $t \in S$ such that $\mathscr{Z}_G(t) = \mathscr{Z}_G(S)$. Since the centralizer of S is connected (11.12), it follows that $\mathscr{Z}_G(S) = \mathscr{Z}_G(s)^o$ if and only if $s \in S$ is regular. In particular G^S is then a Cartan subgroup, and is nilpotent. The proposition below implies therefore that regular tori are semi-regular.

If $\lambda: \mathbf{GL}_1 \to G$ is a one-parameter subgroup then we shall call λ a *regular*, *semi-regular*, or *singular parameter* if the torus $S = \text{im}(\lambda)$ has the corresponding property.

In the next proposition S is a torus in G and $X = G/B$ for some $B \in \mathscr{B}$. We know then (see 11.18) that there is a natural bijection between $\mathscr{B}^S$ and X^S.

Proposition. *The following conditions are equivalent*:

(1) *S is semi-regular.*
(2) *S has an isolated fixed point in X* (i.e. *X^S has a connected component with one point*).
(3) *G^S is solvable.*
(4) *$G^S \subset I(S)$.*

Proof. (1)$\Rightarrow$(2) is obvious since X^S is finite and non-empty.

(2)$\Rightarrow$(3). G^S is connected (11.12) and leaves X^S stable, so it stabilizes the connected components of X^S. If one of these components is reduced to a point then that point is fixed by G^S. The corresponding Borel subgroup contains G^S, so G^S is solvable.

(3)$\Rightarrow$(1) and (4) follows from (11.15, Cor.).

(4)$\Rightarrow$(3) is clear because $I(S)$ is solvable.

Corollary. *Let H be a connected subgroup of G containing S. If S is regular* (*resp., semi-regular*) *in G then it is likewise in H.*

Proof. H^S is nilpotent (resp., solvable) as soon as the larger group G^S is nilpotent (resp., solvable).

13.2 *Singular subtori, and roots.* We fix a *semi-regular torus T.* If $\alpha \in X(T)$ is not zero, then $T_\alpha = (\ker \alpha)^o$ is a subtorus of condimension 1.

We shall denote the roots of G relative to T by Φ in place of the usual $\Phi(T, G)$. Thus

$$\mathfrak{g} = \mathfrak{g}^T \oplus \coprod_{\alpha \in \Phi} \mathfrak{g}_\alpha.$$

Consider also the subset $\Psi = \Phi(T, G/I(T))$. Recall from (8.17) that, if one writes $\mathfrak{g}_\alpha = L(I(T))_\alpha \oplus \mathfrak{g}'_\alpha$, then Ψ is the set of α for which $\mathfrak{g}'_\alpha \neq 0$. These are the "roots of G outside of $I(T)$." Moreover, since $G^T \subset I(T)$, and hence $\mathfrak{g}^T \subset L(I(T))$, we have

$$\mathfrak{g} = L(I(T)) \oplus \coprod_{\alpha \in \Psi} \mathfrak{g}'_\alpha.$$

Proposition. (1) *The following conditions on a subtorus S of T are equivalent*: (a) *S is singular*; (b) *$S \subset T_\alpha$ for some $\alpha \in \Psi$*; (c) *$G^S \not\subset I(T)$*.

(2) *If $\lambda \in X_*(T)$, then λ is semi-regular if and only if $< \alpha, \lambda > \neq 0$ for all $\alpha \in \Psi$.*

An immediate consequence of (1) is:

Corollary. *A singular subtorus of T is contained in a singular subtorus of codimension* 1.

Proof. (1) (a)$\Leftrightarrow$(c). If S is semi-regular then $G^S \subset I(S)$ (condition (4) of (13.1) and clearly $I(S) \subset I(T)$. Conversely if $G^S \subset I(T)$ then, G^S is solvable (condition (3) of (13.1)), so S is semi-regular.

Now the equivalence of (b) and (c) is just the equivalence of 2(a) and 2(c) is Proposition 9.4. We take $I(T)$ for the H in that proposition, and use the fact that G^S is connected (11.12).

(2) is just the equivalence of (a) and (b) applied to $S = \mathrm{im}(\lambda)$.

13.3 *Actions of one-parameter groups at* 0 *and* ∞. We shall write

$$\mathbf{P}_1 = \mathbf{GL}_1 \cup \{0\} \cup \{\infty\}, \quad \text{(disjoint union)}$$

with the following convention: The coordinate ring of $\mathbf{GL}_1$ is $K[\chi, \chi^{-1}]$, and $\mathbf{P}_1$ is covered by the affine lines with coordinate rings $K[\chi]$ and $K[\chi^{-1}]$. The points 0 and ∞ correspond to the loci "$\chi = 0$" and "$\chi^{-1} = 0$," respectively, in these open sets. The character χ is the identity map of $\mathbf{GL}_1 = K^*$.

Suppose $f: \mathbf{GL}_1 \to Y$ is a morphism into a complete variety. Then it follows from AG.18.5(f) that f extends uniquely to a morphism $f: \mathbf{P}_1 \to Y$. Thus we may speak of $f(0)$ and $f(\infty)$.

Now suppose we have a linear representation of G on a vector space V, and let $\lambda: \mathbf{GL}_1 \to T$ be a one-parameter group in a torus T in G. Then G, and hence $\mathbf{GL}_1$, operates on the projective space $\mathbf{P}(V)$. If $x \in \mathbf{P}(V)$ then $f: \mathbf{GL}_1 \to \mathbf{P}(V)$, $f(t) = \lambda(t)x$, extends as above to $\mathbf{P}_1$. In place of $f(0)$ and $f(\infty)$, in this case, we shall write

$$\lambda(0)x \quad \text{and} \quad \lambda(\infty)x.$$

To determine these points, choose a basis $e_1, \ldots, e_n$ of V such that e_i is an eigenvector, say with character α_i, for T; let $\langle \alpha_i, \lambda \rangle = m_i$ (see 8.6). Then if $v = \sum a_i e_i \in V$ and if $t \in \mathbf{GL}_1$ we have

$$\lambda(t)v = \sum a_i t^{m_i} e_i.$$

Assume $v \neq 0$, let $x = [v]$ and $I_v = \{i | a_i \neq 0\}$. Let J_m be the set of $i \in I_v$ such that m_i takes the minimal value, $m = \min m_i (i \in I)$. Similarly let J_M be the set of $i \in I_v$ such that m_i takes the maximum value $M = \max m_i (i \in I_v)$. If $[v] \in \mathbf{P}(V)$ denotes the image of v under the projection $\pi: V - \{0\} \to \mathbf{P}(V)$ then we have, for any $t \in \mathbf{GL}_1$, $[v] = [t^{-m}v] = [t^{-M}v]$. Define morphisms

$$g_m : \mathbf{GL}_1 \cup \{0\} \to V - \{0\}$$

by

$$g_m(t) = \sum_i a_i t^{m_i - m} e_i,$$

$$g_M : \mathbf{GL}_1 \cup \{\infty\} \to V - \{0\}$$

by

$$g_M(t) = \sum_{i \in I} a_i t^{m_i - M} e_i.$$

Since $m_i - M \leqq 0 \leqq m_i - m$ for all $i \in I$, both formulas make sense (at 0 and ∞, resp.) and they give non-zero vectors of V. If $t \in \mathbf{GL}_1$, then $g_m(t) = t^{-m}\lambda(t)v$ and $g_M(t) = t^{-M}\lambda(t)v$. Thus the morphism $f: t \mapsto [\lambda(t)v]$ coincides with both $\pi \circ g_m$ and $\pi \circ g_M$. The former of these two gives the extension of f to $t = 0$, and the latter to $t = \infty$. Explicitly, we have

$$\lambda(0)[v] = \left[\sum_{j \in J_m} a_j e_j \right]$$

$$\lambda(\infty)[v] = \left[\sum_{j \in J_M} a_j e_j \right].$$

It is clear from these formulas that:

$$\lambda(0)[v] = \lambda(\infty)[v]$$

$\Leftrightarrow I_m = I_M$
$\Leftrightarrow m = M$ (i.e. all $m_i (i \in I)$ are equal)
$\Leftrightarrow v$ is an eigenvector of $\mathbf{GL}_1$ under λ
$\Leftrightarrow [v]$ is a fixed point of $\mathbf{GL}_1$ under the action induced by λ.

In case λ is such that $m_i = \langle \alpha_i, \lambda \rangle$ are distinct for distinct $\alpha_i (1 \leqq i \leqq n)$, then the eigenvectors of $\mathbf{GL}_1$ (under λ) coincide with the eigenvectors of T. In this case, therefore, $\lambda(0)[v] = \lambda(\infty)[v]$ if and only if $[v]$ is a fixed point of T.

13.4 Lemma. *Let W be a hyperplane in a vector space V, let Y be a closed subvariety of $\mathbf{P}(V)$, and let $H = \mathbf{P}(W) \subset \mathbf{P}(V)$.*

(a) *If* $\dim Y \geqq 1$ *then* $Y \cap H \neq \phi$.
(b) *If Y is irreducible and not contained in H then each irreducible component of $Y \cap H$ has dimension* $\dim Y - 1$.

Proof. (a) If $Y \cap H = \phi$, then Y is a complete variety in the affine variety $\mathbf{P}(V) - H$, so Y is finite (see 10.1(2)).

(b) Locally on $\mathbf{P}(V)$, the hyperplane H is defined by a single (linear) equation, and hence likewise for $Y \cap H$ on Y. Therefore part (b) follows from AG.9.2.

13.5 Proposition. *Let T be a torus with a given linear representation on a vector space V. Let Y be a non-empty closed subset of $\mathbf{P}(V)$ stable under T. Then T has at least* $\dim Y + 1$ *fixed points on Y.*

Proof. Being connected, T leaves every irreducible component of Y stable, therefore we may assume Y to be irreducible. If it is of dimension 0, then it consists of one point, which is necessarily fixed under T. We may therefore argue by induction, assume $\dim Y \geqq 1$ and the proposition to be true for any Y' of strictly smaller dimension.

Let $\alpha_1, \ldots, \alpha_n$ be the distinct characters of T in V. We can choose $\lambda \in X_*(T) \cong \mathrm{Hom}(X(T), \mathbb{Z})$ (see 8.6) so that the $m_i = \langle \alpha_i, \lambda \rangle$ are all distinct. Then $\mathbf{GL}_1$ (via λ) and T have the same eigenvectors in V, and hence the same fixed points in $\mathbf{P}(V)$. Thus we may, without loss, assume that $T = \mathbf{GL}_1$.

If $\dim Y^T \geqq 1$, then Y^T is infinite. From now on, assume that $\dim Y^T \leqq 0$. Also, replacing V by the intersection of all the hyperplanes containing Y, which is obviously stable under T, we may (and do) assume that no hyperplane in $\mathbf{P}(V)$ contains Y. We use the conventions and notation of 13.3 and assume moreover, for convenience, that $m_i \leqq m_j$ if $i \leqq j$. Since Y is not contained in any hyperplane, there exists $v \in V$ such that $a_1 \neq 0$ and $[v] \in Y$. Then (see 13.3) $[\lambda(0) \cdot v]$ is defined, belongs to Y and is fixed under T. It does not belong to the intersection of Y and the hyperplane $a_1 = 0$. This intersection has dimension equal to $\dim Y - 1$, (see 13.4), and is stable under T. By induction assumption, it contains at least $\dim Y$ fixed points under T. Altogether, we get at least $\dim Y + 1$ fixed points, as asserted.

13.6 Corollary. *If $P \subsetneqq G$ is a parabolic subgroup, and if T is a torus in G, then T has at least two fixed points on G/P.*

Proof. According to 5.1 (cf. also the proof of 6.8) we can choose a linear representation $G \to GL(V)$ and an $x \in \mathbf{P}(V)$ so that $g \mapsto gx$ induces an isomorphism of G/P onto the orbit $Y = Gx$. The hypotheses imply that Y is closed and that $\dim Y \geqq 1$, so the corollary follows from 13.5.

13.7 Proposition. *Let T be a maximal torus of G. Then G is generated by all $B \in \mathscr{B}^T$.*

Proof. Let P be the subgroup generated by the $B \in \mathscr{B}^T$. It is closed, connected (2.2). Fix $B \in \mathscr{B}^T$, and consider the quotient morphisms

$$G \xrightarrow{\pi} G/B \xrightarrow{p} G/P.$$

Suppose $P \neq G$. Since P is clearly parabolic, 13.6 implies that T has a fixed point in G/P distinct from $p(\pi(e))$. The inverse image Y in G/B of that fixed point is closed and stable under T. Thus T has a fixed point $\pi(y) \in Y$, and, by construction,

$$p(\pi(y)) \neq p(\pi(e)).$$

Now $Ty \subset yB$, i.e. $y^{-1} \cdot T \cdot y \subset B$. By the conjugacy of maximal tori in B we can write ${}^{y^{-1}}T = {}^{b}T$ for some $b \in B$, in which case $yb \in \mathcal{N}_G(T)$. But it is clear from the definition of P that $\mathcal{N}_G(T)$ normalizes P. By the normalizer theorem 11.15 we therefore have $\mathcal{N}_G(T) \subset \mathcal{N}_G(P) = P$, so $y = (yb)b^{-1} \in PB = P$, and hence $p(\pi(y)) = p(\pi(e))$. Contradiction.

Remark. We shall see later (14.1, Cor. 1) that any connected k-group is in fact generated by *two* suitably chosen Borel subgroups.

13.8 We fix a *semi-regular torus* T in G and write $X_*(T)_{sr}$ for the set of semi-regular one-parameter subgroups in T. According to 13.2(2) this set is not empty. More precisely (see (13.2)),

$$X_*(T)_{sr} = \{\lambda \in X_*(T) \mid \langle \alpha, \lambda \rangle \neq 0 \text{ for all } \alpha \in \Psi\}.$$

Fix a $B_0 \in \mathscr{B}$ and put $X = G/B_0$. If $\lambda \in X_*(T)_{sr}$ then the set of fixed points of $\operatorname{im}(\lambda)$ in X is finite, and hence coincides with X^T.

Proposition. *Let* $\lambda \in X_*(T)_{sr}$.

(1) *There is a unique point* $x(\lambda) \in X$ *such that* $\lambda(\infty)x = x(\lambda)$ *for all* x *in some neighbourhood of* $x(\lambda)$. *The corresponding Borel subgroup,* $B(\lambda)$, *contains* T.
(2) $U = \{x \in X \mid \lambda(\infty)x = x(\lambda)\}$ *is the complement of a* T*-invariant hyperplane section (in some* $\mathbf{P}(V)$*) of* X. *In particular,* $\dim(X - U) = \dim X - 1$.
(3) *There is a set* $\{\beta_i\}$ *of non-trivial characters of* T, *with trivial restrictions to* $T \cap \mathscr{R}(G)$, *such that, for* $\lambda' \in X_*(T)_{sr}$, *we have* $B(\lambda) = B(\lambda')$ *if and only if* $\langle \beta_i, \lambda' \rangle > 0$ *for each* i.

Remark. Conversely, the existence of an $x(\lambda)$ as above implies that λ is semi-regular, because (see 13.1(2)) $x(\lambda)$ must then be an isolated fixed point of $\operatorname{im}(\lambda)$. The group $B(\lambda)$ will be said to *be associated to* λ.

Proof. Since X is irreducible any two non-empty open sets meet, and this clearly implies the uniqueness of $x(\lambda)$.

Since $X = G/B_0 \cong (G/\mathscr{R}(G))/(B_0/\mathscr{R}(G))$ we can choose a linear representation of $G/\mathscr{R}(G)$, and hence of G, on a vector space V so that X can be identified with the G-orbit of a point in $\mathbf{P}(V)$. Let $\pi: V - \{0\} \to \mathbf{P}(V)$, $v \mapsto [v]$ be the canonical morphism. Replacing V by the subspace spanned by $\pi^{-1}(X)$, if necessary, we can further arrange that *X lies in no hyperplane in* $\mathbf{P}(V)$.

Let $e_1, \ldots, e_n$ be a basis of V such that each e_i is an eigenvector, say with character α_i, of T. Put $m_i = \langle \alpha_i, \lambda \rangle$ and assume the basis ordered so that $m_1 \geqq \cdots \geqq m_n$. Say $m_1 = \cdots = m_r$ and $m_r > m_i$ for $i > r$. Let W be the hyperplane in V spanned by $e_2, \ldots, e_n$, and suppose $v = \sum a_i e_i \notin W$ (i.e. $a_1 \neq 0$). Then the calculation of 13.3 shows that:

$$(*) \qquad \lambda(\infty)[v] = [a_1 e_1 + \cdots + a_r e_r].$$

Let H be the hyperplane $\mathbf{P}(W)$ in $\mathbf{P}(V)$. We propose to show that $r = 1$, and that $x(\lambda) = [e_1]$ has the property required by part (1). Moreover we will prove that the U of part (2) is $X - (X \cap H)$, and this, by virtue of 13.4, will yield (2).

Suppose $r > 1$. We can find infinitely many $b \in K$ and v of the form $v = e_1 + be_2 + \cdots$ such that $[v] \in X$. Otherwise X would lie in the union of H and of a finite number of hyperplanes, "$a_2 = ba_1$." This is impossible since X is irreducible and lies in no hyperplane. Since $r > 1$ it follows from (*) that, for $v = e_i + be_2 + \cdots$ as above, the $\lambda(\infty)[v]$ are distinct for distinct b. Thus we obtain infinitely many fixed points of $\mathrm{im}(\lambda)$ in X, contradicting the assumption that λ is semi-regular. Thus indeed $r = 1$.

Now that $r = 1$ it follows further from (*) that $\lambda(\infty)[v] = [e_1]$ if and only if $v \notin W$. Thus

$$\{x \in X \mid \lambda(\infty)x = [e_1]\} = X - (X \cap H).$$

This completes the proof of (1) and (2).

To prove (3), suppose we are given $\lambda' \in X_*(T)_{sr}$. Let $m'_i = \langle \alpha_i, \lambda' \rangle$. The proof above shows that, for some j, m'_j is strictly larger than m'_i for all $i \neq j$, and that $x(\lambda') = [e_j]$. Thus $x(\lambda') = x(\lambda) \Leftrightarrow m'_1 > m'_i$ for all $i > 1 \Leftrightarrow \langle \alpha_1, \lambda' \rangle > \langle \alpha_i, \lambda' \rangle$ for all $i > 1 \Leftrightarrow \langle \beta_i, \lambda' \rangle > 0$ for all $i > 1$, where $\beta_i = \alpha_1 - \alpha_i$. Since α_i are characters of T, trivial on $T \cap \mathscr{R}(G)$, this proves (3).

Let $f: G \to G'$ be an *isomorphism* of algebraic groups, and put $T' = f(T)$. If $\lambda \in X_*(T)_{sr}$ then $f \circ \lambda \in X_*(T')_{sr}$, clearly, and $B(f \circ \lambda) = f(B(\lambda))$. Now suppose $n \in \mathscr{N}_G(T)$ and $f = \mathrm{Int}(n)$. Then $T' = T$ and, if we write ${}^n\lambda = \mathrm{Int}(n) \circ \lambda$, we have

$$B({}^n\lambda) = {}^nB(\lambda) \quad \text{for } n \in \mathscr{N}_G(T).$$

13.9 Lemma. *Let T be a maximal torus, $W = W(T, G)$ and $X = G/B_o$ as above.*

(1) *If* $\dim X \geqq 1$ *then* $\operatorname{card} W \geqq 2$.
(2) *If* $\dim X \geqq 2$ *then* $\operatorname{card} W \geqq 3$.

Proof. Recall (11.16, 11.19) that W acts simply transitively on X^T; hence $\operatorname{card} W = \operatorname{card} X^T$. Thus the lemma follows from 13.5.

13.10 *Weyl chambers.* Let T be a *maximal* torus, and $W = W(T, G)$. To $\lambda \in X_*(T)_{sr}$ we have the associated $B(\lambda) \in \mathscr{B}^T$ constructed above in 13.8. Given $B \in \mathscr{B}^T$.

$$WC(B) = \{\lambda \in X_*(T)_{sr} \mid B(\lambda) = B\}$$

is called the *Weyl chamber* of B (with respect to T in G).

Proposition. (1) *There is a set $\{\beta_i\}$ of non-trivial characters of T, which are trivial on $T \cap \mathscr{R}(G)$, such that*

$$WC(B) = \{\lambda \in X_*(T)_{sr} \mid \langle \beta_i, \lambda \rangle > 0 \textit{ for each } i\}.$$

(2) *The Weyl group W acts simply transitively on the set of Weyl chambers $WC(B)$ of $B \in \mathscr{B}^T$.*

Proof. Part (2) will imply each Weyl chamber is non-empty, whereupon (1) reduces to part (3) of 13.8.

If $n \in \mathscr{N}_G(T)$ and $\lambda \in WC(B)$ then (see end of 13.8) we have $B({}^n\lambda) = {}^nB(\lambda)$, so ${}^nWC(B) = WC({}^nB)$. If $n \in G^T$ then ${}^n\lambda = \lambda$ so $W = \mathscr{N}_G(T)/G^T$ acts on the set of Weyl chambers in such a way that $f: B \mapsto WC(B)$ is W-equivariant. Since W is transitive on the B's it follows that each $WC(B)$ is not empty (since at least one of them is not empty). In that case $WC(B)$ determines B, clearly, so f is bijective. The simple transitivity of W on the $WC(B)$'s now follows from the simple transitivity of W on $\mathscr{B}^T$ (11.19).

13.11 *Centralizers of singular subtori of codimension* 1. Let S be a singular subtorus of codimension 1 in the maximal torus T, and let $B \in \mathscr{B}^T$. Then T is a maximal torus in G^S, and we know from 11.15 that B^S is a Borel subgroup of G^S. Indeed, the map $B \mapsto B^S$ is a surjection from $\mathscr{B}(G)^T$ to $\mathscr{B}(G^S)^T$. When we write $WC(B^S)$, it is understood with reference to B^S as an element of $\mathscr{B}(G^S)^T$.

Proposition.

(1) *The Weyl group $W(T, G^S)$ has order* 2.
(2) *In $X_*(T)$ we have $WC(B) \subset WC(B^S)$.*
(3) *If C is one of the two elements of $\mathscr{B}(G^S)^T$, then there is a non-trivial $\alpha \in X(T/S) \subset X(T)$ such that, for any $B \in \mathscr{B}^T$, we have*

$$B^S = C \Leftrightarrow \langle \alpha, \lambda \rangle > 0 \text{ for all } \lambda \in WC(B).$$

If C' is the other element of $\mathscr{B}(G^S)^T$ then

$$B^S = C' \Leftrightarrow \langle \alpha, \lambda \rangle < 0 \text{ for all } \lambda \in WC(B).$$

Proof. (1) Put $H = G^S$ and let $\pi: H \to H' = H/\mathscr{R}(H)$ be the quotient morphism. Then $T' = \pi(T)$ is a maximal torus in H', and 11.20 implies that $W(T, H) \to W' = W(T', H')$ is an isomorphism. We shall show that $\dim T' = 1$, and then deduce from this that card $W' = 2$, thus proving (1).

Since $S \subset \mathscr{R}(H)$, and S has codimension 1 in T, we have $\dim T' = \dim(T/T \cap \mathscr{R}(H)) \leqq 1$. On the other hand, since S is singular, $H = G^S$ is not solvable, hence H' is not solvable, and $T' \neq \{e\}$ by 11.5. Thus $\dim T' = 1$.

We just observed that H' is not solvable, so 13.9(1) implies that card $W' \geqq 2$. On the other hand $W' = \mathscr{N}_{H'}(T')/\mathscr{Z}_{H'}(T')$ acts faithfully on T', and, since $T' \cong \mathbf{GL}_1$, it follows from 8.3 and 8.4 that $\mathrm{Aut}(T') \cong \mathbf{GL}_1(\mathbb{Z})$ has order 2. Thus card $W' \leqq 2$.

(2) Consider the commutative square

$$\begin{array}{ccc} G & \xrightarrow{\pi} & G/B \\ \cup & & \uparrow j \\ G^S & \xrightarrow[\pi^S]{} & G^S/B^S, \end{array}$$

where π and π^S are the quotient morphisms, and $j(\pi^S(g)) = g\cdot\pi(e)$. Then j is injective. If $\lambda\in WC(B)$, then $\pi(e) = x(\lambda)$, so $\lambda(\infty)$ projects an open set onto $\pi(e)$. Hence, with respect to the corresponding action in G^S/B^S, $\lambda(\infty)$ projects an open set onto $\pi^S(e)$, i.e. B^S is associated to λ in G^S, as was to be shown.

(3) It follows from (1) and 13.10(2) that there are two Weyl chambers of T as a torus in G^S. According to 13.10(1), each one is of the form $\{\lambda\in X_*(T)_{\mathrm{sr}} | \langle\beta_i,\lambda\rangle > 0$ for each $i\}$. Here $X_*(T)_{\mathrm{sr}}$ refers to the set of λ which are semi-regular in G^S, and the β_i are non-trivial characters of T/S. Since $\dim T/S = 1$ we have $X(T/S)\cong\mathbb{Z}$. Since there are two Weyl chambers, each non-empty, they must be of the form

$$\{\lambda\in X_*(T)_{\mathrm{sr}} | \langle\alpha,\lambda\rangle > 0\}$$

where α varies over the two generators of $X(T/S)$. In particular $WC(C)$ has this form for some α, so that (3) follows immediately from (2).

13.12 Corollary. *Let Q be a singular subtorus of codimension 1 in T, distinct from S. Then there is a $B'\in\mathscr{B}^T$ such that*

$$B'^S = B^S \quad \textit{and} \quad B'^Q \neq B^Q.$$

Proof. Using part (3) above, we can find non-trivial $\alpha\in X(T/S)$ and $\beta\in X(T/Q)$ such that, for $B'\in\mathscr{B}^T$ and $\lambda'\in WC(B')$, we have

$$B'^S = B^S \Leftrightarrow \langle\alpha,\lambda'\rangle > 0$$
$$B'^Q = B^Q \Leftrightarrow \langle\beta,\lambda'\rangle > 0.$$

Since $S\neq Q$ it follows that α and β are linearly independent in $X(T)$ (their kernels have distinct connected components) so we can find a $\lambda'\in X_*(T)_{\mathrm{sr}}$ such that $\langle\alpha,\lambda'\rangle > 0$ and $\langle\beta,\lambda'\rangle < 0$. Then $B' = B(\lambda')$ solves our problem.

13.13 *Groups of semi-simple rank* 1, *and* $\mathbf{PGL}_2$. In the following T denotes a maximal torus in G, and $W = W(T, G)$. The *semi-simple rank* of G is defined to be $\dim(T/(T\cap\mathscr{R}(G)))$, i.e. the dimension of a maximal torus in $G/\mathscr{R}(G)$. The conjugacy of maximal tori shows that this depends only on G. For example $\mathbf{PGL}_2$ is not solvable and it has a maximal torus of dimension 1 (see 10.8); hence its semi-simple rank is 1.

Proposition. *The following conditions are equivalent*:

(1) *G has semi-simple rank* 1.

(2) card $W = 2$.
(3) $\dim G/B = 1$ (*where* $B \in \mathscr{B}$).
(4) *G/B is isomorphic to* $\mathbf{P}_1$.
(5) *There is a surjective morphism* $\varphi: G \to \mathbf{PGL}_2$ *such that* $\ker \varphi = \bigcap_{B' \in \mathscr{B}} B'$ (*and so* $(\ker \varphi)^o = \mathscr{R}(G)$).

Proof. The implication (1)$\Rightarrow$(2) is contained in the proof of part (1) of 13.11.

(2)$\Rightarrow$(3). Since card $W = \text{card}\, \mathscr{B}^T > 1$, G is not solvable, so $\dim G/B \geq 1$. Since card $W < 3$ it follows from 13.9(2) that $\dim G/B < 2$.

(3)$\Rightarrow$(4). Let $\lambda \in X_*(T)$ be a regular one-parameter subgroup. Then $\mathbf{GL}_1$ (via λ) does not act trivially on G/B because T does not. Since G/B is irreducible of dimension 1 it follows that, for $x \in G/B$ and not fixed by T, the orbit map $\mathbf{GL}_1 \to G/B$, $t \mapsto \lambda(t)x$, is dominant. Therefore we obtain an inclusion of function fields $K(G/B) \subset K(\mathbf{GL}_1)$. Since $K(\mathbf{GL}_1)$ is a pure function field in one variable it follows from Lüroth's theorem that $K(G/B)$ is likewise. Since G/B is a complete non-singular curve, it must therefore be isomorphic to $\mathbf{P}_1$.

(4)$\Rightarrow$(5). Since $G/B \cong \mathbf{P}_1$ it follows from 10.8 that the action of G on G/B is given by a morphism $\varphi: G \to \mathbf{PGL}_2$ ($= \text{Aut}(\mathbf{P}_1)$). Clearly the kernel is $\bigcap B' (B' \in \mathscr{B})$, so $(\ker \varphi)^o = \mathscr{R}(G)$. Since G is not solvable (because $B \neq G$) it follows from 11.6 that $\dim \varphi(G) > 2$. However $\mathbf{PGL}_2$ is connected and has dimension 3. Hence φ is surjective.

(5)$\Rightarrow$(1). The existence of φ clearly implies that the semi-simple rank of G coincides with that of $\mathbf{PGL}_2$, which is 1.

Corollary. *Suppose G has semi-simple rank* 1, *and let B_0, B_1, B_∞ be distinct Borel subgroups of G. Put $I = \bigcap B$ $(B \in \mathscr{B})$. Then $(B_0 \cap B_1)/I \cong \mathbf{GL}_1$ and $B_0 \cap B_1 \cap B_\infty = I$.*

Proof. We have a surjection $\varphi: G \to \mathbf{PGL}_2$, with kernel I, such that the B_i are stability groups of distinct points of $\mathbf{P}_1$. (With the aid of an automorphism of $\mathbf{P}_1$ we can even assume B_i is the stability group of i $(i = 0, 1, \infty)$ (see 10.8).) The assertions of the corollary follow from the fact that $\mathbf{PGL}_2$ acts simply transitively on triples of distinct points in $\mathbf{P}_1$, and that the subgroup fixing a pair of points is isomorphic to $\mathbf{GL}_1$ (see 10.8).

13.14 *Reductive groups of semi-simple rank* 1. In the following proposition we assume G to be a reductive group of semi-simple rank 1, and T a maximal torus in G. We put

$$I = \bigcap_{B \in \mathscr{B}} B \quad \text{and} \quad \mathscr{B}^T = \{B, B'\}.$$

The Lie algebras are denoted:

$$\mathfrak{g} = L(G), \quad \mathfrak{b} = L(B), \quad \mathfrak{b}' = L(B').$$

Proposition.

(1) *$I(T) = B \cap B' = T$, and $I = \mathscr{C}(G)$.*

(2) *$B_u \cong \mathbf{G}_a$ and the action of T on B_u is given by a generator, α, of $X(T/(T \cap I))$. $\Phi(T, B) = \{\alpha\}$, and $\mathfrak{b} = L(T) \oplus \mathfrak{g}_\alpha$. Moreover B_u is the unique T-invariant connected subgroup of G such that $L(B_u) = \mathfrak{g}_\alpha$. Similar conclusions apply to B' with $-\alpha$ in place of α.*

(3) *$\mathfrak{b} \cap \mathfrak{b}' = L(T)$.*

$$\mathfrak{b} + \mathfrak{b}' = \mathfrak{g} = L(T) \oplus \mathfrak{g}_\alpha \oplus \mathfrak{g}_{-\alpha}.$$

$$\Phi(T, G) = \{\alpha, -\alpha\}.$$

(4) $WC(B) = \{\lambda \in X_*(T) | \langle \alpha, \lambda \rangle > 0\}$

$$WC(B') = \{\lambda \in X_*(T) | \langle \alpha, \lambda \rangle < 0\}.$$

Proof. From 13.13, we have a surjective morphism $\varphi: G \to \mathbf{PGL}_2$ with kernel I. Since $\mathscr{R}_u(G) = \{e\}$ it follows that $\varphi: B_u \to \varphi(B_u) \cong \mathbf{G}_a$ has finite kernel. Thus B_u is connected, unipotent, and one dimensional, so 10.9 implies there is an isomorphism $\theta: \mathbf{G}_a \to B_u$. Similarly we have an isomorphism $\theta': \mathbf{G}_a \to B'_u$. If $t \in T$ and $b \in \mathbf{G}_a$ we have

$$t\theta(b)t^{-1} = \theta(t^\alpha b)$$

for some $\alpha \in X(T)$ (see 10.10). Passing to $\mathbf{PGL}_2$, we see that α generates $X(\varphi(T)) = X(T/(T \cap I))$ and that the action of T on B'_u is given by $-\alpha$ (see 10.8).

Now $B_u \cap B'_u$ corresponds to a proper subgroup of $\mathbf{G}_a$ stable under the non-trivial linear action of T given by α, so $B_u \cap B'_u = \{e\}$. Since $B = T \cdot B_u$ it follows that $B \cap B' = T \cdot (B_u \cap B') = T \cdot (B_u \cap B'_u) = T$. Since $T \subset I(T) = (B \cap B')^o$ this proves the first part of (1). It further implies that $I \subset T$, so I is a normal and diagonalizable subgroup of G. By rigidity it follows that $I \subset \mathscr{Z}(G)$. The reverse inclusion follows from 11.11, thus proving (1).

We have $B = T \cdot B_u$ so that $\mathfrak{b} = L(T) \oplus L(B_u)$, and the remarks above show that $L(B_u)$ is a one dimensional subspace of $\mathfrak{g}_\alpha$. Similarly $\mathfrak{b}' = L(T) \oplus L(B'_u)$ with $L(B'_u)$ a one dimensional subspace of $\mathfrak{g}_{-\alpha}$. Hence $\mathfrak{b} \cap \mathfrak{b}' = L(T)$, and $\mathfrak{b} + \mathfrak{b}' = \mathfrak{g}$ by dimension count, for $\dim(\mathfrak{b} + \mathfrak{b}') = 2 + \dim T$, $\dim G = \dim I + \dim \mathbf{PGL}_2 = \dim I + 3$, and $\dim T = \dim I + \dim \varphi(T) = \dim I + 1$.

This proves all of (2) and (3) except for the assertion: If H is a connected T-invariant subgroup such that $L(H) = \mathfrak{g}_\alpha$, then $H = B_u$.

Since $\dim H = 1$ it follows that H is either a torus or unipotent. If H were a torus T would have to centralize H, by rigidity. But T acts non-trivially on $L(H)$ so we must have $H = H_u$. Next note that $T \cdot H$ is a connected solvable subgroup containing T, hence contained in B or B'. Since $\Phi(T, B') = \{-\alpha\}$ and $\Phi(T, T \cdot H) = \{\alpha\}$ we must have $T \cdot H \subset B$, and hence $H \subset B_u$. Dimension count now implies that $H = B_u$.

Finally we prove (4). Let $\pi: G \to G/B$ be the quotient morphism and put $x_0 = \pi(e)$. Since B is the stability group of x_0 and $B'_u \cap B = \{e\}$, it follows that the points $\theta'(c)x_0(c\in\mathbf{G}_a)$ cover a neighborhood of x_0 in $G/B\cong\mathbf{P}_1$. Suppose $\lambda\in X_*(T)$ is such that $m=\langle\alpha,\lambda\rangle>0$. For $c\in\mathbf{G}_a$ and $t\in\mathbf{GL}_1$ we have

$$\begin{aligned}\lambda(t)\theta'(c)x_0&=\lambda(t)\theta'(c)\lambda(t)^{-1}x_0 \quad (T \text{ fixes } x_0)\\ &=\theta'(t^{\langle-\alpha,\lambda\rangle}c)x_0=\theta'(t^{-m}c)x_0.\end{aligned}$$

Specializing t^{-1} to 0 (i.e. t to ∞) we obtain

$$\lambda(\infty)\theta'(c)x_0=\theta'(0)x_0=x_0.$$

If follows that λ is semi-regular and that $x(\lambda)=x_0$. This proves:

$$\langle\alpha,\lambda\rangle>0\Rightarrow\begin{cases}\lambda \text{ is semi-regular}\\ \text{and } x(\lambda)=x_0 \ (\text{i.e. } B(\lambda)=B).\end{cases}$$

According to 13.11(3) the condition, "$\langle\alpha,\lambda\rangle>0$," defines a Weyl chamber in $X_*(T)_{sr}$, so we must have

$$WC(B)=\{\lambda\in X_*(T)|\langle\alpha,\lambda\rangle>0\}.$$

The analogue for B' follows similarly.

13.15 From now on we return to the general setting, i.e. G is no longer assumed to be of semi-simple rank 1, unless otherwise stated.

Lemma. *Let S and Q be distinct singular tori of codimension 1 in the maximal torus T, and let $B\in\mathscr{B}^T$.*

(1) $\dim(B_u^S/(B_u^S\cap I(T)_u))\leqq 1$.
(2) $I(B^S)^Q\subset I(T)$.

Proof. (1) Thanks to 13.11(1) and 13.13, we have a surjective morphism $\varphi: G^S\to\mathbf{PGL}_2$ such that $(\ker\varphi)^o=\mathscr{R}(G^S)$ and $\varphi(B_u^S)\cong\mathbf{G}_a$. It follows that $\dim(B_u^S/(B_u^S\cap\mathscr{R}_u(G^S)))=1$. Now (1) follows because $\mathscr{R}(G^S)\subset I(S)\subset I(T)$ (see 11.18).

(2) Choose B' as in 13.12. Then since $B^Q\neq B'^Q$ it follows from the Corollary of 13.13 that $B^Q\cap B'^Q=T\cdot\mathscr{R}_u(G^Q)$, and 11.18 again implies this lies in $I(T)$. Since $B'^S=B^S$ we have $I(B^S)\subset B\cap B'$, and hence $I(B^S)^Q\subset B^Q\cap B'^Q\subset I(T)$.

13.16 Theorem. *Let T be a maximal torus in G. Then*

$$I(T)_u=\mathscr{R}_u(G).$$

Proof. Clearly $\mathscr{R}_u(G)\subset I(T)_u$, and the latter is connected and unipotent. Hence we need only show that $I(T)_u$ is a normal subgroup of G.

According to 13.7, G is generated by the $B\in\mathscr{B}^T$. Combined with 9.5(2), this shows that G is generated by groups B^S, for $B\in\mathscr{B}^T$, and variable subtori S of codimension 1 in T. Since $B^S=T\cdot B_u^S$, it suffices to show that B_u^S normalizes $I(T)_u$. Note first:

(*) If S is semi-regular, then, by 13.1, we have $B^S \subset G^S \subset I(S) \subset I(T)$, so $B_u^S \subset I(T)_u$.

Now assume S is singular. Then B_u^S is contained in

$$H = (I(B^S) \cap B_u)^0,$$

so it suffices to show that H normalizes $I(T)_u$. Note that $I(T)_u \subset H$ and that H is a connected unipotent group, which is normalized by T (because B^S and B_u are). Hence it will follow from 12.1 Lemma that H normalizes $I(T)_u$ if we show that $\dim H \leqq \dim I(T)_u + 1$.

As above, we know from 9.5(2) that H is generated by the H^Q where Q varies over subtori of codimension 1 in T. If $Q = S$ then evidently $H^S = B_u^S$, and it follows from 13.15(1) that $\dim(H^S/(H^S \cap I(T)_u)) \leqq 1$. If $Q \neq S$ and Q is semi-regular then we have $H^Q \subset B_u^Q \subset I(T)_u$, as pointed out in (*) above. Finally suppose Q is singular and $\neq S$. Then $H^Q \subset I(B^S)^Q$, clearly, and the latter lies in $I(T)$, by 13.15(2). Hence $H^Q \subset I(T)_u$ for all $Q \neq S$, so the natural morphism

$$H^S/(H^S \cap I(T)_u) \to H/I(T)_u$$

is surjective. Since we observed already that the left side has dimension $\leqq 1$, the proof is now complete.

13.17 In the following important corollaries T denotes a maximal torus in G and S a subtorus of G.

Corollary 1.

(a) $\mathscr{R}_u(G^S) = \mathscr{R}_u(G)^S$.
(b) *If S is semi-regular then* $(G^S)_u = \mathscr{R}_u(G)^S$.
(c) $G^T = T \cdot \mathscr{R}_u(G)^T$.

Proof. (c) is a special case of (b), and (b) follows from (a) because $\mathscr{R}_u(H) = H_u$ when H is a connected solvable group (see 10.6).

To prove (a) we may assume $S \subset T$ (11.3). The group $\mathscr{R}_u(G)^S$ is a connected unipotent normal subgroup of G^S, hence contained in $\mathscr{R}_u(G^S)$. On the other hand, $\mathscr{R}_u(G^S)$ lies in every Borel subgroup of G^S, among which are all B^S, $B \in \mathscr{B}^T$. In particular $\mathscr{R}_u(G^S) \subset I(T) = T \cdot \mathscr{R}_u(G)$, by (13.16) so $\mathscr{R}_u(G^S) \subset \mathscr{R}_u(G)^S$.

Corollary 2. *Suppose G is reductive, $S \subset T$.*

(a) *G^S is reductive.*
(b) *If S is semi-regular, then $G^S = T$. In particular, S is regular.*
(c) *$G^T = T$. The Cartan subgroups of G coincide with the maximal tori.*
(d) *The intersection, Z, of all maximal tori is $\mathscr{C}(G)$.*

Parts (a), (b), and (c) follow immediately from the corresponding parts of Corollary 1. Since Z is a normal diagonalizable subgroup of G, it is central by rigidity. Conversely part (c) implies $\mathscr{C}(G)$ lies in every maximal torus. This proves (d).

13.18 *Roots in reductive groups.* T still denotes a maximal torus in G. An automorphism of $X(T)$ will be called a *reflection* if it has order 2 and induces the identity on a subgroup of corank 1.

Recall from 13.2 that $\Psi = \Phi(T, G/I(T))$; we also write Φ for $\Phi(T, G)$.

The next theorem summarizes much of the information we have accumulated about reductive groups.

Theorem. *Suppose G is reductive.*

(1) $\Psi = \Phi$, $L(T) = \mathfrak{g}^T$, *and* $\mathfrak{g} = \mathfrak{g}^T \oplus \coprod_{\alpha \in \Phi} \mathfrak{g}_\alpha$.

(2) *The singular tori of codimension* 1 *in* T *are the* $T_\alpha = (\ker \alpha)^o$ $(\alpha \in \Phi)$, *and* $\left(\bigcap_{\alpha \in \Phi} T_\alpha\right)^o = \mathscr{C}(G)^o$.

(3) Φ *generates a subgroup of finite index in* $X(T/\mathscr{C}(G)^o) \subset X(T)$. *If* α *and* β *in* Φ *are linearly dependent, then* $\beta = \pm\alpha$.

(4) *Let* $\alpha \in \Phi$, *and put* $G_\alpha = \mathscr{Z}_G(T_\alpha)$. *Then* G_α *is a reductive group of semisimple rank* 1, *and*:

- (a) $-\alpha \in \Phi$, *and* $L(G_\alpha) = \mathfrak{g}^T \oplus \mathfrak{g}_\alpha \oplus \mathfrak{g}_{-\alpha}$.
- (b) $\dim \mathfrak{g}_\alpha = 1$.
- (c) *The subgroup* $W(T, G_\alpha)$ *of* $W(T, G)$ *is generated by a reflection,* r_α, *such that* $r_\alpha(\alpha) = -\alpha$.
- (d) *There is a unique connected T-stable subgroup* U_α *of* G *such that* $L(U_\alpha) = \mathfrak{g}_\alpha$. *It is the unipotent part of a Borel subgroup of* G_α *containing* T.

(5) *Let* $B \in \mathscr{B}^T$.

- (a) *For each* $\alpha \in \Phi$, $\Phi(T, B^{T_\alpha}) = \Phi(B) \cap \{\alpha, -\alpha\}$ *has precisely one element. Hence* Φ *is the disjoint union of* $\Phi(B)$ *and* $-\Phi(B)$.
- (b) $WC(B) = \{\lambda \in X_*(T) \mid \langle \alpha, \lambda \rangle > 0$ *for all* $\alpha \in \Phi(B)\}$.
- (c) *If* $\lambda \in WC(B)$ *then* $\Phi(B) = \{\alpha \in \Phi \mid \langle \alpha, \lambda \rangle > 0\}$.
- (d) *One can give* $X(T)$ *the structure of a totally ordered abelian group so that* $\Phi(B)$ *is the set of positive elements in* Φ.
- (e) $L(B) = \mathfrak{g}^T \oplus \coprod_{\alpha \in \Phi(B)} \mathfrak{g}_\alpha$.

Proof. (1) It follows from 13.16 and from the assumption G reductive, that $I(T) = T$. Clearly $\Phi(T, G/T) = \Phi$. One always has

$$\mathfrak{g} = \mathfrak{g}^T \oplus \coprod_{\alpha \in \Phi} \mathfrak{g}_\alpha,$$

(see 8.16). By 9.4, $\mathfrak{g}^T = L(G^T)$, and Corollary 2 of 13.17 tells us that $G^T = T$.

(2) The first assertion follows from 13.2(1) since $\Psi = \Phi$. We know from

13.17, Corollary 2, that $\mathscr{C}(G)^o$ is a subtorus of T. If S is a subtorus of T, then we know, by 9.4, that $G^S = G \Leftrightarrow \mathfrak{g}^S = \mathfrak{g} \Leftrightarrow S \subset T_\alpha$ for each $\alpha \in \Phi$. Thus $\mathscr{C}(G)^o = (\bigcap T_\alpha)^o$.

(3) The equality just proved is equivalent to the condition that Φ generate a subgroup of finite index in $X(T/\mathscr{C}(G)^o)$, clearly. If $\alpha, \beta \in X(T)$, then it is clear that α and β are linearly dependent, i.e. $n\alpha = m\beta$ for some $n, m \in \mathbb{Z}$, not both zero, if and only if $T_\alpha = (\ker \alpha)^o$ and $T_\beta = (\ker \beta)^o$ coincide. In this case we have $\beta \in \Phi(G_\alpha)$ so the last assertion of (3) will follow from (4)(a), which implies that $\Phi(T, G_\alpha) = \{\alpha, -\alpha\}$.

(4) We have $L(G_\alpha) = \mathfrak{g}^T \oplus \coprod_{T_\alpha \subset T_\beta} \mathfrak{g}_\beta$. From 13.11 we know that G_α has semi-simple rank 1, and from 13.17 that it is reductive; hence we can apply 13.14. Since $\alpha \in \Phi(G_\alpha)$, it follows from 13.14 that $\Phi(G_\alpha) = \{\alpha, -\alpha\}$ and that $\dim \mathfrak{g}_\alpha = 1$. In particular $\mathfrak{g}^{T_\alpha} = \mathfrak{g}^T \oplus \mathfrak{g}_\alpha \oplus \mathfrak{g}_{-\alpha}$ so that $T_\alpha \subset T_\beta$ (for $\beta \in \Phi) \Leftrightarrow \beta = \pm\alpha$. This proves (a) and (b).

As already remarked, $W(T, G_\alpha)$ has order 2, say with generator r_α represented by $n \in \mathscr{N}_{G_\alpha}(T)$. The automorphism of T induced by n has order 2 and fixes pointwise the subtorus T_α of codimension 1. Hence the set of commutators (n, T) is a subtorus of dimension 1 (being a non-trivial image of $T/T_\alpha \cong \mathbf{GL}_1$). If $\beta \in X(T)$, then $\beta(ntn^{-1}) = \beta(t)$ for all $t \in T \Leftrightarrow \beta((n, T)) = \{1\} \Leftrightarrow (n, T) \subset \ker \beta$. The set of such β is a subgroup of corank 1 in $X(T)$, not containing α. Since $\Phi(G_\alpha) = \{\alpha, -\alpha\}$ is stable under r_α, we must have $r_\alpha(\alpha) = -\alpha$, for otherwise r_α would fix a subgroup of finite index in $X(T)$, and hence be the identity. This proves (c).

Finally, to prove (d), let H be a connected T-stable subgroup of G such that $L(H) = \mathfrak{g}_\alpha$. Since $\dim H = 1, H$ is either a torus or unipotent. If it were a torus it would be centralized by T, by rigidity of tori, contradicting the non-triviality of the action of T on $L(H)$. Hence, by 10.9, there is an isomorphism $\theta: \mathbf{G}_a \to H$, and we have $t\theta(b)t^{-1} = \theta(t^\alpha b)$ for $t \in T, b \in \mathbf{G}_a$. It follows that $H \subset G_\alpha$. At this point the uniqueness, as well as the existence, of U_α follow from the corresponding assertion in G_α (see 13.14).

(5) (a) follows from 13.14 because $B^{T_\alpha} \in \mathscr{B}(G_\alpha)^T$ and $\Phi(G_\alpha) = \{\alpha, -\alpha\}$, as noted above.

Moreover, it follows from 13.11 that

$$WC(B) \subset WC(B^{T_\alpha}) = \{\lambda \in X_*(T) | \langle \alpha, \lambda \rangle > 0\}$$

for each $\alpha \in \Phi(B)$. To prove (b) we claim, conversely, that any $\lambda \in \bigcap_{\alpha \in \Phi(B)} WC(B^{T_\alpha})$ lies in $WC(B)$.

If λ were not semi-regular then, by (2), we would have $S = \operatorname{im}(\lambda) \subset T_\alpha$ for some α, so that $G_\alpha \subset G^S$, contradicting the fact that λ is semi-regular in G_α. Thus $B' = B(\lambda)$ is defined, and we must show that $B' = B$. The hypothesis implies that $B'^{T_\alpha} = B^{T_\alpha}$ for all $\alpha \in \Phi$. But 9.5(2) says B is generated by these B^{T_α}, and similarly for B'; hence $B = B'$ as claimed.

Part (c) is clearly a consequence of parts (a) and (b).

To prove (d), let $\lambda_1, \ldots, \lambda_r$ be a basis for $X_*(T)$ such that $B = B(\lambda_1)$. For a non-zero character α write $\alpha > 0$ if the first non-zero term among the $\langle \alpha, \lambda_i \rangle (1 \leqq i \leqq r)$ is positive. This defines a total ordering of $X(T)$. If $\alpha \in \Phi(B)$ then $\langle \alpha, \lambda_1 \rangle > 0$ so $\alpha > 0$. If $\alpha \in \Phi, \alpha \notin \Phi(B)$ then, by part (a), $-\alpha \in \Phi(B)$, so $\alpha < 0$.

(e) By definition, if $\mathfrak{b} = L(B)$, we have $\mathfrak{b} = \mathfrak{b}^T \oplus \coprod_{\alpha \in \Phi(B)} \mathfrak{b}_\alpha$. But clearly $\mathfrak{b}^T = L(T) = \mathfrak{g}^T$, and, if $\alpha \in \Phi(B)$, $\mathfrak{b}_\alpha = \mathfrak{g}_\alpha$ because $\dim \mathfrak{g}_\alpha = 1$ (part (4)(b)).

13.19 Proposition. *Let G be connected, reductive, $X \in \mathfrak{g}$ and $A \in G$ semi-simple elements. Then $\mathscr{Z}_G(X)^o$ and $\mathscr{Z}_G(A)^o$ are reductive.*

Proof. We keep the notation of 13.18, and let $H = \mathscr{Z}_G(X)^o$. By 11.8, we may assume $X \in \mathfrak{t}$. We have

$$\mathfrak{z}(X) = \mathfrak{t} + \coprod_{\alpha \in \Psi} \mathfrak{g}_\alpha \quad (\Psi = \{\alpha \in \Phi, d\alpha(X) = 0\}).$$

By 9.1, $\mathfrak{z}(X) = L(H)$. In particular, if $\alpha \in \Psi$, then $G_\alpha \subset H$. We have to show that $\mathscr{R}_u(H)$ is reduced to $\{e\}$. By 9.5, $\mathscr{R}_u(H)$ is generated by centralizers of singular tori in T, i.e. by its intersections with some G_α, i.e. finally by some U_α, with α necessarily in Ψ. But if $U_\alpha \subset \mathscr{R}_u(H)$, then U_α belongs to the unipotent radical of G_α. The latter group being reductive (13.17, Cor. 2) this is a contradiction. Same proof for A, with $A \in T$ and $\Psi = \{\alpha \mid A^\alpha = 1\}$.

13.20 Proposition. *Let G be connected reductive, T a maximal torus of G and H a closed connected subgroup normalized by T. Then:*

$$(1) \qquad L(H) = L(T \cap H) \oplus \bigoplus_{\alpha \in \Phi(T,H)} \mathfrak{g}_\alpha, \quad H = \langle (T \cap H)^o, U_\alpha \mid \alpha \in \Phi(T,H) \rangle.$$

Proof. By full reducibility of the representation of T in $\mathfrak{g}$, we have

$$(2) \qquad L(H) = L(H) \cap L(T) \oplus \bigoplus_\alpha (L(H) \cap \mathfrak{g}_\alpha).$$

But $\mathfrak{g}_\alpha$ is one-dimensional. Therefore either $\mathfrak{g}_\alpha \subset L(H)$ and $\alpha \in \Phi(T;H)$ or $\mathfrak{g}_\alpha \cap L(H) = \{0\}$ and $\alpha \notin \Phi(T,H)$. Since T is its own centralizer in G, we also have $L(H) \cap L(T) = L(T \cap H)$ (9.2, Cor.). This proves the first part of (1). If $U_\alpha \subset H$, then $\mathfrak{g}_\alpha \subset L(H)$ obviously. The main point is to prove the converse, namely:

$$(3) \qquad \mathfrak{g}_\alpha \subset L(H) \Rightarrow U_\alpha \subset H, \quad (\alpha \in \Phi(T,H)).$$

We claim that it suffices to prove (3) when $G = \mathscr{Z}_G(T_\alpha)$ and H is replaced by $(H \cap \mathscr{Z}_G(T_\alpha))^o$. Assume this has been done. Recall that $\mathscr{Z}_G(T_\alpha) = \langle T, U_\alpha, U_{-\alpha} \rangle$ (13.18). Then in particular $\mathfrak{g}_\alpha \subset L(\mathscr{Z}_G(T_\alpha))$. Then $\mathfrak{g}_\alpha \subset L(H)$ implies $\mathfrak{g}_\alpha \subset L(H) \cap L(\mathscr{Z}_G(T_\alpha))$. But the latter is equal to $L(H \cap \mathscr{Z}_G(T_\alpha))$ by 9.2, Cor. Therefore $U_\alpha \subset (H \cap \mathscr{Z}_G(T_\alpha))^o \subset H$.

So let now $G = \mathscr{Z}_G(T_\alpha)$. The group H is not a torus, since $L(H)$ contains $\mathfrak{g}_\alpha$. Assume it is solvable. Then $T \cdot U_\alpha = T \cdot H$, since $T \cdot U_\alpha$ and $T \cdot U_{-\alpha}$ are the only Borel subgroups of $\mathscr{Z}_G(T_\alpha)$ which are invariant under T. Hence $U_\alpha \subset H$. Assume now H is not solvable, then it has a non-trivial semisimple quotient and contains at least two distinct Borel subgroups. Consequently, $T \cdot H$ contain at least two Borel subgroups. This forces U_α, $U_{-\alpha} \subset H$.

Let M be the subgroup generated by $(T \cap H)^o$ and the U_α $(\alpha \in \Phi(T, H))$. It is contained in H and its Lie algebra contains $L(H)$, hence M and H have the same dimension and $M = H$.

13.21 Corollary. *Let H' be a closed connected subgroup normalized by T. Then $L(H \cap H') = L(H) \cap L(H')$.*

This follows from 13.20(1), applied to H, H' and $(H \cap H')^o$.

Bibliographical Note

The results of this section up to 13.18 are almost all due to Chevalley (see [13], in particular Exp. 10, 11, 12). Instead of 13.5, it is proved there that if $\dim Y \geqq 1$, then T has at least two fixed points. A few years after I had noticed the easy generalization of Chevalley's lemma provided by 13.5, I saw that in fact, over $\mathbf{C}$, it had already been established in 1896 by Guido Fano [16]. 13.20, 13.21 will be made more precise with regard to fields of definition in §20. These results are borrowed from [4].

§14. Root Systems and the Bruhat Decomposition in Reductive Groups

In this section the connected affine group G is assumed to be *reductive*. T denotes a maximal torus in G, and we shall write Φ for $\Phi(T, G)$. For each $\alpha \in \Phi$, we put $T_\alpha = (\ker \alpha)^o$ and $G_\alpha = \mathscr{Z}_G(T_\alpha)$.

The Weyl group $W = W(T, G)$ operates on $X(T)$ and leaves Φ stable. We propose to show that Φ is a reduced root system in a suitable subspace of $X(T)_{\mathbf{Q}} = X(T) \bigotimes_{\mathbf{Z}} \mathbf{Q}$, with Weyl group W (14.8). In view of §13, what remains to be shown is mainly the integrality condition $r_\alpha(\beta) - \beta \in \mathbf{Z} \cdot \alpha$, to be proved in 14.6, after some preliminary work in 14.3 to 14.5.

Let $B \in \mathscr{B}^T$, let $\pi: G \to G/B$ be the quotient morphism and put $o = \pi(e)$. The Bruhat decomposition (14.11) refers to the following: Let $U = B_u$. Then $w \mapsto Uw(o)$ is a bijection from W to the set of U-orbits in G/B. Moreover, for $w \in W$, $Uw(o)$ is isomorphic to an affine space (a cell) and $w(o)$ is the unique fixed point of T in $Uw(o)$.

14.1 Theorem. *Let $\lambda \in WC(B)$. For B, $B' \in \mathscr{B}^T$, the following conditions are equivalent:*

(I) $B \cap B' = T$; (i) $\mathfrak{b} \cap \mathfrak{b}' = L(T)$ $(\mathfrak{b} = L(B), \mathfrak{b}' = L(B'))$

(II) *The product morphism* $B \times B' \to G$ *is dominant and separable*; (ii) $\mathfrak{b} + \mathfrak{b}' = \mathfrak{g}$.

(III) $B' = B(-\lambda)$; (iii) $\Phi(B') = -\Phi(B)$.

In view of condition (III) we see that there exists a unique B' satisfying the above conditions. It is called the *opposite* Borel subgroup to B. The "big cell" associated with T and B is $B \cdot B'$, which contains a dense open set in G, by (II).

Proof. We know from 13.18 that:

$$\mathfrak{g} = \mathfrak{g}^T \oplus \coprod_{\alpha \in \Phi} \mathfrak{g}_\alpha, \quad \mathfrak{g}^T = L(T),$$

$$\mathfrak{b} = \mathfrak{g}^T \oplus \coprod_{\alpha \in \Phi(B)} \mathfrak{g}_\alpha,$$

$$\mathfrak{b}' = \mathfrak{g}^T \oplus \coprod_{\alpha \in \Phi(B')} \mathfrak{g}_\alpha,$$

and Φ is the disjoint union of $\Phi(B)$ and $-\Phi(B) = \Phi(B(-\lambda))$. From these facts the equivalence of (i), (ii), (iii), and (III) is clear.

The equivalence of (II) and (ii) is also clear (see AG. 17.3).

(I)$\Rightarrow$(iii). Suppose, on the contrary, that $B \cap B' = T$ but that there is an $\alpha \in \Phi(B) \cap \Phi(B')$. Then we must have $B^{T_\alpha} = B'^{T_\alpha} \subset B \cap B'$; contradiction.

(i)$\Rightarrow$(I). Since $T \subset B \cap B'$ and $L(B \cap B') \subset \mathfrak{b} \cap \mathfrak{b}'$ we see that (i) implies $T = (B \cap B')^0$. Since $B = T \cdot B_u$ we have $B \cap B' = T \cdot C$ where $C = B_u \cap B' = B_u \cap B'_u$. Now C is a finite group normalized, and hence centralized by (the connected group) T. Thus $C \subset G^T \cap B_u = T \cap B_u = \{e\}$ (see 13.17 Corollary 2).

Corollary 1. *Let* H *be a connected k-group,* T *a maximal torus of* H, *and* B *a Borel subgroup of* H *containing* T. *There exists one and only one Borel subgroup* B' *of* H *verifying the following three conditions, which are equivalent*:

$$B \cap B' = T \cdot R_u(H); \quad \mathfrak{b} \cap \mathfrak{b}' = \mathfrak{t} + L(R_u(H)); \quad \mathfrak{b} + \mathfrak{b}' = \mathfrak{h}.$$

Let $\pi: H \to H' = H/R_u(H)$ be the canonical projection. The group H' is reductive (11.21) and $\pi(B)$ (resp. $\pi(T)$) is a Borel subgroup (resp. a maximal torus) of H' (11.14). Moreover, the Borel subgroups of H all contain $R_u(H)$ and are the inverse images of the Borel subgroups of H'. This reduces the corollary to the theorem.

Corollary 2. *Let* H *be a connected k-group and* P *a parabolic subgroup of* H. *Then* P *is equal to the normalizer in* H *of* $L(P)$.

Let $Q = \mathcal{N}_H(L(P))$. Then Q contains P, hence is parabolic, and therefore connected (11.15). Let B be a Borel subgroup of H contained in P. If B' is a Borel subgroup of Q, then it is conjugate to B by an element of Q, hence $L(B') \subset L(P)$. On the other hand, by Cor. 1, there exists such a B' verifying $L(B') + L(B) = L(Q)$. Therefore $L(Q) = L(P)$ and $Q = P$.

Corollary 3. *Let B, B' and C, C' be two pairs of opposite Borel subgroups. Then there exists $g \in G$ such that ${}^gB = C$ and ${}^gB' = C'$.*

In view of the theorem, this follows from the fact that Int G is transitive on the pairs Q, T consisting of a Borel subgroup Q and a maximal torus contained in Q (11.19c)).

14.2 *The center and derived group.* Write $C = \mathscr{C}(G)^o$ for the "connected center" of G. Since G is reductive, C coincides with $\mathscr{R}(G)$ (see 11.21), so that G/C is semisimple.

Proposition.

(1) $C = \left(\bigcap_{\alpha \in \Phi} T_\alpha \right)^o = (T^W)^o$.

(2) $\mathscr{D}G$ *is semi-simple*

(3) $G = C \cdot \mathscr{D}G$ *and* $C \cap \mathscr{D}G$ *is finite.*

Proof. (1) That $C = \left(\bigcap_{\alpha \in \Phi} T_\alpha \right)^o$ follows from (13.18)(2), and clearly $C \subset T^W$. It remains to be shown that $(T^W)^o \subset T_\alpha$ for each $\alpha \in \Phi$. According to 13.18 (4)(c) there is a $w \in W$ such that $w(\alpha) = -\alpha$. Now if $t \in T^W$ we have ${}^wt = t$ so $t^\alpha = ({}^wt)^\alpha = t^{w(\alpha)} = t^{-\alpha}$. Thus $T^W \subset \ker(2\alpha)$, and $(\ker(2\alpha))^o = (\ker \alpha)^o = T_\alpha$.

(3) To prove that $G = C \cdot \mathscr{D}G$ we first recall (10.8) that $\mathbf{PGL}_2$ is its own derived group. Since G_α/T_α is isogenous to $\mathbf{PGL}_2$ it follows that $G_\alpha = T_\alpha \cdot \mathscr{D}G_\alpha$ for each $\alpha \in \Phi$. We know (cf. 9.4(4)) that the $G_\alpha (\alpha \in \Phi)$ generate G, so it remains to show that $T \subset C \cdot \mathscr{D}G$. Put $D = (\mathscr{N}_G(T), T)^o \subset (\mathscr{D}G \cap T)$. It will suffice to prove that $T = C \cdot D$, or that $T'_{\mathbb{Q}} = C'_{\mathbb{Q}} + D'_{\mathbb{Q}}$, where $T' = X_*(T)$, $C' = X_*(C)$, $D' = X_*(D)$. But $C' = T'^W$, by (1), and by construction all $(1 - w)\lambda$ $(w \in W, \lambda \in T')$ lie in D'. Now the subspace these span in $T'_{\mathbb{Q}}$ has complement $(T'_{\mathbb{Q}})^W = (T'^W)_{\mathbb{Q}}$, since the group algebra $\mathbb{Q}[W]$ is semi-simple. This proves that $G = C \cdot \mathscr{D}G$.

Once we show that $C \cdot \mathscr{D}G$ is finite the semi-simplicity of $\mathscr{D}G$ will follow because $\mathscr{D}G \to G/C = G/\mathscr{R}(G)$ is surjective with finite kernel. Thus the proof is completed by the:

Lemma. *Let C be a central torus in a connected group H. Then $C \cap \mathscr{D}H$ is finite.*

Proof. Using a faithful linear representation we may assume $H \subset GL(V)$. Write $V = \oplus V_i (1 \leqq i \leqq n)$, where $V_i = V_{\alpha_i}$, α_i ranging over the weights of C in V. Then $H \subset GL(V)^C = GL(V_1) \times \cdots \times GL(V_n)$. If $t \in C$ then $t = (t^{\alpha_1} Id., \ldots, t^{\alpha_n} Id.)$ in these coordinates. If further $t \in \mathscr{D}H$ then each $t^{\alpha_i} Id.$ has determinant 1, so $(t^{\alpha_i})^{m_i} = 1$, where $m_i = \dim V_i$. Thus $C \cap \mathscr{D}H$ lies in a group of the form $C_1 \times \cdots \times C_n$, where C_i is cyclic of order dividing m_i.

Corollary. (a) *The following three conditions are equivalent: G is semi-simple; $G = \mathscr{D}G$; and $\mathscr{C}(G)$ is finite.* (b) *Let H be a closed connected normal subgroup of G. Then H is reductive, $\mathscr{C}(H)^o = (\mathscr{C}(G) \cap H)^o$, and $\mathscr{D}H = (\mathscr{D}G \cap H)^o$,*

The first assertion is an obvious consequence of the proposition. Let H be as in the statement. Then $\mathscr{R}_u(H) \subset \mathscr{R}_u(G)$, hence $\mathscr{R}_u(H) = \{e\}$ and H is reductive. The group $\mathscr{C}(H)^o$ is a torus, normal in G, hence central (8.10, Cor.) and contained in $(\mathscr{C}(G)^o \cap H)^o$. The other inclusion is obvious. That $\mathscr{D}H \subset (\mathscr{D}G \cap H)^o$ is clear. That it is no smaller follows from $\mathscr{C}(H)^o \subset \mathscr{C}(G)$ and the proposition.

14.3 *Direct spanning.* Let $(H_i)_{i \in I}$ be a finite family of closed connected subgroups of a connected group H. We shall say that H is *directly spanned* by the H_i if, for some ordering $i_1, \ldots, i_n$ of I, the product morphism

$$H_{i_1} \times \cdots \times H_{i_n} \to H$$

is an isomorphism of varieties. We shall denote this circumstance by writing

$$H = H_{i_1} \cdot H_{i_2} \cdot \ldots \cdot H_{i_n}.$$

In case $n = 2$ and one of the groups normalizes the other we have, as a special case, just a semi-direct product decomposition.

14.4 *Certain actions of T on unipotent groups.* We consider an action of T on a connected unipotent group U, subject to the following assumptions, were $\Phi(U)$ stands for $\Phi(T, U)$:

(i) Each weight α of T in $\mathfrak{u} = L(U)$ is not zero, so that

$$\mathfrak{u} = \coprod_{\alpha \in \Phi(U)} \mathfrak{u}_\alpha,$$

and $\dim \mathfrak{u}_\alpha = 1$ for each $\alpha \in \Phi(U)$.

(ii) If $\alpha, \beta \in \Phi(U)$ are distinct, then they are linearly independent, i.e. the subtori $T_\alpha = (\ker \alpha)^o$ and $T_\beta = (\ker \beta)^o$ are distinct.

Proposition. (1) *If $\alpha \in \Phi(U)$ then $U_\alpha = U^{T_\alpha}$ is the unique T-stable closed subgroup of U with Lie algebra $\mathfrak{u}_\alpha$.*

(2) *Let Λ denote the set of T-stable closed subgroups of U.*

(a) *If $H \in \Lambda$ then H is connected and H is directly spanned by*

$$\{U_\alpha | \alpha \in \Phi(H)\} = \{U_\alpha | \mathfrak{u}_\alpha \subset \mathfrak{h}\},$$

in any order.

(b) *$H \mapsto \mathfrak{h}$ is a lattice monomorphism from Λ to the lattice of T-stable subalgebras of $\mathfrak{u}$.*

(c) *If H, ${}^xH \in \Lambda$ for some $x \in U$, then $H = {}^xH$.*

Proof. We know from 9.4 that for a subtorus S of T, the group U^S is connected, and $L(U^S) = \mathfrak{u}^S = \coprod \mathfrak{u}_\beta (S \subset T_\beta)$. Taking $S = T$ we see that $U^T = \{e\}$. Taking $S = T_\alpha$ for some $\alpha \in \Phi(U)$, we see that U_α is connected and, thanks to assumption (ii), that $L(U_\alpha) = \mathfrak{u}_\alpha$. The uniqueness of U_α follows from (2)(a), which we now prove.

Let $H \in \Lambda$. We assume first H to be *connected*. From 9.4 again we know that H is generated by the subgroups $H^{T_\alpha} (\alpha \in \Phi(U))$. Now $H^{T_\alpha} \subset U_\alpha$ and, if $\mathfrak{h} = L(H)$, we have $L(H^{T_\alpha}) = \mathfrak{h}^{T_\alpha} = \mathfrak{h}_\alpha \subset \mathfrak{u}_\alpha$. From assumption (i) we have $\dim U_\alpha = \dim \mathfrak{u}_\alpha = 1$. Thus either

$$H^{T_\alpha} = \{e\}, \quad \mathfrak{h}_\alpha = 0, \quad \text{and} \quad \alpha \notin \Phi(H)$$

or

$$H^{T_\alpha} = U_\alpha, \quad \mathfrak{h}_\alpha = \mathfrak{u}_\alpha, \quad \text{and} \quad \alpha \in \Phi(H).$$

Let $\alpha_1, \ldots, \alpha_n$ by some ordering of $\Phi(H)$ and let

$$f : P = U_{\alpha_1} \times \cdots \times U_{\alpha_n} \to H$$

be the product map. To prove (2)(a) we must show that f is an isomorphism of varieties. Clearly $(df)_e$ is an isomorphism, so f is dominant and separable.

There is no loss in generality in assuming that the ordering is chosen so that the U_{α_i} which lie in $\mathscr{C}(H)$ occur last; say $U_{\alpha_1}, \ldots, U_{\alpha_m}$ are those not contained in $\mathscr{C}(H)$. We distinguish two cases:

(i) $m = 0$; i.e. H is commutative. Then P is a group on which T acts subject to the analogues of the assumptions made on U, and f is a dominant T-equivariant homomorphism with finite kernel. But then, since T is connected, $\ker f \subset P^T$, and we saw above that the assumptions (i) and (ii) imply $P^T = \{e\}$. Thus f is an isomorphism.

(ii) General case. Let $\pi : H \to H/\mathscr{C}(H)^o$ be the quotient morphism. If $i \leq m$ then $U_{\alpha_i} \to \pi(U_{\alpha_i})$ is bijective, and we have $\pi(H) = \pi(U_{\alpha_1}) \cdot \ldots \cdot \pi(U_{\alpha_m})$, by induction on $\dim H$. Therefore $H = U_{\alpha_1} \cdot \ldots \cdot U_{\alpha_m} \cdot \mathscr{C}(H)^o$. By (i) $\mathscr{C}(H)^o = U_{\alpha_{m+1}} \cdot \ldots \cdot U_{\alpha_n}$.

In case $H \in \Lambda$ is not connected we apply the conclusion above to H^o and to U to write U in the form $U = H^o \cdot V$, where say $V = U_{\beta_1} \cdot \ldots \cdot U_{\beta_q}$. Then H is the set theoretic cartesian product of H^o and $F = H \cap V$. Since F is a finite T-stable subset of U we have $F \subset U^T = \{e\}$.

This completes the proof of (1) and of (2)(a). Part (2)(b) is an immediate consequence of (2)(a).

There remains the proof of (2)(c), so suppose $H, {}^xH \in \Lambda$ for some $x \in U$. We will show, by induction on $\dim U$, that $H = {}^xH$.

Choose a $U_\gamma \subset \mathscr{C}(U)$ and let $\pi : U \to U' = U/U_\gamma$ be the quotient morphism. By induction we have $\pi(H) = \pi({}^xH)$. If $U_\gamma \subset H$ then $U_\gamma = {}^xU_\gamma \subset {}^xH$ also, and we see that $H = {}^xH$. If not, then at least $M = H \cdot U_\gamma$ coincides with ${}^xH \cdot U_\gamma$. Thus $L(M) = L(H) \oplus \mathfrak{u}_\gamma = L({}^xH) \oplus \mathfrak{u}_\gamma$. Since $L(H)$ and $L({}^xH)$ are T-stable, and since the weights of T in $\mathfrak{u}$ have multiplicity 1, it follows that $L(H) = L({}^xH)$. Hence, by (2)(a) (or (2)(b)) we have $H = {}^xH$.

Remark. Each U_α above is isomorphic to $\mathbf{G}_a$. Thus the product map

$$f : U_{\alpha_1} \times \cdots \times U_{\alpha_n} \to H$$

in the proof above gives rise to a T-equivariant isomorphism of H (as a variety) with the affine space K^n on which T acts diagonally via $\alpha_1, \ldots, \alpha_n$. This also shows the existence of a T-equivariant isomorphism of varieties of $L(H)$ onto H. In characteristic zero, it is given by the exponential map.

14.5 *Special sets of roots.* Recall first that, if $\alpha \in \Phi$ then (see 13.18 (4)(d)) there is a unique connected T-stable subgroup U_α with Lie algebra $\mathfrak{g}_\alpha$.

If $\alpha, \beta \in \Phi$ we denote by (α, β) the set of roots $\gamma \in \Phi$ of the form $\gamma = r\alpha + s\beta$, where r, s are strictly positive integers. If Ψ and Ψ' are subsets of Φ write

$$(\Psi, \Psi') = \cup(\alpha, \beta) \quad (\alpha \in \Psi, \beta \in \Psi').$$

We shall call Ψ *special* if

(a) $(\Psi, \Psi) \subset \Psi$, and

(b) there is a $\lambda \in X_*(T)$ such that $\langle \alpha, \lambda \rangle > 0$ for all $\alpha \in \Psi$.

There is no loss, in (b), in assuming that λ is regular, i.e. that $\langle \alpha, \lambda \rangle \neq 0$ for all $\alpha \in \Phi$. For let λ' be any regular one-parameter subgroup. Since Φ is finite we can choose a large positive integer N so that, if $\lambda'' = N\lambda + \lambda'$, we have $\langle \alpha, \lambda'' \rangle > 0$ whenever $\alpha \in \Phi$ and $\langle \alpha, \lambda \rangle > 0$. Then λ'' is evidently regular, and serves as well as λ in (b).

The terminology "special" is provisional. Once it is established that Φ is a root system (14.8), then we shall see that $\Psi \subset \Phi$ is special if and only if it is closed and belongs to a positive set of roots for some ordering on Φ (see 14.7).

Proposition.

(1) *If $\alpha, \beta \in \Phi$, and if $\beta \neq \pm\alpha$, then*

$$[\alpha, \beta) = \{\gamma \in \Phi \mid \gamma = r\alpha + s\beta \text{ for } r, s \in \mathbb{Z} \text{ with } s > 0\}$$

is special.

Let $\Psi \subset \Phi$ be special.

(2) *The set $\{U_\alpha \mid \alpha \in \Psi\}$ directly spans, in any order, a T-stable subgroup U_Ψ of G.*

(3) *If $\alpha \in \Phi$ and $(\alpha, \Psi) \subset \Psi$ then U_α normalizes U_Ψ.*

Proof. Suppose $\alpha, \beta \in \Phi$ and $\beta \neq \pm\alpha$. Condition (a) above is obviously satisfied by $[\alpha, \beta)$. To establish (b) recall from 13.18 (3) that $\beta \neq \pm\alpha$ implies α and β to be linearly independent. Hence there is a $\lambda \in X_*(T)$ such that $\langle \alpha, \lambda \rangle = 0$ and $\langle \beta, \lambda \rangle > 0$. This λ is clearly positive on $[\alpha, \beta)$, so we have proved (1).

Note that $(\alpha, \beta) \subset [\alpha, \beta)$. Since condition (a) is obvious for (α, β) we see that (α, β) is also special. We claim the following.

(*) *Let $U_{(\alpha,\beta)}$ denote the product, in some order, of $\{U_\gamma \mid \gamma \in (\alpha, \beta)\}$. Then $(U_\alpha, U_\beta) \subset U_{(\alpha,\beta)}$.*

(In case $(\alpha, \beta) = \phi$ we take $U_{(\alpha,\beta)} = \{e\}$).

We shall now give the proof of (2) and (3), using (*), and then prove (*) at the end.

We are given a special $\Psi \subset \Phi$. The remark preceding the proposition shows that we can choose a regular $\lambda \in X_*(T)$ such that $\langle \alpha, \lambda \rangle > 0$ for all $\alpha \in \Psi$. Put $B = B(\lambda)$, $U = B_u$, and $\Phi^+ = \Phi(U) = \Phi(B)$. Then it follows from 13.18 that the action of T on U satisfies hypotheses (i) and (ii) of 14.4. If $\alpha \in \Phi^+$, moreover, the group U_α here coincides with the group so denoted (w.r.t. T and U) in 14.4. Suppose that the product, in some order, of $\{U_\alpha | \alpha \in \Psi\}$ is a subgroup, call it U_Ψ. Then U_Ψ is clearly a closed T-stable subgroup of U with Lie algebra $\sum_{\alpha \in \Psi} \mathfrak{g}_\alpha$. It follows therefore from 14.4 that $\Psi = \Phi(U_\Psi)$ and that U_Ψ is directly spanned, in any order, by $\{U_\alpha | \alpha \in \Psi\}$.

Now we shall prove (2) and (3) by induction on card Ψ. If card $\Psi = 0$ both assertions are clear, with $U_\Psi = \{e\}$. Otherwise write $\Psi = \{\beta\} \cup \Psi'$ where $\beta \notin \Psi'$ and $\langle \beta, \lambda \rangle \leqq \langle \gamma, \lambda \rangle$ for all $\gamma \in \Psi'$. Then it is easy to see that Ψ' is special and that $(\beta, \Psi') \subset \Psi'$. By induction, therefore, we have the group $U_{\Psi'}$ directly spanned by $\{U_\gamma | \gamma \in \Psi'\}$. In view of (*) $U_{\Psi'}$ is normalized by U_β (see the proof of (3) below). In particular $U_\beta \cdot U_{\Psi'}$ is a subgroup, U_Ω of U, and the paragraph above shows that U_Ψ is directly spanned, in any order, by $\{U_\gamma | \gamma \in \Psi\}$. This proves (2).

To prove (3), suppose $(\alpha, \Psi) \subset \Psi$. To show that U_α normalizes U_Ψ it suffices to show that, for $x \in U_\alpha$ and $\gamma \in \Psi$, we have ${}^xU_\gamma \subset U_\Psi$. Suppose $y \in U_\gamma$. Then $x_y = (xyx^{-1})(y^{-1}y) = (x, y)y \in (U_\alpha, U_\gamma)U_\gamma$. Thus it suffices to see that $(U_\alpha, U_\gamma) \subset U_\Psi$. But, according to (*), $(u_\alpha, U_\gamma) \subset U_{(\alpha,\gamma)}$, where $U_{(\alpha,\gamma)}$ is the product, in some order, of $\{U_\delta | \delta \in (\alpha, \gamma)\}$. Since $(\alpha, \gamma) \subset \Psi$ we have $U_{(\alpha,\gamma)} \subset U_\Psi$, and this completes the proof, modulo the:

Proof of (*). Put $\Psi = (\alpha, \beta) \cup \{\alpha, \beta\}$. It is clear (see proof of (1) above) that Ψ is special. Hence we can choose λ, $B = B(\lambda)$, $U = B_u$, and $\Phi^+ = \Phi(U)$, as above, so that $\Psi \subset \Phi^+$.

If $\gamma \in \Phi^+$, let $\theta_\gamma : \mathbf{G}_a \to U_\gamma$ be an isomorphism. Then,

$$ {}^t\theta_\gamma(x) = \theta_\gamma(t^\gamma x) \quad (\text{for } t \in T,\ x \in \mathbf{G}_a). $$

Let $\alpha_1, \ldots, \alpha_n$ be the elements of Φ^+, in any fixed order. According to 14.4 the product morphism

$$ U_{\alpha_1} \times \cdots \times U_{\alpha_n} \to U $$

is an isomorphism of varieties. Define

$$ f : \mathbf{G}_a \times \mathbf{G}_a \to U, \quad f(x, y) = (\theta_\alpha(x), \theta_\beta(y)). $$

Then the isomorphism above shows that

$$ f(x, y) = \prod_{1 \leqq i \leqq n} \theta_{\alpha_i}(P_i(x, y)) $$

(product in ascending order), where the P_i are polynomials in two variables.

Say

$$P_i(x, y) = \sum_{r,s \geqq 0} c_{i,r,s} x^r y^s.$$

Since $f(0, y) = e = f(x, 0)$ we see that each monomial in P_i involves both x and y, i.e. the summation is actually over r, $s > 0$.

For $t \in T$ and $x, y \in \mathbf{G}_a$ we see that ${}^t f(x, y)$ is equal to

$$(\theta_\alpha(t^\alpha x), \theta_\beta(t^\beta y)) = \prod_i \theta_{\alpha_i}(P_i(t^\alpha x, t^\beta y))$$

as well as to

$$\prod_i \theta_{\alpha_i}(t^{\alpha_i} P_i(x, y)).$$

This yields, for each $i = 1, \dots, n$,

$$\sum_{r,s>0} c_{i,r,s}(t^\alpha x)^r (t^\beta y)^s = \sum_{r,s>0} c_{i,r,s} t^{\alpha_i} x^r y^s.$$

It follows that

$$c_{i,r,s} = 0 \text{ unless } \alpha_i = r\alpha + s\beta.$$

Since, as already observed, $c_{i,r,s} = 0$ unless $r, s > 0$, we therefore have $c_{i,r,s} = 0$ unless $\alpha_i \in (\alpha, \beta)$. Thus,

$$P_i = 0 \quad \text{unless} \quad \alpha_i \in (\alpha, \beta).$$

The latter is precisely what we sought to prove. It asserts, for x, $y \in \mathbf{G}_a$, that $(\theta_\alpha(x), \theta_\beta(y))$ lies in the product (in the above order) of those U_{α_i} for which $\alpha_i \in (\alpha, \beta)$.

Remarks.

(1) Since α and β are linearly independent there is at most one expression for an α_i in the form $r\alpha + s\beta$. Hence the proof above shows that *each P_i is a monomial*, zero unless $\alpha_i \in (\alpha, \beta)$.

(2) We have shown the existence of a smallest set of roots $\Phi(\alpha, \beta)$ such that $(U_\alpha, U_\beta) \subset U_{\Phi(\alpha,\beta)}$, which is contained in (α, β). In characteristic zero, the rule $[\mathfrak{g}_\alpha, \mathfrak{g}_\beta] = \mathfrak{g}_{\alpha+\beta}$ implies readily that $\Phi(\alpha, \beta) = (\alpha, \beta)$. In positive characteristic, there may be a strict inclusion, though only in small characteristics. In particular, it may happen that $(U_\alpha, U_\beta) = \{1\}$; even though $\alpha + \beta$ is a root. For a list of the cases in which $\Phi(\alpha, \beta) \neq (\alpha, \beta)$ see [6:4.3]. Obviously, $\Phi(\alpha, \beta) = \Phi(\beta, \alpha)$. We shall show (14.8, Cor. 2) that $\Phi(-\alpha, -\beta) = -\Phi(\alpha, \beta)$.

14.6 Corollary. *Let $\alpha \in \Phi$, and let $r_\alpha \in W$ be the generator of the subgroup $W(T, G_\alpha)$. Then if $\beta \in \Phi$ we have*

$$r_\alpha(\beta) = \beta - n_{\beta,\alpha}\alpha$$

with $n_{\beta,\alpha} \in \mathbb{Z}$. Moreover $n_{\alpha,\alpha} = 2$.

Proof. We know from 13.18 (4)(c) that $r_\alpha(\alpha) = -\alpha = \alpha - 2\alpha$ and that r_α fixes the elements of a subgroup of corank 1 in $X(T)$. Passing to $X(T)_{\mathbb{Q}}$, and extending α to a basis whose remaining members are in the fixed hyperplane of r_α (extended to $X(T)_{\mathbb{Q}}$), we see that, for any $\gamma \in X(T)_{\mathbb{Q}}$, $\gamma - r_\alpha(\gamma)$ is a (rational) multiple of α. In particular, $r_\alpha(\beta) = \beta - n_{\beta,\alpha}\alpha$ for some $n_{\beta,\alpha} \in \mathbb{Q}$, moreover $n_{\alpha,\alpha} = 2$ and $n_{\alpha,-\alpha} = -2$. Therefore, to prove $n_{\beta,\alpha} \in \mathbb{Z}$, we may assume $\beta \neq \pm\alpha$.

We apply the proposition above, which says that $[\alpha, \beta)$ is special, and hence that the $U_\gamma(\gamma \in [\alpha, \beta))$ directly span a T-stable subgroup $H = U_{[\alpha,\beta)}$. Evidently $(\alpha, [\alpha, \beta)) \subset [\alpha, \beta)$ and $(-\alpha, [\alpha, \beta)) \subset [\alpha, \beta)$, so the proposition implies that U_α and $U_{-\alpha}$ normalize H. Since $U_\alpha, U_{-\alpha}$ and T generate G_α it follows that G_α normalizes H. Now r_α arises from conjugation by an $n \in \mathcal{N}_{G_\alpha}(T)$, and we have just seen that this n normalizes H. Since $\beta \in [\alpha, \beta)$ and $nU_\beta n^{-1} = {}^{r_\alpha}U_\beta = U_{r_\alpha(\beta)}$ it follows that $r_\alpha(\beta) \in [\alpha, \beta)$, i.e. that $r_\alpha(\beta) = r\alpha + s\beta$ for suitable r, $s \in \mathbb{Z}$, $s > 0$. Thus $s = 1$ and $n_{\beta,\alpha} = -r \in \mathbb{Z}$. Q.E.D.

14.7 *Review of root systems.* The facts to be reviewed here can all be found in ([31], Chap. V and *p.* VII-13), or in [9:VI].

Let R be a subfield of $\mathbb{R}$. If V is a vector space over R we write $V^* = \mathrm{Hom}_R(V, R)$. Let α be a non-zero vector in V. We call $r \in GL(V)$ a *reflection* with respect to α if $r(\alpha) = -\alpha$ and if r fixes the points of a hyperplane H in V. Thus $r(\beta) = \beta - \langle\beta, \lambda\rangle\alpha$, for $\beta \in V$, where $\lambda \in V^*$ has kernel H, and $\langle\alpha, \lambda\rangle = 2$.

If Φ is a finite spanning set of V, there is at most one reflection with respect to α leaving Φ stable.

A *root system* is a pair (V, Φ) where V is a vector space over R, and where Φ is a subset of V satisfying:

(1) Φ is finite, spans V, and does not contain zero.

(2) For each $\alpha \in \Phi$ there is a reflection r_α with respect to α which leaves Φ stable (and which is therefore unique, by the remark above).

(3) If α, $\beta \in \Phi$ then $r_\alpha(\beta) = \beta - n_{\beta,\alpha}\alpha$ with $n_{\beta,\alpha} \in \mathbb{Z}$. The elements of Φ are called *roots.*

The notion of isomorphism of root systems is evident. We will usually denote the root system by Φ, and say that "Φ is a root system in V." In particular we have $\mathrm{Aut}(\Phi) \subset GL(V)$. The subgroup $W(\Phi)$ of $\mathrm{Aut}(\Phi)$ generated by the $r_\alpha(\alpha \in \Phi)$ is called the *Weyl group* of Φ.

Let $\alpha \in \Phi$ be such that the only roots, $a\alpha$, proportional to α are such that $|a| \leqq 1$. If $a\alpha$ is one such then $-a\alpha = r_\alpha(a\alpha) = a\alpha - n_{a\alpha,\alpha}\alpha$, so that $2a = n_{a\alpha,\alpha} \in \mathbb{Z}$. Thus the roots proportional to α are either $\{-\alpha, \alpha\}$ or $\{-\alpha, -\alpha/2, \alpha/2, \alpha\}$. If the latter case never occurs, i.e. if, whenever α and β are proportional roots we have $\beta = \pm\alpha$, then the root system Φ is said to be *reduced.*

Fix a root system Φ in V. A *basis* of Φ is a subset Δ of Φ, which is a basis of V such that each root β is a linear combination, $\beta = \sum_{\alpha \in \Delta} m_\alpha \alpha$, with the m_α integers all of the same sign. We then define the *positive roots* Φ^+ (with respect to Δ) to be those β for which all m_α are $\geqq 0$. Thus Φ is the disjoint union of Φ^+ and $\Phi^- = -\Phi^+$. We call

$$WC(\Delta) = \{\lambda \in V^* | \langle \alpha, \lambda \rangle > 0 \text{ for all } \alpha \in \Delta\}$$

the *Weyl chamber* of Δ (or of Φ^+. One can clearly replace Δ by Φ^+ in the definition without essentially altering it.)

Call $\lambda \in V^*$ *regular* if $\langle \alpha, \lambda \rangle \neq 0$ for all $\alpha \in \Phi$. For example, a Weyl chamber clearly consists of regular elements. If λ is regular we shall write

$$\Phi^+(\lambda) = \{\alpha \in \Phi | \langle \alpha, \lambda \rangle > 0\}$$

and

$$\Delta(\lambda) = \{\alpha \in \Phi^+(\lambda) | \alpha \text{ is not the sum of two elements of } \Phi^+(\lambda)\}.$$

Theorem. *Let Φ be a root system in V.*

(1) *If $\lambda \in V^*$ is regular then $\Delta(\lambda)$ is a basis of Φ. It is the unique basis contained in $\Phi^+(\lambda)$. Thus, $\Delta \mapsto WC(\Delta)$ is a bijection from the set of bases to the set of Weyl chambers.*

Now suppose Φ is reduced.

(2) *$W(\Phi)$ acts simply transitively on the set of bases of Φ, and (equivalently) on the set of Weyl chambers.*

Let Δ be a basis of Φ.

(3) *The $r_\alpha (\alpha \in \Delta)$ generate $W(\Phi)$.*

(4) $\Phi = \bigcup_{w \in W(\Phi)} w\Delta.$

One associates to a basis Δ of Φ the so-called *Dynkin diagram* $\mathrm{Dyn}(\Phi, \Delta)$ which is a finite graph having Δ as its set of vertices, supplied with suitable "weights," and in which α and β in Δ are joined by $n_{\alpha,\beta} n_{\beta,\alpha}$ edges. The Dynkin diagrams give a complete classification of root systems. Moreover $\mathrm{Dyn}(\Phi, \Delta)$ is functorial in (Φ, Δ), and the automorphism group of Φ is the semi-direct product of W and of $\mathrm{Aut}(\mathrm{Dyn}(\Phi, \Delta))$, the latter being the stability group of Δ in $\mathrm{Aut}(\Phi)$.

The root system (V, Φ) is said to be *irreducible* if one cannot write $V = V_1 \oplus V_2$ as a non-trivial direct sum so that $\Phi = (\Phi \cap V_1) \cup (\Phi \cap V_2)$.

A subset Ψ of Φ is *closed* if $\alpha, \beta \in \Psi$ and $\alpha + \beta \in \Phi$ imply $\alpha + \beta \in \Psi$. We claim that Ψ is special (see 14.5) if and only if it is closed and belongs to Φ^+ for some ordering. Let Ψ be special. Then condition (a) implies that is closed and (b) shows that it belongs to some $\Phi^+(\lambda)$, hence to a positive set of roots for some ordering. The converse follows from the fact that if $\{\alpha_i\}$ $(i = 1, \ldots, N)$ is a set of positive roots whose sum is a root, then there exists a permutation σ of the indices such that all sums

$$\sum_{j=1}^{j=i} \alpha_{\sigma(j)} \quad (i = 1, \ldots, N)$$

are roots (see VI, 1.6, Prop. 19 in [9]).

A positive nondegenerate scalar product on V or V^*, invariant under W, will be called *admissible*. Since W is finite, admissible scalar products always exist.

14.8 Theorem. *Let $V = (X(T/\mathscr{C}(G)^o)_{\mathbb{Q}}$, identified canonically with a subspace of $X(T)_{\mathbb{Q}}$. Then $\Phi = \Phi(T, G)$ is a reduced root system in V, with Weyl group $W = W(T, G)$.*

Proof. From 13.18(3) we conclude that Φ is a finite set of non-zero vectors spanning V, and that, if α and β in Φ are proportional, then $\beta = \pm\alpha$. For the rest we can, without loss, assume $\mathscr{C}(G)^o = \{e\}$, by passing to $G/\mathscr{C}(G)^o$.

From 13.18 (4)(c) we obtain a reflection r_α of $X(T)$ with respect to α which leaves Φ stable. The extension of r_α to V (which we shall also denote by r_α) verifies condition (2) in the definition of a root system.

The (integrality) condition (3) is established by 14.6.

This shows that Φ is a reduced root system in V, and that $W(\Phi) \subset W$.

If we identify V^* with $X_*(T)_{\mathbb{Q}}$ then it is clear from 14.7 and from 13.18(5) that the Weyl chambers in V^* of bases of Φ coincide with the subsets of V^* obtained from Weyl chambers in $X_*(T)$ of Borel subgroups $B \in \mathscr{B}^T$. (The Weyl chamber in V^* corresponding to $B \in \mathscr{B}^T$ is $\{\lambda \in V^* | \langle \alpha, \lambda \rangle > 0$ for $\alpha \in \Phi(T, B)\}$.) According to 13.10(2), W acts simply transitively on these Weyl chambers. But 14.7 asserts $W(\Phi)$ does likewise, and hence the inclusion $W(\Phi) \subset W$ is an equality.

Corollary 1. *Let $B \in \mathscr{B}^T$ and let $\Delta = \Delta(B)$ be the set of $\alpha \in \Phi(B)$ which are not sums of two elements in $\Phi(B)$.*

(1) Δ is a basis of Φ. (We call Δ the set of simple roots associated with B (and T).)

(2) G is generated by $\{G_\alpha | \alpha \in \Delta\}$.

Proof. (1) follows from 14.6 in view of the fact (see 13.18 (5)) that $\Phi(B) = \{\alpha \in \Phi | \langle \alpha, \lambda \rangle > 0$ for $\lambda \in WC(B)\}$, and, for such a λ, $\langle \alpha, \lambda \rangle \neq 0$ for all $\alpha \in \Phi$, i.e. λ is regular.

(2) We know from 13.7 that G is generated by the set of $B \in \mathscr{B}^T$. If $B \in \mathscr{B}^T$ then $B = T \cdot B_u$ and 14.4 implies that B_u is generated (even directly spanned) by the set of $U_\alpha(\alpha \in \Phi(B))$. Thus G is generated by T together with the $U_\alpha(\alpha \in \Phi)$.

Let H be the subgroup generated by $G_\alpha(\alpha \in \Delta)$. G_α contains U_α as well as a representative, $n_\alpha \in \mathscr{N}_{G_\alpha}(T)$, of $r_\alpha \in W$. Hence it follows from 14.7 that H contains a representative, $n = n(w) \in \mathscr{N}_G(T)$, of each $w \in W$. If $\beta \in \Phi$ then ${}^n U_\beta = U_{w(\beta)}$. Thus H contains all U_β for which β is a W-transform of some $\alpha \in \Delta$. According to 14.7 these β's exhaust Φ. Since, clearly, $T \subset H$, this shows that $H = G$.

Corollary 2. *Let $\alpha, \beta \in \Phi$, $\alpha \neq \pm\beta$ and let $\Phi(\alpha, \beta)$ be the smallest set of roots such that $(U_\alpha, U_\beta) \subset U_{\Phi(\alpha,\beta)}$ (see 14.5, Remark 2). Then $\Phi(-\alpha, -\beta) = -\Phi(\alpha, \beta)$.*

Let $S = (\ker \alpha \cap \ker \beta)^o$. It is of codimension two in T and $H = \mathscr{Z}(S)/S$ is a semisimple group of rank two, with maximal torus $T' = T/S$.

If $\alpha+\beta$ is not a root, then (α, β) is empty, so is $\Phi(\alpha, \beta)$ and $(U_\alpha, U_\beta)=\{1\}$. But then $(-\alpha, -\beta)$ is also empty and so is $\Phi(-\alpha, -\beta)$. From now on, assume that $\alpha+\beta$ is a root. In this case $\Phi(T', H)$ is irreducible. By the classification of roots systems of rank 2, $\Phi(T', H)$ is of one of the types $\mathbf{A}_2, \mathbf{B}_2, \mathbf{G}_2$. Assume first it is of one of the last two. Then $-Id$ belongs to the Weyl group. Therefore there is an inner automorphism of H leaving T' invariant and sending U_γ onto $U_{-\gamma}$ for every $\gamma \in \Phi(T', H)$. It maps (U_α, U_β) onto $(U_{-\alpha}, U_{-\beta})$ and $U_{\Phi(\alpha,\beta)}$ onto $U_{-\Phi(\alpha,\beta)}$ whence our assertion in this case.

Assume now $\Phi(T', H)$ to be of type $\mathbf{A}_2$. Since $\alpha+\beta$ is assumed to be a root, we may find an ordering on $\Phi(T', H)$ for which α and β are the simple roots. Then either $\Phi(\alpha, \beta)=\phi$ or $\Phi(\alpha, \beta)=\{\alpha+\beta\}$. The Weyl group contains an element (the reflection to $\alpha+\beta=0$) which sends α and β onto $-\beta$ and $-\alpha$ respectively. The corresponding automorphism of G maps (U_α, U_β) onto $(U_{-\beta}, U_{-\alpha})$ and $U_{\Phi(\alpha,\beta)}$ onto $U_{\Phi(-\beta,-\alpha)}$ and again $\Phi(-\beta, -\alpha)=\Phi(-\alpha, -\beta)$ is either empty or equal to $\{-\alpha, -\beta\}$. Therefore

$$\Phi(\alpha, \beta)=\phi \Leftrightarrow (U_\alpha, U_\beta)=(1) \Leftrightarrow (U_{-\beta}, U_{-\alpha})=\{1\} \Leftrightarrow \Phi(-\alpha, -\beta)=\phi,$$

$$\Phi(\alpha, \beta)=\{\alpha+\beta\} \Leftrightarrow (U_\alpha, U_\beta) \neq (1)$$

$$\Leftrightarrow (U_{-\beta}, U_{-\alpha}) \neq \{1\} \Leftrightarrow \Phi(-\alpha, -\beta) \neq 0 \Leftrightarrow \Phi(-\alpha, -\beta)=\{-\alpha-\beta\}.$$

14.9. *Automorphisms of semi-simple groups.* Assume G is semi-simple, and fix a $b \in \mathscr{B}^T$. In

$$A=\operatorname{Aut}_{\text{alg.grp.}}(G)$$

let $\operatorname{Int}(G)$ be the group of inner automorphisms. Also write $A_{(B,T)}$ for the subgroup of A stabilizing both B and T.

According to (14.8) $\Phi(B)$ is the set of positive roots with respect to a basis $\Delta(B)$ of Φ. We shall write $\operatorname{Dyn}(\Phi, B)$ for the corresponding Dynkin diagram (see 14.7), and $\operatorname{Aut}(\operatorname{Dyn}(\Phi, B))$ for its automorphism group.

If $a \in A_{(B,T)}$ then, since $a \cdot T=T$, a induces an automorphism of the root system Φ. Since $a \cdot B=B$ it follows that a leaves $\Delta(B)$ stable and hence defines an element $a' \in \operatorname{Aut}(\operatorname{Dyn}(\Phi, B))$.

Proposition.

(1) $A=\operatorname{Int}(G) \cdot A_{(B,T)}$.

(2) $\operatorname{Int}(G) \cap A_{(B,T)}$ *is the kernel of the homomorphism* $A_{(B,T)} \rightarrow \operatorname{Aut}(\operatorname{Dyn}(\Phi, B))$ $(a \mapsto a')$ *described above.*

(3) *There is a natural injection* $A / \operatorname{Int}(G) \rightarrow \operatorname{Aut}(\operatorname{Dyn}(\Phi, B))$. *In particular* $\operatorname{Int}(G)$ *has finite index in* A.

Proof. Clearly (3) follows from (1) and (2).

(1) Let $a \in A$. By the conjugacy of Borel subgroups of G we have $cB=B$ where $c=\operatorname{Int}(g) \circ a$ for some $g \in G$. By the conjugacy of maximal tori in B we

have $dT = T$ where $d = \text{Int}(b)\circ c$ for some $b\in B$. Thus we have $d = \text{Int}(b)\circ\text{Int}(g)\circ a\in A_{(B,T)}$, as required.

(2) Suppose $a\in A_{(B,T)}$. If $a = \text{Int}(g)$ for some $g\in G$ then by 10.6, 11.16,

$$g\in\mathcal{N}_G(B)\cap\mathcal{N}_G(T) = B\cap\mathcal{N}_G(T) = \mathcal{N}_B(T) = T,$$

so α induces the identity automorphism of Φ.

Suppose, conversely, that $a'\in\text{Aut}(\text{Dyn}(\Phi, B))$ is the identity. We must show that a is inner. For each $\alpha\in\Delta(B)$ we have an isomorphism $\theta_\alpha:\mathbf{G}_a\to U_\alpha$. Since a leaves U_α stable we have $a\theta_\alpha(x) = \theta_\alpha(c_\alpha x)$ for some $c_\alpha\in K^*$. Since the elements of $\Delta(B)$ are linearly independent we can find a $t\in T$ such that $t^\alpha = c_\alpha$ for each $\alpha\in\Delta(B)$. Then $\text{Int}(t)$ has the same effect as a on each $U_\alpha(\alpha\in\Delta(B))$ so we can replace a by $\text{Int}(t)^{-1}\circ a$ and assume each $c_\alpha = 1$. In that case a fixes the elements of each $U_\alpha(\alpha\in\Delta(B))$.

We claim a also fixes the elements of T. For if $t\in T$ then $t^\alpha = a(t)^\alpha$ for each $\alpha\in\Delta(B)$. Since G is semisimple, $\Delta(B)$ spans a subgroup of finite index in $X(T)$ (14.8). Hence $t = a(t)$ as claimed.

Evidently a stabilizes $G_\alpha = G^{T_\alpha}$ and it fixes the elements of the Borel subgroup $T\cdot U_\alpha$. Hence 11.4 (1) implies $a|G_\alpha$ is the identity. Finally 14.8 asserts that the $G_\alpha(\alpha\in\Delta(B))$ generate G, so a is the identity.

Remark. In (1), G may be any connected algebraic group.

14.10 Proposition. *Assume G is semi-simple and $\neq\{e\}$.*

(1) *Let H be a connected normal subgroup of G, and let $H' := (G^H)^o$.*

- (a) *H is semi-simple.*
- (b) *$G = H\cdot H'$ and $H\cap H'$ is contained in the finite group $\mathscr{C}(G)$.*
- (c) *If $G\to G'$ is a surjective morphism, then G' is semisimple.*

(2) *Let $\{G_i | i\in I\}$ be the minimal elements among the connected normal subgroups of dimension $\geqq 1$.*

- (a) *If $i\neq j$ then $(G_i, G_j) = \{e\}$.*
- (b) *I is finite; say $I = \{1,\ldots,n\}$. The product morphism*

$$G_1\times\cdots\times G_n\to G$$

is an isogeny.

- (c) *If H is connected normal subgroup of G, then H is generated by $\{G_i | G_i\subset H\}$.*

(3) *G is "almost simple," i.e. $G/\mathscr{C}(G)$ is simple, if and only if the root system Φ is irreducible.*

Proof. (1) Assertion (a) follows from (14.2, Cor.). G^H is the kernel of the conjugation homomorphism:

$$G\xrightarrow{\text{Int}|H}\text{Aut}_{\text{alg.grp.}}(H)$$

and, by 14.9, the image of H is a subgroup of finite index. Hence $H \cdot G^H$ has finite index in G. Therefore, by connectivity, $G = H \cdot (G^H)^o = H \cdot H'$. Moreover $(H \cap H')^o \subset \mathscr{Z}(H)^o \subset \mathscr{R}(H) = \{e\}$ so $H \cap H'$ is a finite normal, and hence central, subgroup of G.

(c) Let H be the identity component of $\ker \pi$, H' be as in (1). Then π defines a surjective morphism with finite (hence central) kernel of H' onto G'. Since H' is semi-simple (by (a)), so is G'.

(2) Let H be as above and let $i \in I$. Then (G_i, H) is a connected normal subgroup of G contained in $G_i \cap H$. Hence, by minimality of G_i, it equals $\{e\}$ or G_i. In other words, $G_i \subset (G^H)^o$ or $G_i \subset H$. In particular $(G_i, G_j) = \{e\}$ for $i \neq j$.

Let $J = \{i_1, \ldots, i_r\} \subset I$ and let G_J denote the image of the morphism

$$f_J : G_{i_1} \times \cdots \times G_{i_r} \to G.$$

With the aid of the remarks above an induction on $r = \operatorname{card} J$ shows that $G_J \cap G_h$ is finite if $h \notin J$, and hence that $\ker(f_J)$ is finite. Therefore $\dim G \geqq \dim G_J = \sum_{j \in J} \dim G_j \geqq \operatorname{card} J$, so I must be finite. Moreover, $f_J : G_{i_1} \times \cdots \times G_{i_r} \to G_J$ is an isogeny.

With H and H' as in (1) we see also that $I = J \cup J'$ (disjoint) where $J = \{ j \in I \mid G_j \subset H\}$ and $J' = \{ j \in I \mid G_j \subset H'\}$. It follows then, since $G = G_J \cdot G_{J'} = H \cdot H'$ and $H \cap H'$ is finite, that $H = G_J$.

(3) If $G = H \cdot H'$ as above then it is clear that the root system Φ of G decomposes into the direct sum of those of H and of H', respectively. Thus, if both H and H' have dimension $\geqq 1$, Φ is reducible.

Conversely, suppose Φ is reducible; say $\Phi = \Phi_1 \cup \Phi_2$ is a non-trivial decomposition into a sum of two root systems. Let G_i denote the subgroup generated by all $U_\alpha (\alpha \in \Phi_i)$. Then, since Φ_1 and Φ_2 are both not empty, $\dim G_i \geqq 1 \, (i = 1, 2)$.

We claim first that G_1 and G_2 generate G. For let H be the group they generate. Then $H_\alpha = H \cap G_\alpha$ projects onto the semi-simple quotient $\mathbf{PGL}_2$ of G_α (see 10.8) so H_α contains a complementary torus T'_α to T_α in T. Since $(\bigcap T_\alpha)^o = \{e\}$ (G is semi-simple) it follows that the tori T'_α are independent and generate T. Thus H contains T, and hence each G_α, and hence $H = G$ (see 14.8, Cor. 1(2)).

Next we claim that G_1 centralizes G_2. For if $\alpha \in \Phi_1$ and $\beta \in \Phi_2$ there are no roots of the form $r\alpha + s\beta$ with $r, s > 0$. Hence the assertion $(*)$ in the proof of 14.5 shows that U_α and U_β commute.

Finally, therefore, $G_1 \cap G_2$ commutes with the group generated by G_1 and G_2, which is G, so $G_1 \cap G_2 (\subset \mathscr{Z}(G))$ is finite. This completes the proof of (3), and hence of the proposition.

14.11 Corollary. *Let H, H' be connected linear algebraic groups and $\pi : H \to H'$ a surjective morphism. Then $\pi \mathscr{R} H = \mathscr{R} H'$ and $\pi(\mathscr{R}_u H) = \mathscr{R}_u H'$.*

It is obvious from the definition that $\pi(\mathscr{R} H) \subset \mathscr{R} H'$. By the universal property of quotients (6.3), π induces a surjective morphism of $H / \mathscr{R} H$ onto

$H'/\pi(\mathscr{R}H)$. By 14.10(1c), $H'/\pi(\mathscr{R}H)$ is semisimple, therefore $\pi(\mathscr{R}H) \supset \mathscr{R}H'$, and the first assertion follows. The second one is then a consequence of 10.6.

14.12 *The Bruhat decomposition.* We fix a $B \in \mathscr{B}^T$ and write $U = B_u$, $\Phi^+ = \Phi(B)$, and Δ for the basis of Φ in Φ^+, the set of "simple roots associated with B."

Let $B^- \in \mathscr{B}^T$ be the opposite Borel subgroup (see 14.1). We put $U^- = B_u^-$ and $\Phi^- = \Phi(B^-) = -\Phi^+$. For $\alpha \in \Phi$ we shall write $\alpha > 0$ if $\alpha \in \Phi^+$ and $\alpha < 0$ if $\alpha \in \Phi^-$.

If $w \in W$ we shall allow ourselves to confuse w with a representing element in $\mathscr{N}_G(T)$, whenever the use is unaffected by the choice of a representative.

We shall consider the groups

$$U_w = U \cap {}^wU \text{ and } U'_w = U \cap {}^wU^-.$$

These are both T-stable closed subgroups of U, so it follows from 14.4 that they are directly spanned, in any order, by the $U_\gamma (\gamma > 0)$ that they contain. The sets of such γ are, respectively,

$$\Phi_w^+ = \Phi(U_w) = \{\gamma > 0 \mid \gamma^w > 0\},$$

and

$$\Phi'_w = \Phi(U'_w) = \{\gamma > 0 \mid \gamma^w < 0\},$$

where $\gamma^w = \gamma \circ \operatorname{Int}(n)$ for any $n \in \mathscr{N}_G(T)$ representing w. Since these sets partition Φ^+ it follows also from 14.4 that

$$U = U_w \cdot U'_w = U'_w \cdot U_w.$$

Let x_o denote the fixed point of B in G/B.

Theorem. (a) (*Bruhat decomposition of* G). *G is the disjoint union of the double cosets $BwB (w \in W)$. If $w \in W$ then the morphism $U'_w \times B \to BwB (x, y) \mapsto xwy$, is an isomorphism of varieties.*

(b) (*Cellular decomposition of* G/B). *G/B is the disjoint union of the U-orbits Uwx_o $(w \in W)$. If $w \in W$ then the morphism $U'_w \to Uwx_o u \mapsto uwx_o$, is an isomorphism of varieties.*

Remarks. (1) The fixed points $(G/B)^T$ correspond to $\mathscr{B}^T$, and we know from 11.19 that W acts simply transitively on this set. In particular, $Wx_0 = (G/B)^T$, and this set has the same cardinality as W. Part (b) above therefore asserts that each U-orbit in G/B meets $(G/B)^T$ in precisely one point.

(2) Since $B = U \cdot T$ and W normalizes T it follows that, for $w \in W$, we have $BwB = UwB$ and $Bwx_o = Uwx_o$. Thus it is clear that (a) and (b) are equivalent.

(3) It follows from 14.4 that each U'_w is isomorphic, as a variety, to an affine space. Thus, if $K = \mathbb{C}$, each of the U-orbits is a cell and (b) gives rise to a cell decomposition of G/B in the sense of algebraic topology. Since these

cells are complex varieties they occur only in even (real) dimensions. Hence the $2i^{\text{th}}$ Betti number of G/B is the number of cells of (complex) dimension i. The latter is the number of $w \in W$ for which $\dim U'_w = \operatorname{card}\{\gamma > 0 \mid w^{-1}(\gamma) < 0\}$ is equal to i.

Proof. In view of remark (2) above, the theorem will follow once we establish the following three assertions:

(1) *If* $w, w' \in W$, *then* $Uwx_o = Uw'x_o \Rightarrow w = w'$.
(2) $G = BWB$.
(3) *If* $w \in W$ *then the map* $U'_w \times B \to BwB$ *given by* $(x, y) \mapsto xwy$ *is an isomorphism of varieties.*

Proof of (1). Say $w'x_o = uwx_o$ with $u \in U$. Then the stability group in U of $w'x_o$, i.e. $U \cap {}^{w'}B = U_{w'}$, coincides with that of uwx_o, i.e. with $U_{uw} = U \cap {}^{uw}B = {}^{u}(U \cap {}^{w}B) = {}^{u}U_w$. Thus U_w and ${}^{u}U_w = U_{w'}$ are each closed T-stable subgroups of U. Therefore 14.4 (2)(c) implies $U_w = U_{w'}$. In particular $\Phi^+_w = \Phi(U_w)$ and $\Phi^+_{w'} = \Phi(U_{w'})$ coincide, where $\Phi^+_w = \{\gamma \in \Phi^+ \mid \gamma^w > 0\}$, and similarly for $\Phi^+_{w'}$. Therefore the proof is completed by the:

Lemma. *If* $w, w' \in W$ *and if* $\Phi^+_w = \Phi^+_{w'}$ *then* $w = w'$.

Proof. Suppose $n \in \mathcal{N}_G(T)$ represents w. We then have the actions of w on $\lambda \in X_*(T)$ and on $\alpha \in X(T)$ given by ${}^{w}\lambda = \operatorname{Int}(n) \circ \lambda$ and $\alpha^w = \alpha \circ \operatorname{Int}(n)$. Thus $\alpha^w \circ \lambda = \alpha \circ {}^{w}\lambda$, or, equivalently, $\langle \alpha^w, \lambda \rangle = \langle \alpha, {}^{w}\lambda \rangle$.

If λ is semi-regular then the Weyl chamber to which λ belongs is determined by the signs of the numbers $\langle \alpha, \lambda \rangle$, where α varies over Φ^+. This follows from 13.18(5). Suppose $\lambda \in WC(B)$, i.e. $\langle \alpha, \lambda \rangle > 0$ for all $\alpha > 0$. Then for $\alpha > 0$ we have $\langle \alpha, {}^{w}\lambda \rangle = \langle \alpha^w, \lambda \rangle$ which is > 0 if $\alpha \in \Phi^+_w$ and < 0 otherwise. It follows from the hypothesis, therefore, that ${}^{w}\lambda$ and ${}^{w'}\lambda$ lie in the same Weyl chamber. Therefore $w = w'$ since W acts simply transitively on the Weyl chambers (13.10).

Proof of (2). It will be carried out in several steps.

(i) *If G has semi-simple rank* 1 *then* (2) *holds.*

In this case W has order 2. So part(1) implies that BWx_o consists of two U-orbits. Hence it suffices to show that G/B consists of at most two U-orbits. Consider the morphism $U \to G/B$ $(u \mapsto uy)$, where y is not a fixed point of U. We can identify $U \cong \mathbf{G}_a$ with $\mathbf{P}_1$ minus a point, and then extend the morphism to $\mathbf{P}_1 \to G/B$. The image is closed and one-dimensional, and hence equals G/B. On the other hand this image consists of a one-dimensional U-orbit together with a single fixed point.

(ii) *If* $\alpha \in \Phi$ *and* $x \in (G/B)^T$, *then*

$$G_\alpha x = (U_\alpha x) \cup (U_\alpha r_\alpha x).$$

Put $C=(G_\alpha)_x=G_\alpha\cap B_x$. This is a Borel subgroup of G_α (11.18), and we have a G_α-equivariant and bijective morphism $G_\alpha/C\to G_\alpha x$. Since G_α has semi-simple rank 1 (13.18) and Weyl group $\{e,r_\alpha\}$ (with respect to T). (ii) now follows from (i).

(iii) *Suppose α is a simple root (i.e. $\alpha\in\Delta$) and let $\Psi=\Phi^+-\{\alpha\}$. Then, in the terminology and notation of (14.5), Ψ is special, and so the U_β ($\beta\in\Psi$) directly span a group U_Ψ. Moreover U_Ψ is normalized by G_α, and $U=U_\alpha U_\Psi=U_\Psi U_\alpha$.*

It is clear from the properties of root systems (see 14.7) that Ψ is special and that $(\alpha,\Psi)\subset\Psi$. Furthermore $(-\alpha,\Psi)\subset\Psi$. For suppose $\gamma=r(-\alpha)+s\beta\in\Phi$, where $\beta\in\Psi$ and $r,s>0$. Then $\beta=\sum_{\delta\in\Delta}m_\delta\delta$ with $m_{\delta_0}>0$ for some $\delta_0\neq\alpha$, because Φ is reduced. Hence the δ_0-coordinate of γ is $sm_{\delta_0}>0$, so $\gamma\in\Phi^+$. Clearly $\gamma\neq\alpha$ so $\gamma\in\Psi$.

Now it follows from 14.5 that $\{U_\beta|\beta\in\Psi\}$ directly span (in any order) a group U_Ψ, and that U_Ψ is normalized by U_α and $U_{-\alpha}$, as well as, of course, by T. Thus U_Ψ is normalized by G_α, the latter being generated by U_α, $U_{-\alpha}$, and T. The equalities $U=U_\alpha U_\Psi=U_\Psi U_\alpha$ are now clear.

(iv) *If $\alpha\in\Delta$ and $x\in(G/B)^T$ then $G_\alpha Bx=(Ux)\cup(Ur_\alpha x)$.*

We have $B=UT=U_\alpha U_\Psi T$ (as in (iii)), so

$$\begin{aligned}
G_\alpha Bx &= G_\alpha U_\alpha U_\Psi Tx \\
&= G_\alpha U_\Psi x && (Tx=x \text{ and } U_\alpha\subset G_\alpha)\\
&= U_\Psi G_\alpha x && (G_\alpha \text{ normalizes } U_\Psi;\text{ (iii)})\\
&= U_\Psi((U_\alpha x)\cup(U_\alpha r_\alpha x)) && \text{(part (ii))}\\
&= (Ux)\cup(Ur_\alpha x)
\end{aligned}$$

(v) *If $\alpha\in\Delta$ then $G_\alpha(BwB)\subset BwB\cup Br_\alpha wB$.*

For if $w\in W$ then, by (iv), we have

$$G_\alpha BwB=(UwB)\cup(Ur_\alpha wB)\subset(BwB)\cup(Br_\alpha wB).$$

According to the corollary of 14.8, G is generated by the G_α ($\alpha\in\Delta$). Hence (v) implies $G(BWB)\subset(BWB)$, thus proving (2).

Proof of (3). Since $U_w=U\cap{}^wB=U\cap wBw^{-1}=U\cap wUw^{-1}$ we have $U_w w\subset U$. Similarly, $U'_w w\subset wU^-$. Writing $B=UT=U'_wU_wT$ we see that $BwB=U'_wU_wwB=U'_wwB$, so

$$f:U'_w\times B\to BwB,\quad (x,y)\mapsto xwy$$

is surjective. Since $U'_w w\subset wU^-$ and $U^-\cap B=\{e\}$ it follows that f is injective also. Moreover since $L(U^-)=\sum_{\alpha<0}\mathfrak{g}_\alpha$ has trivial intersection with $L(B)=\mathfrak{g}^T\oplus\sum_{\alpha>0}\mathfrak{g}_\alpha$, it follows that f is separable, and hence an isomorphism.

14.13 Corollary. *If B, B', $B'' \in \mathscr{B}$ then $B \cap B'$ contains a maximal torus of G. If B' and B'' are opposite to B, then they are conjugate by an element of B.*

Proof. B' has a fixed point on $G/B = \bigcup_{w \in W} Uwx_o$ (in the notation of 14.12). Say B' fixes $x = uwx_o$, where $w \in W$ and $u \in U$. Then $B' = {}^{uw}B$. Since $T \subset {}^{w}B$ we have ${}^{u}T \subset {}^{uw}B \cap B = B' \cap B$.

Let T', T'' be maximal tori in $B \cap B'$ and $B \cap B''$ respectively. If B' (resp. B'') is opposite to B, then it is the unique Borel subgroup of G opposite to B containing T' (resp. T'') by 14.1. Then an element $b \in B$ such that ${}^{b}T' = T''$ (see 10.6) will conjugate B' onto B''.

14.14 Corollary. *Let B, $B' \in \mathscr{B}^T$ be opposite Borel subgroups and $U = B_u$, $U' = B'_u$. Then the product map $U' \times B \to G$ is an isomorphism of $U' \times B$ onto an open subset of G. The group G is a rational variety.*

By 14.13, we may assume that B, B' are the B and B^- of 14.12. Let w_o be the element of W which maps Φ^+ onto Φ^-. Then, left translation by w_o is an isomorphism of $U \cdot w_o \cdot B$ onto $w_o \cdot U \cdot w_o \cdot B = U^- \cdot B$, and the first assertion follows from 14.12.

T is isomorphic to a product of $\mathbf{GL}_1$'s over K. In view of 13.18, the remark in 14.4 applies to U, U', hence both are isomorphic, as varieties, to affine spaces. Since B is isomorphic, as a variety, to $T \times U$ by 10.6, it follows that $U^- \cdot B$ is a rational variety, hence so is G.

Remark. We shall see in § 18 that if G is defined over k, then G is unirational over k, rational over a separable extension of k. It follows from 14.14 and 15.1(2) and 15.8 that any connected affine k-group is a rational variety over $\bar{k}$.

14.15 *The Tits system* $\mathscr{T} = (G, B, N, S)$. Let $N = \mathscr{N}_G(T)$ and $S = \{r_\alpha\}$ $(\alpha \in \Delta)$. Then (W, S) is a Coxeter system [9:IV, 2.1]. These data satisfy the following conditions:

T1 G is generated by B, N and $B \cap N = T$ is normal in N.

T2 $W = N/T$ is generated by S.

T3 $sBw \subset BwB \cup Bs \cdot wB \quad (s \in S, w \in W)$.

T4 $sBs \neq B$ for $s \in S$.

In fact, T1 follows from 14.8 and 10.6, T2 from the fact that (W, S) is a Coxeter group, T3 from 14.12(v) and T4 from the fact that if $s = r_\alpha$ $(\alpha \in \Delta)$, then sBs contains $U_{-\alpha}$.

This means that $\mathscr{T}$ is a Tits system [9:IV, 2.1]. By general arguments (*loc. cit.*) these conditions already imply the Bruhat decomposition $G = \coprod_{w} BwB$.

14.16 *Parabolic subgroups.* In a Tits system $\mathscr{T}' = (G', B', N', S')$ the parabolic subgroups are by definition the subgroups which contain a conjugate of B'. For a subset I of S', let W'_I be the subgroup of $W' = N'/(B' \cap N')$ generated by I. Then $P'_I = B' \cdot W'_I \cdot B'$ is a group and every parabolic subgroup of G' is conjugate to exactly one of those [9:IV, 2.5, 2.6].

Coming back to our Tits system $\mathcal{T}$, we note first that $B \cdot W_I \cdot B$ is a closed subgroup. In fact, for each $\alpha \in I$, it contains T, U_α and $r_\alpha U_\alpha r_\alpha = U_{-\alpha}$, hence $G_\alpha = \mathcal{Z}_G(T_\alpha)$. Since I generates W_I, it follows that $B \cdot W_I \cdot B$ is generated by B and the G_α ($\alpha \in I$), which are all irreducible subvarieties containing 1, so that our assertion follows from 2.2. In particular we see that any subgroup containing a conjugate of B is closed. As a consequence, the parabolic subgroups as defined in 11.2 are the same as the parabolic subgroups of the Tits system $\mathcal{T}$.

It follows from general facts about Tits systems that for I, $J \subset \Delta$, there is a canonical bijection of between $W_I \backslash W / W_J$ and $P_I \backslash G / P_J$ [9:IV, 2.5, Rem. 2]. In particular, G is a disjoint union of double cosets $B \cdot w \cdot P_J$, where w runs through W/W_J. We shall come back to this in a more general case in §21.

14.17 *Standard parabolic subgroups.* Once B is chosen, a parabolic subgroup P containing B is called standard. For later use, we rephrase here slightly the description of standard parabolic subgroups. For $I \subset \Delta$, let

$$T_I = \left(\bigcap_{\alpha \in I} \ker \alpha \right)^o.$$

Let us write B_I for $B \cap \mathcal{Z}_G(T_I)$. It is a Borel subgroup of $\mathcal{Z}_G(T_I)$ by 11.15. It is a well known fact on finite euclidean reflection groups that a subgroup of such a group fixing pointwise a vector subspace is generated by the reflections in the group fixing that subspace [9:V, §4, no. 6]. Therefore $W(T, \mathcal{Z}_G(T_I)) = \langle r_\alpha \rangle_{\alpha \in I}$. Moreover, in view of 9.2, $\Phi(T, \mathcal{Z}_G(T_I)) = [I]$, where $[I]$ denotes the set of roots which are linear combinations of elements in I. Let $\Phi(I)^+ = \Phi^+ - [I]$. It is the set of roots

$$\beta = \sum_{\alpha \in \Delta} c_\alpha(\beta) \alpha \tag{1}$$

where at least one of the $c_\alpha(\beta)$, $\alpha \notin I$, is > 0. As a consequence, in the terminology of 14.5, $\Phi(I)^+$ is special and

$$(\alpha, \Phi(I)^+) \subset \Phi(I)^+ \quad \text{for } \alpha \in [I]. \tag{2}$$

14.18 Proposition. *We keep the notation and assumptions of 14.17. The standard parabolic subgroups are the groups $P_I = B \cdot W_I \cdot B$ ($I \subset \Delta$). The group P_I is the semi-direct product of $\mathcal{Z}_G(T_I)$ and of its unipotent radical, which is equal to $U_{\Phi(I)^+}$. The group T_I is the identity component of the center of $\mathcal{Z}_G(T_I)$, is equal to $\mathcal{Z}_G(T_I) \cap \mathcal{R}P_I$ and is a maximal torus and a Cartan subgroup of $\mathcal{R}P_I$.*

Proof. By 14.5, the U_α, for $\alpha \in \Phi(I)^+$ directly span a unipotent subgroup $U_{\Phi(I)^+}$. Moreover, this group is normalized by $U_{\pm\beta}$ for $\beta \in I$, hence by $\mathscr{Z}_G(T_I)$. The intersection of $U_{\Phi(I)^+}$ and $\mathscr{Z}_G(T_I)$ is unipotent on the one hand, central in the latter group, hence consists of semi-simple elements on the other hand, therefore is reduced to $\{1\}$. We have

$$L(U_{\Phi(I)^+}) = \bigoplus_{\alpha \in \Phi(I)^+} \mathfrak{g}_\alpha, \quad L(\mathscr{Z}_G(T_I)) = L(T) \oplus \bigoplus_{\alpha \in [I]} \mathfrak{g}_\alpha.$$

These Lie algebras intersect only at the origin and $\mathscr{Z}_G(T_I) \cdot U_{\Phi(I)^+}$ is a semi-direct product with unipotent radical $U_{\Phi(I)^+}$. It contains B, hence is parabolic. The Bruhat decomposition of $\mathscr{Z}_G(T_I)$ shows that it is equal to P_I. The center of $\mathscr{Z}_G(T_I)$ contains T_I and is contained in T. But T_I is the identity component of the fixed point set of $W_I = W(T, \mathscr{Z}_G(T_I))$, hence, *a fortiori*, the identity component of the center of $\mathscr{Z}_G(T_I)$. The subgroup $T_I \cdot U_{\Phi(I)^+}$ is solvable, normal in P_I, hence belongs to $\mathscr{R}P_I$. But the identity component of $\mathscr{R}P_I \cap \mathscr{Z}_G(T_I)$ is the radical of $\mathscr{Z}_G(T_I)$, hence is contained in T_I. Therefore $\mathscr{R}P_I = T_I \cdot U_{\Phi(I)^+}$ and T_I is a maximal torus of $\mathscr{R}P_I$. By construction, no element in $\Phi(I)^+$ is trivial on T_I, therefore the centralizer of T_I in the Lie algebra of $U_{\Phi(I)^+}$ is reduced to $L(T_I)$ and, by 9.3, 10.6, T_I is its own centralizer in $\mathscr{R}P_I$, hence is a Cartan subgroup of $\mathscr{R}P_I$.

For $\beta \in X(T)$, let $\bar{\beta}$ be the restriction of β to T_I. If $\beta \in \Phi(T, G)$, then $\bar{\beta} = 0$ if and only if $\beta \in [I]$. Otherwise, $\bar{\beta}$ is an integral linear combination with coefficients all of the same sign of the $\bar{\alpha}$, where $\alpha \in I' = \Delta - I$. The $\bar{\alpha}$ $(\alpha \in I')$ are often called the simple roots of $\mathscr{R}P$ or of P with respect to T_I. They are linearly independent. Of course they are trivial on the identity component Z of the center of G. They may also be viewed as characters of $T'_I = (T_I \cap \mathscr{D}G)^o$ or of T_I/Z. Both tori are of dimension equal to Card I', hence the $\bar{\alpha}$ $(\alpha \in I)$ span a Q-basis of $X(T'_I)_{\mathbf{Q}}$ or $X(T_I/Z)_{\mathbf{Q}}$. The group $X_*(T_I)$ contains an element λ on which the $\bar{\alpha}$ $(\alpha \in I')$ are equal to one another and > 0. Then $\Phi(T, P_I)$ (resp. $\Phi(T, \mathscr{R}_u P_I)$, resp. $\Phi(T, \mathscr{Z}(T_I))$) is the set of $\alpha \in \Phi(T, G)$ such that $\langle \bar{\alpha}, \lambda \rangle \geqq 0$ (resp. $\langle \bar{\alpha}, \lambda \rangle > 0$, resp. $\langle \bar{\alpha}, \lambda \rangle = 0$).

14.19 Corollary. *Let P be a parabolic subgroup of G. Then P contains Levi subgroups. The maximal tori of $\mathscr{R}P$ are also Cartan subgroups of $\mathscr{R}P$. The Levi subgroups of P are the centralizers of the maximal tori of $\mathscr{R}P$. Any two are conjugate by a unique element of $\mathscr{R}_u P$.*

The group P is conjugate to a standard parabolic subgroup. The corollary then follows from the proposition and 11.23(ii).

14.20 *Opposite parabolic subgroups.* Two parabolic subgroups of G are said to be opposite if their intersection is a common Levi subgroup.

If P and P' are Borel subgroups, then the Levi subgroups are maximal tori and P, P' are opposite in the sense of 14.1.

14.21 Proposition. *Let P be parabolic subgroup of G and L a Levi subgroup of P.*

(i) *There exists one and only one parabolic subgroup P' opposite to P and containing L. Any two parabolic subgroups opposite to P are conjugate by a unique element of $\mathscr{R}_u P$.*

(ii) *Two parabolic subgroups P, Q contain opposite Borel subgroups if and only if their unipotent radicals intersect only at the identity.*

(iii) *If P' is opposite to P, the product map $\mu:(x, y) \mapsto x \cdot y$ induces an isomorphism of varieties of $\mathscr{R}_u P' \times P$ onto an open subset of G, equal to $P' \cdot P$.*

Proof. We may assume P and L to be in the standard situation of 14.17, whose notation is kept. Then the argument of 14.17 also shows that $U_{-\Phi(I)^+}$ is normalized by $\mathscr{Z}_G(T_I)$, hence that $P_I = U_{-\Phi(I)^+} \cdot \mathscr{Z}_G(T_I)$ is opposite to P. The uniqueness follows from 14.18. Thus the parabolic subgroups opposite to P correspond bijectively to the Levi subgroups of P (by the map $Q \mapsto Q \cap P$), hence also to the maximal tori of $\mathscr{R}P$, whence the conjugacy assertion.

(ii) In this proof we use repeatedly 14.18, 14.19, without explicit mention. Assume P and Q contain opposite Borel subgroups B and B^-. Then $P \cap Q$ contains $T = B \cap B^-$. For each root $\alpha \in \Phi(T, G)$, the group U_α is either in B or in B^-, hence either in P or Q. But if $U_\alpha \subset \mathscr{R}_u P \cap \mathscr{R}_u Q$, then $U_{-\alpha}$ is neither in P nor in Q, whence a contradiction. Therefore $\mathscr{R}_u P \cap \mathscr{R}_u Q = 1$. Assume now this last condition to hold and choose a maximal torus T of G contained in $P \cap Q$ (14.13). From the discussion in 14.17, we see that P (resp. Q) has a Levi subgroup L (resp. M) generated by T and the U_α such that $\alpha, -\alpha \in \Phi(T, P)$ (resp. $\alpha, -\alpha \in \Phi(T, Q)$). Then $R = (L \cap M)^o$ is generated by T and the U_α for which $\alpha \in \Phi(T, L) \cap \Phi(T, M)$. Thus a root α belongs to $\Phi(T, R)$ if and only if neither α nor $-\alpha$ is contained in $\Phi(T, \mathscr{R}_u P) \cup \Phi(T, \mathscr{R}_u Q)$. Let us write I for $\Phi(T, \mathscr{R}_u P)$ and J for $\Phi(T, \mathscr{R}_u Q)$. Then

$$\Phi(T, G) = \Phi(T, R) \cup I \cup J \cup -I \cup -J, \tag{1}$$

(disjoint union). Repeated application of Cor. 2 to 14.8 shows that the $U_{-\alpha}$ $(\alpha \in I)$ (resp. $U_{-\beta} (\beta \in J)$) directly span a unipotent group U_{-I} (resp. U_{-J}) normalized by R. Let $\alpha \in I$. Then $\alpha \notin J$ by assumption, therefore $-\alpha \in \Phi(T, Q)$. As a consequence, $U_{-I} \subset Q$. Similarly $U_{-J} \subset P$. The group U_{-I} (resp. U_{-J}) normalizes $\mathscr{R}_u Q$ (resp. $\mathscr{R}_u P$), therefore generates with it a unipotent group U_A (resp. U_B), where $A = -I \cup J$ and $B = I \cup -J = -A$. By (1), we see that

$$\Phi(T, G) = \Phi(T, R) \cup A \cup -A.$$

Choose two opposite Borel subgroups C, C^- of R containing T. Then $C \cdot U_A$ and $C^- \cdot U_{-A}$ are two opposite Borel subgroups of G contained in P and Q respectively.

(iii) Since $\mathscr{R}_u P' \cap P = \{1\}$, the map is injective. The image contains $B^- \cdot B$, which is open (14.14), and is acted upon transitively by $\mathscr{R}_u P' \times P$ where the first (second) factor acts by left (right) translations, hence $\mathscr{R}_u P' \cdot P$ is open in G. The differential of μ at the origin is an isomorphism, since $L(\mathscr{R}_u P')$ and $L(P)$ are transverse and span $L(G)$. Therefore μ is an isomorphism of varieties.

14.22 Proposition. *Let P and Q be parabolic subgroups of G.*

(i) *$P \cap Q$ is connected and $(P \cap Q) \cdot \mathscr{R}_u P$ is a parabolic subgroup.*
(ii) *Let L be a Levi subgroup of P. Then $H \mapsto H \cdot \mathscr{R}_u P$ is a bijection of the set of parabolic subgroups of L onto the set of parabolic subgroups of P.*
(iii) *If Q is conjugate to P and contains $\mathscr{R}_u P$, then $Q = P$.*

Proof. By 14.13, $P \cap Q$ contains a maximal torus T of G. Fix a Borel subgroup of P containing T and use the setup of 14.17. Then $P = P_\theta$ for some set θ of simple roots. For any $\alpha \in \theta$, the group P_θ contains both U_α and $U_{-\alpha}$. But at least one of them is contained in any Borel subgroup of Q containing T. Thus any Borel subgroup B' of $P \cap Q$ containing T contains at least one of U_α, $U_{-\alpha}$ for every $\alpha \in [\theta]$. Then $B' \cdot \mathscr{R}_u P$ contains T and one of $U_\alpha, U_{-\alpha}$ for every $\alpha \in \Phi(T, G)$. Therefore it is a Borel subgroup and $(P \cap Q) \cdot \mathscr{R}_u P$ is parabolic. It is then necessarily connected, hence $P \cap Q$ is connected.

(ii) We just saw that if Q is parabolic in L, then $Q \cdot \mathscr{R}_u P$ is parabolic in P (or G). If now R is a parabolic subgroup of P, it contains $\mathscr{R}_u P$ and can be written in the form $= (L \cap R) \cdot \mathscr{R}_u P$. We have $P/R = L/(L \cap R)$, hence $L/(L \cap R)$ is complete and $L \cap R$ is parabolic. It is clear that this correspondence is bijective.

(iii) The group $(Q \cap P) \cdot \mathscr{R}_u P$ is parabolic by (i) and contained in $P \cap Q$ by assumption. In particular $P \cap Q$ contains a Borel subgroup; but a parabolic subgroup is conjugate to a unique parabolic subgroup containing a given Borel subgroup (11.17) whence $P = Q$.

14.23 Lemma. *We keep the notation of* 14.12. *Let $X_\alpha (\alpha \in \Delta)$ be a non-zero element of $\mathfrak{g}_\alpha$ and $X = \sum_{\alpha \in S} X_\alpha$. Then* $\mathrm{Tr}(X, \mathfrak{b}) = \{g \in G, \mathrm{Ad}\, g(X) \in \mathfrak{b}\} = B$.

Let $g \in \mathrm{Tr}(X, \mathfrak{b})$. By 14.12, we may write $g = b' \cdot w \cdot b$ $(b, b' \in B, w \in W)$. Since B normalizes $\mathfrak{b}$, we may assume $b' = e$. By (3.17), $\mathrm{Ad}\, b(X) - X$ lies in the Lie algebra of the derived group (U, U) of U. It follows from $(*)$ in 14.5 that (U, U) is contained in the direct span of the U_γ $(\gamma \in \Phi^+, \gamma \notin \Delta)$. Since w permutes the $\mathfrak{g}_\alpha$, we may then write

$$\mathrm{Ad}\, g(X) = \sum_{\alpha \in \Phi} c_\alpha X_{w(\alpha)}, \quad (X_{w(\alpha)} \in \mathfrak{g}_{w(\alpha)}).$$

The set of α's for which $c_\alpha \neq 0$ contains Δ, and the corresponding $X_{w(\alpha)}$ are linearly independent. Since $\mathrm{Ad}\, g(X) \in \mathfrak{b}$, it follows that $w(\Delta) \subset \Phi^+$. By 14.7, 14.8, this yields $w = e$, $g \in B$.

14.24 Lemma. *Let H be a connected group and M a closed subgroup of H. Assume that there exists $X \in \mathfrak{m} = L(M)$ such that the set* $\mathrm{Tr}(X, \mathfrak{m})$ *of $h \in H$ for which* $\mathrm{Ad}\, h(X) \in \mathfrak{m}$ *consists of finitely many left classes* mod M. *Then*

$\mathcal{N}_H(\mathfrak{m})^o = M^o$, *and* $V = \bigcup_{h \in H} \mathrm{Ad}\, \mathfrak{h}(\mathfrak{m})$ *contains a dense open set of* $\mathfrak{h}$. *If* H/M *is complete, then* $V = \mathfrak{h}$.

$\mathcal{N}_H(\mathfrak{m})$ is a closed subgroup of H contained in $\mathrm{Tr}(X, \mathfrak{m})$, hence its identity component is equal to M^o. The proof of the remaining assertions is quite similar to that of 11.9. We consider the morphisms

$$H \times \mathfrak{h} \xrightarrow{\alpha} H \times \mathfrak{h} \xrightarrow{\beta} (H/M) \times \mathfrak{h},$$

where $\alpha(x, Y) = (x, \mathrm{Ad}\, x(Y))$ and $\beta = \pi \times Id$, with $\pi: H \to H/M$ the canonical morphism. Let $Q = \beta\alpha(H \times \mathfrak{m})$. By the same argument as in 11.9, it is seen that Q is closed. By definition, $V = pr_2(Q)$, where pr_2 is the projection on the second factor, hence V is closed if H/M is complete. The fibre of pr_2 over an element Z of V is $\pi(\mathrm{Tr}(Z, \mathfrak{m})^{-1})$. In particular, the fibre over X is finite. On the other hand, by using the projection pr_1 on the first factor, one sees again that $\dim Q = \dim H$, hence pr_2 is dominant.

14.25 Proposition. *Let H be a k-group. Then $\mathfrak{h}$ is the union of its Borel subalgebras.*

(By definition, a Borel subalgebra of $\mathfrak{h}$ is the Lie algebra of a Borel subgroup of H^o.)

We may assume H to be connected. Let R be its radical. The canonical projection $H \to H/R$ defines a bijection between Borel subgroups (11.14). This reduces us to the case where H is semi-simple. Let B be a Borel subgroup of H. In view of 14.23 and 14.24, the set V of conjugates of $\mathfrak{b}$ contains a dense open set of $\mathfrak{h}$. But, since H/B is complete, it is closed by 14.24, whence the proposition.

14.26 Proposition. *Let H be a k-group and $X \in \mathfrak{h}$. Then X is nilpotent if and only if it belongs to the Lie algebra of a closed unipotent subgroup.*

Proof. For the "if" part, see 4.8. Let now X be nilpotent. By 14.25, it belongs to the Lie algebra of a Borel subgroup of H, which reduces us to the case where H is connected, solvable. But then, 10.6(4) yields $X \in L(H_u)$.

Bibliographical Note

Up to 14.14, the results of this paragraph are due to Chevalley [13]. In particular, see Exp. 13 for 14.11, Exp. 16 for 14.8, and Exp. 17 for 14.9, 14.10. The main deviation here from [13] is that the integrality condition 14.6 is proved more directly, without recourse to representation theory.

In 14.12 it is proved that G/B (B a Borel subgroup) admits a partition into finitely many locally closed subvarieties, each isomorphic to an affine space. We also noticed that these are parametrized by the fixed points of a maximal torus T. The existence of such a decomposition has been since proved much more generally for any smooth projective variety in which $\mathbf{GL}_1$ operates with only finitely many fixed points by A. Bialynicki-Birula, Annals of Math. **98** (1973), 480–497.

14.25 is due to Grothendieck ([15], Exp. XIV, Thm. 4.11, p. 33). The proof given here is taken from [3].

Chapter V

Rationality Questions

In this chapter, all algebraic groups are affine. G is a k-group.

§15. Split Solvable Groups and Subgroups

15.1 Definition. *Let G be connected solvable. G splits over k, or is k-split, if it has a composition series $G = G_0 \supset G_1 \supset \cdots \supset G_s = \{e\}$ consisting of connected k-subgroups such that G_i/G_{i+1} is k-isomorphic to $\mathbf{G}_a$ or $\mathbf{GL}_1$ $(0 \leqq i < s)$.*

Examples. (1) The group $\mathbf{D}_n$ of invertible diagonal matrices of degree n splits over the prime field. More generally, if a k-torus splits over k in the sense of 8.2, then it is k-isomorphic to a product of $\mathbf{GL}_1$ (8.2, 8.3), hence is k-split in the present sense. The converse then follows from 8.14, Cor.

(2) Since a connected one-dimensional k-group is $\bar{k}$-isomorphic to $\mathbf{G}_a$ or $\mathbf{GL}_1$ (10.9), it follows from 10.6 that if k is algebraically closed, then any connected solvable k-group is k-split.

15.2 Proposition. *Let G be connected, solvable and k-split, and V a complete k-variety on which G acts k-morphically. If $V(k) \neq \phi$, then G has a fixed point in $V(k)$.*

Proof by induction on $\dim G$. Let N be a normal connected k-subgroup of G such that G/N is isomorphic to $\mathbf{G}_a$ or $\mathbf{GL}_1$. By induction, there exists $x \in V(k)$ fixed under N. The orbit map $g \mapsto g \cdot x$ is defined over k, and induces a k-morphism $f: G/N \to V$, whose image is the orbit $G(x)$ of x. By assumption, G/N is k-isomorphic, as a variety, to $\mathbf{P}_1 - A$ where A consists of one or two points rational over k. Since V is complete, f extends to a k-morphism of $\mathbf{P}_1$ into V. Then $f(\mathbf{P}_1) = G(x) \cup f(A)$ is complete, hence is the Zariski-closure of $G(x)$, hence is stable under G. The set $f(A)$ consists of one or two points rational over k, each of which is fixed under G since otherwise its orbit would meet $G(x)$. Q.E.D.

15.3 Definition. *A k-subgroup H of $\mathbf{GL}_n$ is trigonalizable over k if there exists $x \in \mathbf{GL}(n, k)$ such that $x \cdot H \cdot x^{-1}$ consists of upper triangular matrices.*

A flag F in K^n is rational over k if it consists of subspaces defined over k. This is the case if and only if F is the transform by an element of $\mathbf{GL}(n, k)$ of the standard flag $F_0:[e_1] \subset [e_1, e_2] \subset \cdots$. Thus, H is trigonalizable over k if and only if it leaves stable a flag rational over k. More intrinsically, we may say therefore that if V is a k-vector space, a k-subgroup H of $GL(V)$ is trigonalizable over k if and only if it leaves stable a point rational over k of the flag manifold $\mathscr{F}(V)$.

A trigonalizable group is necessarily solvable. If k is algebraically closed, any connected solvable k-subgroup of $\mathbf{GL}_n$ is trigonalizable over k by the Lie-Kolchin theorem (10.5).

15.4 Theorem. *Let G be connected, solvable.*

(i) *If G splits over k, then every image of G under a k-morphism f (resp. under a k-morphism into GL(V)) splits over k (resp. is trigonalizable over k) and $\mathscr{R}_u G$ is k-split.*

Assume G to be linear.

(ii) *The following conditions are equivalent*: (a) *G is trigonalizable over k*; (b) *G_u is defined over k and G/G_u splits over k*; (c) *$X(G) = X(G)_k$.*

(iii) *If k is perfect, G splits over k if and only if it is trigonalizable over k.*

(i) We show first that $G' = f(G)$ is k-split. To start with, assume G to be of dimension one, and $G' \neq \{e\}$, hence of dimension one, too. If $G = \mathbf{GL}_1$, then G' is k-isomorphic to $\mathbf{GL}_1$ by 8.2. Let $G = \mathbf{G}_a$. The group G' is then unipotent; it acts faithfully and k-morphically on the projective line $\mathbf{P}_1$; it has exactly one fixed point, say P, which is rational over $k^{p^{-\infty}}$, and one open orbit (10.9, Ramark). Then G acts k-morphically on $\mathbf{P}_1$ via f, with P as its only fixed point. By 15.2, P is then rational over k, hence (10.9, Remark), G' is k-isomorphic to $\mathbf{G}_a$.

In the general case, we have a composition series (G_i) of G as in 15.1. Then $(f(G_i))$ is a composition series for G', and f induces a surjective k-morphism of G_i/G_{i+1} onto $f(G_i)/f(G_{i+1})$, $(i = 0, \ldots, s-1)$. Our assertion now follows from the one-dimensional case.

Let now G' be a k-subgroup of $\mathbf{GL}_n$ and $\mathscr{F}_n$ the flag manifold of K^n. Since G' splits over k, and $\mathscr{F}_n(k) \neq \phi$, the group G' has a fixed point in $\mathscr{F}_n(k)$ by 15.2, hence is trigonalizable over k (15.3).

To prove that $\mathscr{R}_u G$ is k-split, we argue by induction on dim G. By definition, G contains a k-split normal subgroup N of codimension one. By induction, $\mathscr{R}_u N$ is k-split. If G/N is a torus, then $\mathscr{R}_u G = \mathscr{R}_u N$. So assume $G/N \tilde{\to} \mathbf{G}_a$. Let $\pi: G \to G' = G/\mathscr{R}_u N$ be the canonical projection. $\mathscr{R}_u G'$ has dimension one and $\mathscr{R}_u G = \pi^{-1}(\mathscr{R}_u G')$. It suffices therefore to show that the latter is isomorphic to $\mathbf{G}_a$. The group $N' = \pi(N)$ is a torus, which is normal, hence central, in G'. Therefore G' is nilpotent and is the direct product of N' and $\mathscr{R}_u G'$ (10.6). The latter is defined over k by (ii) below, hence k-isomorphic to $G'/N' \doteq G/N \doteq \mathbf{G}_a$.

(ii) We prove first that (a)$\Rightarrow$(b). Let G be contained in the group $\mathbf{T}_n$ of upper triangular matrices of degree n, and let $\mathbf{U}_n$ be the unipotent part of $\mathbf{T}_n$. Then $G_u = G \cap \mathbf{U}_n$. By 10.6(4), the Lie algebra of G_u consists of all the nilpotent elements in $L(G)$, hence $L(G_u) = L(G) \cap L(\mathbf{U}_n)$, and G_u is defined over k by 6.12. The k-morphism of $\mathbf{T}_n$ onto $\mathbf{D}_n$ with kernel $\mathbf{U}_n$ induces an injective k-morphism of G/G_u into $\mathbf{D}_n$, hence G/G_u is k-isomorphic to a direct product of $\mathbf{GL}_1$ (8.4).

(b)$\Rightarrow$(c) Since G is $\bar{k}$-isomorphic to the semidirect product of G/G_u and G_u (10.6), and $X(G_u) = \{1\}$, it is clear that the map $\pi^*: X(G/G_u) \to X(G)$ induced by the projection $\pi: G \to G/G_u$ is an isomorphism. If G_u is defined over k, then so is π, hence π^* maps $X(G/G_u)_k$ into $X(G)_k$. Therefore $X(G/G_u) = X(G/G_u)_k$ implies $X(G) = X(G)_k$.

(c)$\Rightarrow$(a) Let $\lambda: G \to GL(V)$ be a k-morphism. By the Lie-Kolchin theorem, there exists $\chi \in X(G)$ such that the eigenspace V_χ is $\neq 0$. Since, by assumption, χ is defined over k, the space V_χ is defined over k (5.2). Using induction on dim V, we see then that $\lambda(G)$ is trigonalizable over k, whence our contention.

(iii) Let now k be perfect. In view of (i), there remains to show that if G is trigonalizable over k, then it splits over k. Let $G \subset \mathbf{T}_n$. By taking the identity components of the intersections of G with the standard normal series of $\mathbf{T}_n$ (see 10.2), we get a normal series (G_i) consisting of connected k-groups, whose successive quotients are either k-isomorphic to one-dimension images of subgroups of $\mathbf{D}_n$, hence are k-isomorphic to $\mathbf{GL}_1$ (8.2, 15.1) or are unipotent, one-dimensional. Since k is perfect, the latter quotients are k-isomorphic to $\mathbf{G}_a$ by (10.9, Remark). Thus G splits over k.

Remark. By 15.4(ii), a linear k-torus is k-split if and only if it is trigonalizable over k. On the other hand, by the same result, a connected unipotent k-group is always trigonalizable over k, while it need not be k-split. In fact, [26, p. 46] gives an example of a one-dimensional such group, over a field of characteristic >2 (which is necessarily imperfect in view of 15.4(iii)).

15.5 Corollary. (i) *Let G be linear and trigonalizable over k. Then the image of G under a k-morphism $f: G \to \mathrm{GL}(V)$ is trigonalizable over k.*

(ii) *Let G be unipotent. Then G is trigonalizable over k. If k is perfect, G splits over k.*

(i) Let $G' = f(G)$. Then $G'_u = f(G_u)$ and f induces a surjective k-morphism of G/G_u onto G'/G'_u. Our assertion follows from 15.4(i), (ii).

(ii) This follows from 15.4(ii), (iii).

15.6 Proposition. *Let $G = \mathbf{G}_a, \mathbf{GL}_1$. Let X be a (non-empty) k-variety on which G acts k-morphically and transitively. Then $X(k) \neq \phi$.*

The variety X is irreducible. If $\dim X = 0$, then X is reduced to a point, necessarily rational over k. Otherwise, $\dim X = 1$, and for $x \in X$, the orbit map $f_x: g \mapsto g \cdot x$ is surjective (with finite fibres), hence its comorphism is an injective homomorphism of $K(X)$ into $K(G)$. But, here, $K(G) = K(T)$, where T is an indeterminate, hence, by Luroth's theorem, $K(X)$ is also a purely transcendental extension of K, of dimension one. In other words, X is a rational curve; it is obviously smooth. There exists therefore a k-isomorphism of X onto a k-open subset of a smooth complete curve Y of genus 0. The action of G on X extends to a k-morphic action of G on Y, and $Y-X$ consists of finitely many fixed points of G.

We have $\mathbf{P}_1 = G \cup A$ where either $A = \{0\}$ or $A = \{0\} \cup \{\infty\}$. The orbit map f_x extends to a morphism of $\mathbf{P}_1$ into Y, which is then surjective, since its image is closed, one-dimensional. It follows that $Y - X = f_x(A)$. The morphism f_x is defined over $k(x)$, so $f_x(A) \subset Y(k(x))$. This is true for any point $x \in X$. But we may find two points $x, y \in X(K)$ such that $k(x) \cap k(y) = k$, e.g. two "independent generic points," or a generic point x and an algebraic point y. Therefore $f(A) \subset Y(k)$ and the latter set is not empty. Since Y is of genus zero, it is then k-isomorphic to $\mathbf{P}_1$ (10.9, Remark). As a consequence, Y has at least three rational points (corresponding to 0, 1, ∞). Since $f(A)$ consists of at most two points, this proves that $X(k) \neq \phi$.

15.7 Corollary. *Let H be a k-group, L a connected solvable k-split subgroup, and $\pi: H \to H/L$ the canonical projection. Then $\pi(k): H(k) \to (H/L)(k)$ is surjective.*

Proof by induction on $\dim L$. Let N be the first non-trivial term of a composition series splitting L. Thus N is k-split, of codimension one, and L/N is k-isomorphic to $\mathbf{GL}_1$ or $\mathbf{G}_a$. The map π is the composition of the canonical projections

$$H \xrightarrow{\alpha} H/N \xrightarrow{\beta} H/L.$$

Let $x \in (H/L)(k)$ and $X = \beta^{-1}(x)$. Since β is separable, X is defined over k. The group L normalizes N, hence the right translations on H/N define a k-morphic action of L/N on H/N (6.11). Obviously, its orbits are the fibres of β. Therefore, 15.6 shows that $X(k) \neq \phi$. Since, by induction assumption $\alpha(k)$ is surjective, the corollary is proved.

15.8 Corollary. *Assume H to be connected. Then H is birationally isomorphic over k to $(H/L) \times L$. If H is a k-split solvable group, then it is a rational variety over k.*

By 15.7 and AG, 13.6, there exists a k-open subset U of H/L and a k-morphism $\sigma: U \to H$ such that $\pi \circ \sigma$ is the identity. The map $\varphi: (u, g) \mapsto \sigma(u) \cdot g$ is then a k-isomorphism of $U \times L$ onto $\pi^{-1}(U)$ (6.14, note that $U \times L$ and $\pi^{-1}(U)$ are smooth). Assume now H to be itself k-split solvable. If it is one-dimensional, then it is k-isomorphic to $\mathbf{G}_a$ of $\mathbf{GL}_1$, hence rational over k. Let N be a normal k-split subgroup of codimension 1 of H. Then H/N is rational and, by the first assertion, G is birationally k-isomorphic to $N \times G/N$. Arguing by induction on the dimension, we may assume that N is rational over k, and the second assertion is proved.

Our next goal is to extend 15.6 to connected solvable k-split groups. This requires some preparation.

15.9 Lemma. *Let G operate morphically on the irreducible affine variety V with closed orbits. Assume that G^o is either a torus or unipotent. Then $K(V)^G$ is the quotient field of $I = K[V]^G$.*

Proof. Let $N = G^o$ and $f \in K(V)^N$. If N is a torus, then f is in the quotient ring of $K[V]^N$ by 8.19. Let now N be unipotent. The ideal J of elements $q \in K[V]$ such that $f \cdot q \in K[V]$ is not zero and N-invariant. Let E be a non-zero finite dimensional N-invariant subspace of J (1.9). The space D of fixed points of N in E is at least one-dimensional, as follows from 10.5. There exists therefore $q \neq 0$, such that $r = f \cdot q$ is regular, hence in $K[V]^N$. Thus $f = r/q$, with r, $q \in K[V]^N$ and $q \neq 0$. Let now finally f be invariant under G. We have just seen that $f = a/b$, with a, $b \in K[V]$ fixed under G^o and $b \neq 0$. The finite group G/G^o operates on $K[V]^{G^o}$. Multiply a and b by the product of the $g \cdot b$ ($g \in G/H$, where H is the isotropy group of b in G) to get $f = a'/b'$ with $b' \neq 0$, invariant under G, hence a' invariant under G, too.

15.10 Lemma. *Let V be an irreducible affine k-variety on which G acts k-morphically and transitively. Let N be a normal k-subgroup of G whose identity component is unipotent or a torus. Assume that the quotient G/NH, where H is an isotropy group on V, is affine. Then V/N exists over k and is affine.*

Proof. Let $v \in V$, H its isotropy group and $o_v : g \mapsto g \cdot v$ the orbit map. Then o_v induces a bijective morphism φ of G/H onto V, hence G/H is affine, too. We note that, by construction of quotient spaces, we have

$$K[G/H] = K[G]^H, \quad K[G/HN] = K[G]^{NH}, \tag{1}$$

hence

$$K[G/NH] = K[G/H]^N.$$

The comorphism $\varphi^o : K[V] \to K[G/H]$ is injective and $K[G/H]$ is a finitely generated $K[V]$-module. There exists then $n \in \mathbb{N}$ such that

$$K[G/H]^{p^n} \subset K[V] \subset K[G/H], \tag{2}$$

(here p is the characteristic exponent). This implies

$$(K[G/H]^N)^{p^n} \subset K[V]^N \subset K[G/H]^N, \tag{3}$$

hence $K[V]^N$ is finitely generated. We claim that the affine variety Y with coordinate ring $K[V]^N$ is the desired quotient V/N.

The commutative diagram of inclusions

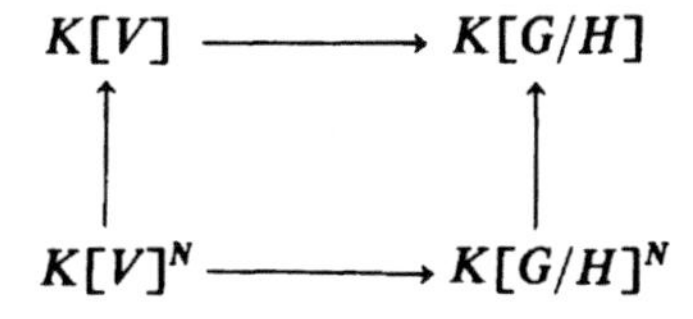

gives rise to a commutative diagram of G-equivariant dominant morphisms

$$\begin{array}{ccc} G/H & \xrightarrow{\varphi} & V \\ \downarrow{\scriptstyle \mu} & & \downarrow{\scriptstyle \nu} \\ G/NH & \xrightarrow{\psi} & Y \end{array}$$

where φ, ψ are bijective, μ is open, surjective, separable. Note that $K[V]^N$ is defined over k (8.19), hence Y and ν are defined over k. Since μ, φ, ψ are surjective, so is ν. The orbits of N are the fibres of μ, and φ is bijective, therefore the orbits of N on V are the fibres of ν. Thus ν is an orbit map. Since μ is open, and a purely inseparable morphism is a homeomorphism for the Zariski topology (AG, 12.1), ν is also open. By 15.9, $K(Y)$, which is by definition the quotient field of $K[Y]$, is also equal to $K(V)^N$, hence ν is separable (AG, 2.4). The group G is transitive on Y, hence Y is smooth, and in particular normal. It now follows from 6.2 that ν is a quotient morphism, hence $Y = V/N$.

15.11 Theorem. *Let G be a connected k-split solvable k-group and V a k-variety on which G acts k-morphically and transitively. Then V is an affine variety and $V(k) \neq \varnothing$.*

Proof. We show first that V is affine. Let $v \in V(\bar{k})$ and H its isotropy group. It is defined over $\bar{k}$. The orbit map o_v yields a bijective morphism of G/H on V. By AG, 18.3, it suffices to prove that G/H is affine. So assume that $V = G/H$. The group H/H^o operates on G/H^o and $G/H = (G/H^o)/(H/H^o)$ (6.10). Since the quotient of an affine variety by a finite group is affine (6.15), it suffices to show that G/H^o is affine, so we may assume H to be connected. Assume now in addition that H is unipotent. If $H = \mathscr{R}_u G$, then $G/H =$ is a k-group, hence is affine, so we argue by descending induction on $\dim H$. Let H be a proper connected subgroup of $\mathscr{R}_u G$. By 10.2, there exists a connected subgroup H' of $\mathscr{R}_u G$ normalizing H and such that H'/H has dimension one, and therefore is isomorphic to $\mathbf{G}_a$ (10.9). By induction, G/H' is an affine variety.

The group H' operates on G by right translations whence, by 6.10, a morphic action of H'/H on G/H which is free, whose orbits are the fibres of the projection $\tau: G/H \rightarrow G/H'$. Let k' be an extension of $\bar{k}$ in K. Since τ is separable, the orbit over a point $y \in (G/H')(k')$ is defined over k', hence contains a k'-rational point by 15.6. Thus $\sigma(k'): (G/H)(k') \rightarrow (G/H')(k')$ is surjective for all extensions of $\bar{k}$ in K. As a consequence, τ has local cross sections (AG, 13.6). Translating them by $G(\bar{k})$, we see that their domains of definition cover G/H. The inverse image of such an open set U is isomorphic to $U \times (H'/H)$, as remarked in 6.14, hence is affine. Therefore G/H' is covered by affine open subsets whose inverse images are affine; hence τ is an affine morphism and therefore the inverse image of any open affine subset of G/H' is affine (AG, 6.5). In particular, G/H is affine.

Let now H be any connected subgroup of G. We already know that $G/\mathscr{R}_u H$ is affine. Since $H/\mathscr{R}_u H$ is a torus whose orbits in $G/\mathscr{R}_u H$ are all closed, it follows from 8.21 that $(G/\mathscr{R}_u H)/(H/\mathscr{R}_u H)$ is affine. But it is isomorphic to G/H (6.10).

We now prove the second assertion. If G is one-dimensional, it is the content of 15.6, so we argue by induction on $\dim G$. The group G contains a proper non-trivial connected normal k-split subgroup N which is either a torus or is unipotent: If G is a torus, then it is a direct product of one-dimensional k-split tori; if it is unipotent, then take a k-split normal subgroup of codimension one; if it is neither, then set $N = \mathscr{R}_u G$ (15.4). By 15.10, the quotient V/N exists and is acted upon k-morphically, and obviously transitively, by G/N. By induction, $V/N(k)$ is not empty. We have seen in the proof of 15.10, or could deduce from 6.5, that the projection $V \to V/N$ is separable. Therefore the inverse image F of a point of $(V/N)(k)$ is defined over k. It is acted upon k-morphically and transitively by N hence, again by induction, $F(k)$ is not empty.

15.12 Corollary. *Let $\pi: V \to W$ be a separable surjective k-morphism of irreducible k-varieties. Assume that G operates k-morphically on V and that the fibres of π are the orbits of G. Then $\pi(k'): V(k') \to W(k')$ is surjective for every extension k' of k in K. The map π admits local cross sections.*

If $w \in W(k')$, then $\pi^{-1}(w)$ is defined over k' by separability, and it is acted upon k'-morphically and transitively by G, which is solvable and k'-split. Therefore $\pi^{-1}(w)$ contains a point of $V(k')$ and the first assertion is proved. The second one then follows from AG.13.6.

15.13. For information, let us mention some further properties of k-split solvable groups.

(a) *If V is as in* 15.11, *then V is k-isomorphic, as a variety, to a product of a certain number of copies of $\mathbf{G}_a$ and $\mathbf{GL}_1$, only of $\mathbf{G}_a$'s if the isotropy groups contain maximal tori of G, in particular if G is unipotent* [28: Thm. 5].

(b) *Let G be connected, unipotent. Assume it admits a k-morphic action of a k-torus T by automorphisms such that only the identity is a fixed point of G. Then G is k-split. There is a T-equivariant k-isomorphism of varieties of G onto its Lie algebra.*

The first assertion of (b) is a consequence of Cor. to Thm. 3 in [28], applied to the semi-direct product G' of T and G. The second one follows from [3: 9.12], also applied to G'.

We shall not need these results, except in one case in which a direct proof will be given (see §21).

15.14 Theorem. *Let k be perfect and G be connected. The maximal connected solvable k-split subgroups (resp. maximal connected unipotent k-subgroups, resp. maximal k-split tori) of G are conjugate by elements of $G(k)$. If R is one of them, $(G/R)(k)$ is the set of rational points of a projective k-variety V containing G/R as k-open subset, on which G acts k-morphically.*

Let R be a connected solvable k-split subgroup of G of maximal dimension. Let $\pi: G \to GL(V)$ be a faithful k-morphism such that $d\pi$ is injective, and V contains a line D defined over k whose isotropy group in G (resp. algebra in $L(G)$) is R (resp. $L(R)$) (see 5.1). The image of R in $GL(V/D)$ under the natural representation is trigonalizable over k (15.4). Therefore, there exists a flag P in V rational over k, whose one-dimensional subspace is D, and which is stable under R. Let $\mathscr{F}(V)$ be the flag manifold of V and $f: g \mapsto g(P)$

the orbit map of G into $\mathscr{F}(V)$. Let $X = G(P)$. This is a projective k-variety on which G operates k-morphically, and it is the union of $G(P)$ and of orbits of strictly smaller dimension. Let $Q \in X(k)$, and H its isotropy group. It is defined over k (since k is perfect), trigonalizable over k, since it leaves fixed an element of $\mathscr{F}(V)(k)$, hence split over k (15.4(iii)). Consequently, $\dim H \leqq \dim R$, and $\dim G(Q) \geqq \dim G(P)$. It follows that $Q \in G(P)$, hence $X(k) = G(P)(k)$. In view of the construction of P, the isotropy group (resp. algebra) of P is R (resp. $L(R)$), hence f is separable, and $G(P) = G/R$, whence $(G/R)(k) = X(k)$.

Let now H be a connected solvable k-split subgroup of G. By 15.2, it has a fixed point $x \in X(k)$. By the above, $x \in G(P)(k) = (G/R)(k)$. It follows from 15.7 that x is the image of an element $g \in G(k)$ under the orbit map f. But, then, $g \cdot H \cdot g^{-1} \subset R$, and, if H is unipotent $g \cdot H \cdot g^{-1} \subset R_u$. This shows that any maximal connected solvable k-split subgroup (resp. connected unipotent k-subgroup) of G is conjugate under $G(k)$ to R (resp. R_u), and that a k-split torus H is conjugate under $G(k)$ to a subtorus of R. We already know that the k-tori of R are k-split (15.4). There remains to show that two maximal ones T, T' are conjugate by an element of $R(k)$. We proceed by induction on $\dim R$. Let Q be a connected one-dimensional k-subgroup of R_u normal in R. It is trigonalizable over k and, since k is perfect, it is k-split, k-isomorphic to $\mathbf{G}_a$. Using induction and 15.7, we see that we are reduced to the case where $T' \subset T \cdot Q$, i.e. where $R_u = \mathbf{G}_a$ is one-dimensional. If R_u commutes with T, then $R = T \times Q$ is nilpotent, and $T = T'$. If not, then $\mathscr{Z}(T) = T$ and $R_u = (T, R)$ (see 9.3). Let S be the identity component of the centralizer of R_u in T. It has codimension one, is defined over k, normal in R. The groups T, T' are conjugate under R_u (10.6), therefore $S \subset T'$, and, dividing out by S, we may assume $T = \mathbf{GL}_1$. Let $Y = \{n \in R_u,\ n \cdot T \cdot n^{-1} = T'\}$. This is a closed set, not empty (10.6), defined over k. If x, $y \in Y$, then $y^{-1} \cdot x \in \mathscr{N}(T)$, hence $y^{-1} \cdot x \in \mathscr{Z}(T) = T$ (10.6), and finally $x \in y \cdot T$. Thus T acts transitively on Y by right translations, and $Y(k) \neq \phi$ by (15.6).

Bibliographical Note

Except for 15.14, the results of this section are due to M. Rosenlicht. Up to 15.5, see [26], which is one of the first papers devoted to rationality questions on affine algebraic groups, and [28]. For 15.6, see [25: p. 425] or [28: Theor. 4]. 15.11 and 15.12 are contained in Theorem 10 of [25]. The argument here is quite similar to Rosenlicht's, though expressed in a somewhat different language. The validity of 15.9 for G connected unipotent was noted in [27: p. 220]. For 15.8, see Cor. 1 and 2 to Theorem 10 in [25]. For 15.13, see [28: 8.2].

§16. Groups Over Finite Fields

In this section, k is a finite field, $q = p^s$ the number of elements of k; and $F_q : x \mapsto x^q$ the Frobenius homomorphism of a field of char p.

16.1 Let V be a k-variety. We denote by $v^{(q)}$ the image of $v \in V$ under the "Frobenius morphism" $V \to V$, also to be denoted by F_q. We recall that if $V \subset K^n$ is affine, then the coordinates of $v^{(q)}$ are obtained by applying F_q to those of v, and the comorphism $F_q^0 : k[V] \to k[V]$ is the q-th power homomorphism $f \mapsto f^q$.

The map F_q is a purely inseparable isogeny. It is bijective, and its differential at any point is the zero map. The fixed point set of F_q is $V(k)$, hence is finite. If V is a k-group, then F_q is a homomorphism.

16.2 Let $f : G \times G \to G$ be defined by $f(g, h) = g^{-1} \cdot h \cdot g^{(q)}$, and let f_g be the map $h \mapsto f(g, h)$. Then $f_{gh} = f_h \circ f_g$, and f is defined over k. Hence G operates on itself by means of the f_g's and this is a right k-morphic action.

16.3 Theorem. (Lang). *Let $a \in G$. Then the orbit map $s_a : g \mapsto g^{-1} \cdot a \cdot g^{(q)}$ is separable. Its image is open and closed.*

For the second assertion, it suffices to show that $s_a(G^o)$ is the connected component of a in G. We may therefore assume G to be connected. Let $i : x \mapsto x^{-1}$. Then (3.2):

$$(ds_a)_e(X) = di_e(X)a + a \cdot dF_q(X), \quad (X \in \mathfrak{g}).$$

But $(di) = -Id.$ (3.2) and $dF_q \equiv 0$ (16.2) hence

$$(ds_a)_e(X) = -X \cdot a,$$

which shows that ds_a is an isomorphism. As a consequence, s_a is dominant, separable. The orbit $s_a(G)$ of a then contains a non-empty open set, hence is open by homogeneity. This being true for any $a \in G$, the orbit is also closed.

16.4 Corollary. *Let G be connected. Then the map $g \mapsto g^{-1} \cdot g^{(q)}$ is surjective, separable.*

Apply 16.3 to the case $a = e$.

16.5 Corollary. (i) *Let G be connected, and V be a non-empty k-variety on which G acts k-morphically and transitively. Then $V(k) \neq \phi$.*

(ii) *Let H be a closed connected k-subgroup of G. Then the canonical map $\pi : G(k) \to (G/H)(k)$ is surjective.*

(i) Let $v \in V$. By assumption, there exists $g \in G$ such that $g \cdot v^{(q)} = v$. By 16.4, we may write $g = h^{-1} \cdot h^{(q)}$ for some $h \in G$. We have then $h \cdot v = h^{(q)} \cdot v^{(q)} = (h \cdot v)^{(q)}$, hence $h \cdot v \in V(k)$.

(ii) Let $x \in (G/H)(k)$. Then $\pi^{-1}(x)$ is defined over k. It is acted upon transitively by H. Therefore $\pi^{-1}(x)(k)$ is not empty by (i) and $x \in \pi(G(k))$.

16.6 Proposition. *G^o has a Cartan subgroup (resp. a maximal torus, resp. a Borel subgroup) defined over k. Two Borel subgroups defined over k are conjugate by an element of $G^o(k)$.*

Let H be a Cartan subgroup (resp. maximal torus, resp. Borel subgroup) of G^o. Then so is its transform $H^{(q)}$ under the Frobenius map. Hence there exists $g \in G^o$ such that

$$g \cdot H^{(q)} \cdot g^{-1} = H$$

(11.1, 11.3). By 16.4, we may find $a \in G^o$ such that $g = a^{-1} \cdot a^{(q)}$. Consequently

$$aHa^{-1} = a^{(q)} H^{(q)} a^{(q)-1} = (a \cdot H \cdot a^{-1})^{(q)}$$

hence aHa^{-1} is defined over k.

Let B, B' be two Borel subgroups defined over k. The variety $V = Tr(B, B') = \{x \in G \mid xBx^{-1} = B'\}$ is defined over k (since k is perfect), not empty (11.1). Since B is equal to its normalizer (11.16), B acts transitively by right translations on V. By (16.5), $V(k) \neq \phi$.

Remark. The last assertion is in fact valid over an arbitrary field (20.9).

16.7 Proposition. *Let H be a connected k-group and $f: G^o \to H$ a surjective k-morphism. Then a Cartan subgroup (resp. a maximal torus, resp. a Borel subgroup) of H defined over k is the image of such a subgroup of G^o.*

Let M be a Cartan subgroup of H, defined over k, and $M' = f^{-1}(M)^o$. Then M' is defined over k (since it is k-closed, and k is perfect), and $M = f(M')$, since M is connected. By (16.6), M' has a Cartan subgroup C' defined over k. By 11.14, the group $f(C')$ is a Cartan subgroup of $f(M') = M$. Hence $f(C') = M$. The proof in the other two cases is the same.

16.8 Proposition. *Let G be connected, H a connected k-group and $f: G \to H$ a k-isogeny. Then $G(k)$ and $H(k)$ have the same number of elements.*

Given an isogeny $r: M \to N$ of connected algebraic groups, we let $\deg r$ denote the degree of the field extension $k(M)$ over $r^o k(N)$. The degree of separability of this extension is the order of $\ker r$.

Let a_G be the map $g \mapsto g^{-1} \cdot g^{(q)}$. It is separable, surjective (16.4) of degree equal to the number $[G(k)]$ of elements in $G(k)$. Similarly the analogous map a_H has degree $[H(k)]$. But $f \circ a_G = a_H \circ f$, therefore

$$\deg(f \circ a_G) = \deg f \cdot \deg a_G = \deg f \cdot \deg a_H.$$

Since $\deg f \neq 0$, this proves the proposition.

16.9 Without giving any details, we mention that 16.4 has an interpretation in Galois cohomology. It is equivalent to the following fact: if L is a finite (Galois) extension of k, then

$$H^1(\mathrm{Gal}(L/k)), G(L)) = 0.$$

(For the definition of H^1, see e.g. [30].)

Bibliographical Note

Except for 16.6, 16.7 the results of this section are due to *S.* Lang [21]. That 16.6 was a consequence of Lang's theorem (16.3) was noticed by Serre (and mentioned in [26, footnote, p. 45]).

§17. Quotient of a Group by a Lie Subalgebra

In this paragraph, $\operatorname{char} k = p > 0$. *We put* $A_k = k[G]$ *and* $A_K = K[G]$. We recall 3.3 that $\mathfrak{g}$ may be identified with the algebra of left-invariant K-derivations of A_K by means of the map $X \mapsto *X$, and that $\mathfrak{g}(k) = \mathfrak{g} \cap \operatorname{Der}_k(A_k, A_k)$.

In §6, we introduced the quotient G/H of G by a closed subgroup. Of course, both G and H were "reduced," since only such groups have been dealt with in this book. However, this quotient can be (and has been) defined in the broader category of group schemes [14]. In this paragraph, we discuss another special case of this situation, where H is a restricted subalgebra of $\mathfrak{g}$, i.e. is of the simplest type among non-reduced groups of dimension zero. Our main objective here is 17.8, which will play an essential role in §18.

17.1 Lemma. *Let* $\mathfrak{m}$ *be a restricted Lie subalgebra of* $\mathfrak{g}$, *which is defined over* k. *Let* B *be the set of elements of* A_k *annihilated by* $\mathfrak{m}(k)$. *Then* B *contains* A_k^p *and is a finitely generated* k*-algebra. Moreover,* $\mathfrak{m} = \{X \in \mathfrak{g}, X \cdot B = 0\}$.

Since $\mathfrak{g}$ acts via derivations on A_K, it annihilates A_K^p, hence $B \supset A_k^p$. Therefore A_k is integral over B. The latter is then finitely generated by a known result (see e.g. Lemma 10, p. 58 in [29]).

To prove the second assertion, it suffices to consider the case where G is connected. Let $L = K(G)$ and M be the quotient field of B_K. Then $L \supset M \supset L^p$, hence L is a purely inseparable extension of M, of height one. The given action of $\mathfrak{g}$ on A_K extends in the abvious way to make $\mathfrak{g} \bigotimes_K L$ into a restricted Lie algebra over L of derivations of L. The field M is the field of invariants of $\mathfrak{m} \bigotimes_K L$. (If $D(a/b) = 0$, where D is a derivation, then $D(a \cdot b^{p-1}) = 0$, hence $a/b = a \cdot b^{p-1}/b^p$ is the quotient of two D-invariants.) But then, by Jacobson Galois theory of inseparable extensions of height one (Jacobson, Lectures in Abstract Algebra III, van Nostrand, Chap. IV, Theor. 19, p. 186), $\mathfrak{m} \bigotimes_K L$ is the algebra of *all* derivations of L which are zero on M, whence our second assertion.

17.2 Proposition. *Let* $\mathfrak{m}$ *be a restricted subalgebra of* $\mathfrak{g}$, *which is defined over* k. *There exists an affine* k*-variety* G/m, *and a* k*-morphism* $\pi: G \to G/\mathfrak{m}$ *having the following properties*:

(i) π *is bijective,* $\ker(d\pi)_x = x \cdot \mathfrak{m} (x \in G)$. *The comorphism* π^0 *induces an isomorphism of* $k[G/\mathfrak{m}]$ *onto the algebra* B *of elements in* A_k *annihilated by* $\mathfrak{m}(k)$.

(ii) *If* Z *is an affine k-variety and* $s: G \to Z$ *a k-morphism such that* $\ker(ds)_x \supset x \cdot \mathfrak{m}(x \in G)$, *then there exists a unique k-morphism* $f: G/\mathfrak{m} \to Z$ *such that* $s = f \circ \pi$.

The pair $(G/\mathfrak{m}, \pi)$ *is unique up to a k-isomorphism.*

Let (V_i, π_i) $(i = 1, 2)$ be two pairs satisfying the conditions imposed on $(G/\mathfrak{m}, \pi)$ in (i), (ii). Then we have unique k-morphisms $f: V_1 \to V_2$ and $h: V_2 \to V_1$ such that $\pi_2 = f \circ \pi_1$ and $\pi_1 = h \circ \pi_2$. It follows then immediately that f, h are bijective and that the comorphisms f^o, h^o are isomorphisms. There remains to show the existence.

Let $b_1, \ldots, b_s$ be a generating set for B as a k-algebra (17.1). By 1.9, we can find a finite dimensional subspace W_i of A_K, stable under G under right translations, defined over k, and containing b_i $(1 \leqq i \leqq s)$. Let $E = \amalg W_i$, and $b = b_1 + \cdots + b_s$. We claim that the orbit $V = G \cdot b$ of b, under the natural representation of G in E, and the orbit map $\pi: g \mapsto g \cdot b$ satisfy our conditions.

If $g \cdot b = b$, then $b_i(g) = b_i(e)$ $(1 \leqq i \leqq s)$. Since the b_i's generate B and B contains A_k^p, this implies $f(g) = f(e)$ for any $f \in A_k$, hence $g = e$, and π is bijective. Let $X \in \mathfrak{g}$. Then by (3.11), $X \in \ker(d\pi)_e$ if and only if $b_i * X = 0$, i.e., by 17.1, if and only if $X \in \mathfrak{m}$. Thus $\ker(d\pi)_e = \mathfrak{m}$. The equality $\ker(d\pi)_x = x \cdot \mathfrak{m}$ follows then from the fact that the orbit map is equivariant.

The orbit map being bijective, and V being normal, V is affine (AG.18.3) and π^o is injective. The relation $\ker(d\pi)_e = \mathfrak{m}$ implies that $\pi^o(k[V]) \subset B$. To establish the reverse inclusion, it suffices to prove that $b_i \in \operatorname{Im} \pi^o$ $(1 \leqq i \leqq s)$. Fix i. Let $u_1, \ldots, u_t$ be a basis of $W_i(k)$ and $a_1, \ldots, a_t$ the dual basis of $W_i^*(k)$. The restriction of $a_j \in W_i^* \subset E^*$ to V is an element of $k[V]$. If we denote it also by a_j, we have:

$$b_i(g) = g \cdot b_i(e) = \Sigma_j a_j(g \cdot b_i) \cdot u_j(e) = \Sigma_j \pi^o a_j(g) \cdot u_j(e),$$

which proves our contention.

Let now s and Z be as in (ii). Then $s^o(k[Z])$ is annihilated by $\mathfrak{m}(k)$, hence contained in B. Consequently, the unique k-morphism $f: V \to Z$ whose associated comorphism is $s^o: k[Z] \to B$ satisfies the relation $s = f \circ \pi$.

Remark. To show that V is affine, we have quoted AG.18.3. In fact, this can be avoided. With a little more work, one can show directly that V is closed in E. (For a similar argument, see e.g. A. Borel, Introduction aux groupes arithmétiques, Hermann, Paris, 1969, Prop. 7.7.)

17.3 Lemma. *We keep the notation of* 17.2 *and* §3. *Let* $f \in A_K$, $g \in G$, $X \in \mathfrak{g}$. *Then*

(i) $(i^o f) * X = -i^o(X * f)$;

(ii) $(f * X)(g) = (\operatorname{Ad} g(X) * f)(g)$.

Let $f_i, h_i \in A_K$ be such that $\mu^o f = \Sigma_i f_i \otimes h_i$. We have then

$$i^o f(g \cdot x) = f(x^{-1} \cdot g^{-1}) = \Sigma_i i^o f_i(x) \cdot h_i(g^{-1}).$$

Since $(i^o f * X)(g) = X(\lambda_{g^{-1}} \cdot i^o f)$ (see 3.4), this yields

$$(i^o f * X)(g) = \Sigma_i X(i^o f_i) \cdot h_i(g^{-1}),$$

hence, by 3.19(e),

$$(i^o f * X)(g) = -\Sigma_i X f_i \cdot h_i(g^{-1}) = -(X * f)(g^{-1}),$$

which proves (i). The equality $f(g \cdot x) = f(gxg^{-1} \cdot g)$ gives

$$\Sigma_i f_i(g) \cdot h_i(x) = \Sigma_i f_i(gxg^{-1}) \cdot h_i(g).$$

Therefore

$$(f * X)(g) = \Sigma_i X(f_i \circ \operatorname{Int} g) \cdot h_i(g).$$

Since the differential of Int is Ad by definition 3.13, the definition of the differential of a map, and 3.19, yield

$$(f * X)(g) = \Sigma_i(\operatorname{Ad} g(X) \cdot f_i) \cdot h_i(g) = \operatorname{Ad} g(X) * f)(g).$$

17.4 Proposition. *Let* $\mathfrak{m}$ *be a restricted subalgebra of* $\mathfrak{g}$ *which is defined over* k *and is stable under* Ad G. *Then the variety* $G/\mathfrak{m}$ *of* 17.2 *admits a canonical structure of* k*-group such that* $\pi: G \to G/\mathfrak{m}$ *is a* k*-isogeny. If* G' *is a* k*-group and* $s: G \to G'$ *a* k*-morphism whose differential annihilates* $\mathfrak{m}$, *then there is a unique* k*-morphism* $f: G/\mathfrak{m} \to G'$ *such that* $s = f \circ \pi$.

We keep the previous notation. Since $\mathfrak{m}$ is stable under Ad G, 17.3(ii) shows that B_K is also the set of invariants of the left convolutions $X * (X \in \mathfrak{m})$. In view of 17.3(i), it follows then that $i^o B = B$. Let $f \in B$, and write $\mu^o f = \Sigma f_i \otimes h_i$ (f_i, $h_i \in A_k$). We may assume the f_i's (resp. h_i's) to be linearly independent over k. For $X \in \mathfrak{g}$, we have

$$f * X = \Sigma f_i \cdot X h_i, \quad X * f = \Sigma X f_i \cdot h_i.$$

If $X \in \mathfrak{m}$, both left hand sides are zero. Consequently $Xf_i = Xh_i = 0$ for all i, which shows that $\mu^o B \subset B \otimes B$. Thus, B, endowed with μ, i^o, ε, satisfies all conditions of 1.5 for the coordinate ring over k of an affine k-group. This yields a group structure on $G/\mathfrak{m}$. Since it is compatible with the inclusion $B \subset A_k$, the map π of 17.2 is a k-isogeny. This proves the first assertion of 17.4. The second one follows from 17.2(ii).

17.5 Examples. (1) If $\mathfrak{m} = \mathfrak{g}$, then $k(G/\mathfrak{m}) = A_k^p$. The Frobenius isogeny of 16.1 may be viewed as the k-isogeny π of G onto $G/\mathfrak{g}$.

(2) Let $p = 2$ and $G = \mathbf{SL}_2$. Then $\mathfrak{g} = \mathfrak{sl}_2$ is the Lie algebra of 2×2 matrices with trace zero. It contains the one-dimensional space $\mathfrak{m}$ of multiples of the identity, which is a restricted ideal, normalized (in fact centralized) by $\mathbf{SL}_2$. It can be seen easily that $G/\mathfrak{m} = \mathbf{PGL}_2$. The morphism $\pi: G \to G/\mathfrak{m}$ is realized by the natural action of G on $\mathbf{P}_1$ (see 10.8). Then $\operatorname{Im} d\pi = \mathfrak{n}$ is a two dimensional ideal of $L(G/\mathfrak{m})$, stable under p-power operation and Ad$(G/\mathfrak{m})$. It can be checked that $\mathbf{PGL}_2/\mathfrak{n} \cong \mathbf{SL}_2$ and that the composition of the projections $G \to G/\mathfrak{m} \to (G/\mathfrak{m})/\mathfrak{n}$ is the Frobenius isogeny of G.

17.6 A character of $\mathbf{GL}_1$ is of the form $x \mapsto x^m$ ($m \in \mathbb{Z}$). Its differential is $X \mapsto m \cdot X$. Let now T be a torus. From this remark, it follows that $L(T) \cong X_*(T) \otimes k$, $L(T)^* \cong X(T) \otimes k$. In particular, $a \in X(T)$ has a zero differential if and only if $a \in p \cdot X(T)$.

17.7 Lemma. *Let G be connected. Assume all semi-simple elements of $\mathfrak{g}$ are central. Then the set of semi-simple elements of $\mathfrak{g}$ is a subspace defined over k, and is the Lie algebra of any maximal torus of G.*

Let T be a maximal torus of G. Let $X \in \mathfrak{g}$ be semi-simple. It is tangent to a torus (11.8), say S, and by 11.3, there exists $g \in G$ such that $g \cdot S \cdot g^{-1} \subset T$. By assumption, X is central in $\mathfrak{g}$, and therefore also in G (9.1). Hence $\mathrm{Ad}\, g(X) = X$, and $X \in \mathfrak{t}$. But $\mathfrak{t}$ consists of semi-simple elements (8.2, Cor.), hence $\mathfrak{t}$ is the set of all semi-simple elements of $\mathfrak{g}$. There exists a power $q = p^s$ of p such that the s-th iterate $[q]$ of $[p]$ annihilates all nilpotent elements of $\mathfrak{g}$ (if $G \subset \mathbf{GL}_n$, $s = n$ will do). If $X = X_s + X_n$ is the Jordan decomposition of $X \in \mathfrak{g}$, we have then $X^{[q]} = X_s^{[q]}$, which shows that $[q]$ maps $\mathfrak{g}$ into $\mathfrak{t}$. Since $[q]$ is bijective on $\mathfrak{t}$, as follows from 3.3(2), $\mathfrak{t}$ is in fact the image of $[q]$. But $[q]$ is a k-morphism of varieties (if $G \subset \mathbf{GL}_n$, $X^{[q]}$ is the matrix product of q copies of X, see 3.1), hence $\mathfrak{t} = \mathrm{Im}[q]$ is defined over k.

17.8 Proposition. *Let G be connected, not nilpotent, and assume that every semi-simple element of $\mathfrak{g}$ is central. Let T be a maximal torus of G. Then there exists a k-group G', such that not all semi-simple elements of $\mathfrak{g}'$ are central, and a purely inseparable k-isogeny $\pi: G \to G'$ such that* $\ker d\pi = \mathfrak{t}$ *and* $\mathrm{Im}\, d\pi$ *is a supplementary subspace in $\mathfrak{g}'$ to the Lie algebra of any maximal torus.*

Let $\Phi = \Phi(T, G)$ be the set of roots of G with respect to T (8.17). Since G is not nilpotent, T is not central in G (11.5) and Φ is not empty (9.2, Cor.). Let c be the greatest positive integer such that $\Phi \subset p^c X(T)$. In view of (17.6), $\mathfrak{t}$ is central in $\mathfrak{g}$ if and only if $c \geqq 1$. By 17.7 and the assumption, we have then $c \geqq 1$. The proof is carried out by descending induction on c.

By 9.1 all elements of $\mathfrak{t}$ are centralized by G. Of course, $\mathfrak{t}$ is restricted and invariant under $\mathrm{Ad}\, G$, hence we may apply 17.4, which yields a k-group $G_1 = G/\mathfrak{t}$ and a k-isogeny $\pi_1: G \to G_1$ such that $\ker d\pi_1 = \mathfrak{t}$. Let T' be a maximal torus of G_1. By (11.4), $\dim T = \dim T'$. By (17.7), $d\pi_1$ annihilates all semi-simple elements of $\mathfrak{g}$, hence (4.4), $d\pi_1(\mathfrak{g})$ consists of nilpotent elements. If T' is a maximal torus of G_1, we have then $d\pi_1(\mathfrak{g}) \cap L(T') = \{0\}$, hence

$$L(G_1) = d\pi_1(\mathfrak{g}) \oplus L(T'), \tag{1}$$

for dimensional reasons. Let now $T' = \pi_1(T)$. Since $d\pi_1$ annihilates $\mathfrak{t}$, it follows from 17.6 that the induced homomorphism $\pi_1^*: X(T') \to X(T)$ maps $X(T')$ into $p \cdot X(T)$. On the other hand, $d\pi_1$ is injective on the sum of the root spaces $\mathfrak{g}_\alpha$ ($\alpha \in \Phi$). It follows then from equivariance and (1) that π_1^* induces a bijection of $\Phi(T', G_1)$ onto Φ. Since $\pi_1^* X(T') \subset p \cdot X(T)$, this shows that if d is the greatest integer such that $\Phi(T', G_1) \subset p^d X(T')$, then $d < c$. If $d = 0$, then we take $G_1 = G'$, $\pi_1 = \pi'$. If not, by induction, we choose G' and π_2 which verify our conditions with respect to G_1. It is then clear that $(G', \pi' = \pi_2 \circ \pi_1)$ satisfy our requirements.

17.9 Proposition. *Let $f: G \to G'$ be a surjective k-morphism whose kernel N is defined over k. Then there exist a sequence of k-groups $G_o = G, G_1, \ldots, G_m = G'$ and a factorization $f = f_m \circ \cdots \circ f_o$ of f where $G_1 = G/N$, f_o is the quotient morphism $G \to G/N$, $f_i: G_i \to G_{i+1}$ is a quotient k-isogeny of height one $(i = 1, \ldots, m-1)$ and $f_m: G_m \to G'$ is a k-isomorphism.*

Proof. By the universal property of quotient morphisms (6.3) we have a factorization $f = h_1 \circ f_o$, where $h_1: G_1 \to G'$ is a k-morphism, necessarily bijective, purely inseparable. If it is an isomorphism (as is the case in characteristic zero), we are done. Assume $p \neq 0$. There exists $s \in \mathbb{N}$ such that $k[G_1]^{p^s} \subset h_1^o(k[G'])$. To factor h_1 we argue by induction on the smallest such s. If $s = 1$, then we are in the situation of 17.4 and $G' = G_1/\mathfrak{m}_1$, where $\mathfrak{m}_1$ is the subalgebra of $\mathfrak{g}_1$ which annihilates $h_1^o(k[G'])$, noting that $\mathfrak{m}_1$ is $\operatorname{Ad} G_1$-stable and restricted. Then $h_1 = f_2 \circ f_1$, where $f_1: G_1 \to G_1/\mathfrak{m}_1$ is the quotient morphism and f_2 a k-isomorphism. Assume now that $s > 1$. Let A be the subalgebra of $k[G_1]$ generated by $k[G_1]^p$ and $h_1^o(k[G'])$. Every derivation $X \in \mathfrak{g}_1$ is zero on $k[G_1]^p$ but no $X \neq 0$ is zero on $k[G_1]$, therefore if $A = k[G_1]$, then $h_1^o(k[G']) = k[G_1]$ and h_1 is a k-isomorphism. If not, let $\mathfrak{m}_1$ be the subalgebra of $\mathfrak{g}_1$ which annihilates A. Again, it is restricted and stable under $\operatorname{Ad} G_1$. We can form the k-group $G_2 = G_1/\mathfrak{m}_1$ and, by 17.4, have, over k, a factorization $h_1 = h_2 \circ f_1$, where f_1 is the quotient morphism $G_1 \to G_1/\mathfrak{m}_1$ and $h_2: G_2 \to G'$ is a purely inseparable k-morphism. By construction, $A = k[G_2]$ and

$$f_1^o(k[G_2]^{p^{s-1}}) = A^{p^{s-1}} = h_1^o(k[G']^{p^{s-1}}) \cdot k[G_1]^{p^s} \subset h_1^o(k[G']).$$

Thus $f_1^o(k[G_2]^{p^{s-1}}) \subset f_1^o(h_2^o(k[G']))$ and therefore

$$k[G_2]^{p^{s-1}} \subset h_2^o(k[G'])$$

so that we can apply the induction assumption to h_2.

Bibliographical Note

As was already mentioned, 17.2, 17.4 are contained in much more general results of [14]. 17.3 was first proved by Serre (Am. J.M. **80** (1958), 715–739), and Barsotti. See also P. Cartier (Bull. S.M. France **87** (1959), 191–220, §7). 17.8 is Prop. 2.2 of [2]. For a slightly more general version, see [3, §5.3].

§18. Cartan Subgroups Over the Groundfield. Unirationality. Splitting of Reductive Groups.

18.1 *Regular elements.* Let nil X ($X \in \mathfrak{g}$) be the multiplicity of the eigenvalue zero of ad X, and let $n(\mathfrak{g})$ be the minimum of nil X, as X ranges through $\mathfrak{g}$. The element X is *regular* if nil $X = n(\mathfrak{g})$, *singular* otherwise. Clearly, nil $X =$ nil X_s, hence X is regular if and only if its semi-simple part X_s is. If $p \neq 0$, then X is regular if and only if $X^{[p]}$ is so.

Taking coordinates in $\mathfrak{g}$ with respect to a base of $\mathfrak{g}(k)$, we can write

$$\det(\operatorname{ad} X - T) = T^{n(\mathfrak{g})}(c_0(X) + c_1(X)T + \cdots + c_q(X)\cdot T^q),$$

where the c_i's are homogeneous polynomials with coefficients in k, and $q = \dim \mathfrak{g} - n(\mathfrak{g})$. By the definition of $n(\mathfrak{g})$, the polynomial c_0 is not identically zero. The singular elements are the zeros of c_0, hence they form a proper k-closed algebraic subset of $\mathfrak{g}$, and the regular elements form a dense open set.

Let k be infinite. Then there always exists a semi-simple $X \in \mathfrak{g}(k)$ which is regular. In fact, since $\mathfrak{g}(k)$ is Zariski dense in $\mathfrak{g}$ (k infinite), there exists a regular $X \in \mathfrak{g}(k)$. Let $X = X_s + X_n$ be its Jordan decomposition (4.4). Then X_s is regular. If k is perfect, then $X_s \in \mathfrak{g}(k)$. Let $p \neq 0$. There exists a power $[q]$ of $[p]$ which annihilates X_n. Then $X_s^{[q]} = X^{[q]}$ will do.

18.2 Theorem. *Let G be connected.*

(i) *G contains a maximal torus and a Cartan subgroup defined over k.*
(ii) *If G is reductive or k is perfect, G is unirational over k and its center $\mathscr{C}G$ is defined over k.*

(i) A Cartan subgroup is the centralizer of a maximal torus T, and is defined over k if T is (9.2, Cor.). It suffices therefore to show the existence of a maximal torus defined over k. If G is nilpotent, see 10.6(3). If k is finite, see 16.6. In the general case, we shall use induction on dim G.

Let now k be infinite, and G be not nilpotent.

Let T be a maximal torus of G. Since G is not nilpotent, T is not central (11.5). If $p = 0$, then a non-zero character of T has a non-zero differential, therefore $\mathfrak{t}$ is not central in $\mathfrak{g}$, and $\mathfrak{g}$ has non-central semi-simple elements. We now let G' and $\pi: G \to G'$ be as in 17.8 if all semi-simple elements of $\mathfrak{g}$ are central (and hence $p \neq 0$), and $G' = G$, $\pi = Id$. otherwise. By (18.1), $\mathfrak{g}'(k)$ contains a regular semi-simple element Y. By construction, $n(\mathfrak{g}') \neq \dim \mathfrak{g}'$, hence $\mathfrak{z}(Y) \neq \mathfrak{g}'$. We let G operate on $\mathfrak{g}'$ via Ad$\circ\pi$ and, for $Z \in \mathfrak{g}'$, denote by G_Z the stability group of Z in G. Clearly, $\pi(G_Z) = \mathscr{Z}_{G'}(Z)$. In particular, by (9.1), $\dim G_Y = \dim \mathfrak{z}_{\mathfrak{g}'}(Y) \neq \dim \mathfrak{g}'$ hence $G_Y \neq G$. We claim that G_Y is defined over k. This is clear if k is perfect, or, by 9.1, if π is the identity. In the remaining case, it suffices to show that the orbit map $f: g \mapsto g(Y) = \operatorname{Ad}\pi(g)(Y)$ is separable (6.7). This amounts to proving that df_e is surjective. By 3.16(b) and 9.1, the tangent space at Y to the orbit Ad $G'(Y) = G \cdot Y$ is $Y + [Y, \mathfrak{g}']$.

On the other hand, by 3.16:

$$df_e(X) = Y + [d\pi(X), Y], \quad (X \in \mathfrak{g}).$$

By (17.8), $d\pi(\mathfrak{g})$ is supplementary in $\mathfrak{g}'$ to the Lie algebra of any maximal torus. We know Y is in the Lie algebra of a maximal torus (see 11.8). Since the latter commutes with Y, we get $[Y, d\pi(\mathfrak{g})] = [Y, \mathfrak{g}']$, which proves our contention.

Since Y is tangent to a maximal torus, $\mathscr{Z}_{G'}(Y)$ contains a maximal torus of G', and therefore G_Y^o contains a maximal torus of G. By conjugacy, all maximal tori of G_Y^o are maximal tori of G. By induction assumption, one of them is defined over k, whence (i).

(ii) Recall first (AG, 17.3) that an irreducible k-variety V is unirational over k if its function field $k(V)$ is contained in a purely transcendental extension of k. If (V_i) $(1 \leqq i \leqq m)$ are unirational over k, and $f: V_1 \times \cdots \times V_m \to V$ is a dominant k-morphism, then, clearly, V is unirational over k.

Let k be perfect. Then the unipotent radical $\mathscr{R}_u(G)$ of G and its center, which are always k-closed, are defined over k. The group $\mathscr{R}_u G$ splits over k (15.4), hence is a rational variety over k (15.8). Moreover, also by (15.8), G is birationally k-isomorphic, as a variety, to $G/\mathscr{R}_u(G) \times \mathscr{R}_u(G)$. This reduces the proof of unirationality to the case where G is reductive.

From now on, G is reductive. If G is a torus, see (8.13)(2). We now consider the case where k is infinite, and prove unirationality by induction on $\dim G$. We keep the notation of the proof of (i) and let H be the group generated by the groups G_Y^o, as Y ranges through the regular semi-simple elements of $\mathfrak{g}'(k)$. We claim that $H = G$.

Assume this is not the case. Then $\mathfrak{h}' = L(\pi(H))$ is a proper subalgebra of $\mathfrak{g}'$. Since k is assumed to be infinite, there exists $Z \in \mathfrak{g}'(k)$ which is regular, and not contained in $\mathfrak{h}'$. Let $Z = Z_s + Z_n$ be its Jordan decomposition. For some iterate $[q]$ of $[p]$, we have $Z_s^{[q]} = Z^{[q]}$, and the element $U = Z_s^{[q]}$ is regular, semi-simple, rational over k. It commutes with Z (since Z_s does, and $Z_s^{[q]} = Z_s^q$ in a matrix realization of G'), hence $\mathfrak{z}_{\mathfrak{g}'}(U) \not\subset \mathfrak{h}'$. But (9.1) $\mathfrak{z}_{\mathfrak{g}'}(U)$ is the Lie algebra of $\mathscr{Z}_{G'}(U)^o = \pi(G_U)$. As a consequence, $G_U \not\subset H$, a contradiction. Thus $H = G$. There exists then in the set of G_Y^o's finitely many groups $H_1, \ldots, H_t$ such that the product map $H_1 \times \cdots \times H_t \to G$ is surjective. The groups H_i are $\neq G$, defined over k by (i), and reductive since their images under the isogeny π are so by 13.19. By induction, they are unirational over k, hence so is G.

Let now k be finite. Then (16.6) it has a Borel subgroup B defined over k, and the latter contains a maximal torus T defined over k. Since k is perfect, the unique Borel subgroup B^- opposite to B and containing T is also defined over k, and the unipotent radicals U, U^- of B and B^- are defined over k. By (10.6), B is k-isomorphic to the semi-direct product $T \cdot U$. By (14.14), G, as a variety, is then birationally k-isomorphic to $U^- \times T \times U$. By (15.5), U, U^- split over k, hence 15.8 are rational varieties over k. Since T is unirational over k (8.13)(2), G is unirational over k.

The center $\mathscr{C}$ of G is contained in any maximal torus T of G. By (i) there is one which is defined over k. Therefore $\mathscr{C}G$ is defined over k_s, (8.2, Cor. and 8.11). Since it is k-closed, it is defined over k.

Remark. It can be shown that any connected k-group is generated by its Cartan subgroups defined over k: for finite k, see [3: 2.9]. For infinite k, this follows from the fact that the variety of Cartan subgroups of G is rational over k, due to Grothendieck (over arbitrary fields) ([15], Exp. XIV, Th. 6.1, p. 39; this is also proved in [3], 7.9, 7.10). As a consequence G is unirational over k if its Cartan subgroups are; this includes both cases of (ii).

18.3 Corollary. *Let G be connected, k infinite. If either k is perfect, or G is reductive, $G(k)$ is Zariski-dense in G.*

This follows from unirationality. We note that Rosenlicht has given an example of a one-dimensional unipotent k-group over an infinite k, in which $G(k)$ is finite [26, p. 46].

18.4 Corollary. *Let G be connected, solvable. Then G splits over an algebraic extension of k.*

By (18.2), G has a maximal torus T defined over k. Its unipotent radical is k-closed, hence splits over a finite purely inseparable extension of k, in view of 15.5. Since T splits over a finite (separable) extension of k (8.11), the corollary follows.

18.5 Lemma. *Let $\mathfrak{h}$ be the Lie algebra of a closed subgroup H of G. Assume that $\mathfrak{h}$ is defined over k and $\mathfrak{h} = \mathfrak{n}_{\mathfrak{g}}(\mathfrak{h})$. Then $\mathscr{N}_G(\mathfrak{h})$ is defined over k, the group H is of finite index in $\mathscr{N}_G(\mathfrak{h})$, defined over a finite separable extension of k, and H^o is defined over k.*

Let $N = \mathscr{N}_G(\mathfrak{h})$. Then N contains H, and $L(N)$ contains $\mathfrak{h}$ and normalizes $\mathfrak{h}$. Thus $L(N) = \mathfrak{h}$, $\dim N = \dim H$. Since $N \supset H$, this proves that $H^o = N^o$ and H has finite index in N. Assume N to be defined over k. Then so is $H^o = N^o$ (see 1.2(b)), and H is defined over k_s by AG.12.3. There remains to show that N is defined over k.

Let $d = \dim \mathfrak{h}$, $E = \Lambda^d \mathfrak{g}$ and $\pi = \Lambda^d \mathrm{Ad}$ the natural representation of G in E. Let D be the line representing $\mathfrak{h}$. It follows from the assumptions and the lemma in 5.1 that N is the stability group of D, and $L(N)$ is the stability algebra of D. Let $f: g \mapsto g \cdot D$ be the orbit map in the associated projective space $\mathfrak{P}(E)$. Since π and D are defined over k, the map f is defined over k. The kernel of df_e is the Lie algebra of the stability group of D, hence f is separable (see proof of 6.8). Consequently N is defined over k (6.7).

18.6 *Split reductive groups.* Let G be reductive, connected, T a maximal torus of G, and $\Phi = \Phi(T, G)$ the set of roots of G with respect to T. For each $\alpha \in \Phi$ there exists a connected unipotent subgroup U_α of G, normalized by T, and an isomorphism $\theta_\alpha : \mathbf{G}_a \to U_\alpha$ such that

$$t \cdot \theta_\alpha(x) \cdot t^{-1} = \theta_\alpha(t^\alpha \cdot x) \quad (x \in K, t \in T). \tag{1}$$

The group U_α is the unique one-dimensional subgroup normalized by T satisfying this condition (see 13.18). The group G is said to *split over* k, or to be k-split, if we can choose a T split over k (8.2) and isomorphisms θ_α defined over k.

Note that if θ' is an isomorphism of $\mathbf{G}_a$ onto U_α, then $f = \theta_\alpha^{-1} \circ \theta'$ is an automorphism of $\mathbf{G}_a$, hence there exists $c \in K^*$ such that $f(x) = c \cdot x$ $(x \in K)$. Then θ' also satisfies (1).

18.7 Theorem. *Let G be connected, reductive. Then G splits over k if it has a maximal torus which splits over k.*

Let T be a maximal torus which splits over k. Let U_α $(\alpha \in \Phi)$ be the unipotent one-parameter subgroup normalized by T. In view of the end remark of 18.6 it suffices to show that each U_α is k-isomorphic to $\mathbf{G}_a$.

Let $\alpha \in \Phi(T, G)$ and $T_\alpha = (\ker \alpha)^0$. The group T_α is defined over k (8.4), hence so is $G_\alpha = \mathscr{Z}(T_\alpha)$ (9.2, Cor.). The group G_α is reductive, generated by T, U_α, $U_{-\alpha}$, and $T \cdot U_\alpha$, $T \cdot U_{-\alpha}$ are the Borel subgroups of G_α containing T (see 13.18). This reduces us to the case where G has semi-simple rank equal to one. We shall prove first that U_α, $U_{-\alpha}$ are defined over k. Since they are the derived groups of $T \cdot U_\alpha$ and $T \cdot U_{-\alpha}$, it suffices to show that the two Borel subgroups $B = T \cdot U_\alpha$ and $B^- = T \cdot U_{-\alpha}$ containing T are defined over k. We have

$$\mathfrak{g} = \mathfrak{t} \oplus \mathfrak{g}_\alpha \oplus \mathfrak{g}_{-\alpha} \quad \mathfrak{b} = \mathfrak{t} \oplus \mathfrak{g}_\alpha, \quad \mathfrak{b}^- = \mathfrak{t} \oplus \mathfrak{g}_{-\alpha}.$$

The group T is k-split, hence α is defined over k, and $\mathfrak{g}_\alpha$, $\mathfrak{g}_{-\alpha}$ are defined over k (5.2). Thus $\mathfrak{b}$ and $\mathfrak{b}^-$ are defined over k. Since B and B^- are the normalizers of $\mathfrak{b}$ and $\mathfrak{b}^-$ (14.2, Cor. 2) they are k-closed. This finishes the proof when k is perfect. Let now k be infinite, of non-zero characteristic. Assume first that $\mathfrak{t}$ is not central in $\mathfrak{g}$, which amounts to supposing that $d\alpha \neq 0$. We have then

$$[\mathfrak{t}, \mathfrak{g}_\alpha] = \mathfrak{g}_\alpha \quad [\mathfrak{t}, \mathfrak{g}_{-\alpha}] = \mathfrak{g}_{-\alpha},$$

which implies that $\mathfrak{b}$ (rest. $\mathfrak{b}^-$) is its own normalizer in $\mathfrak{g}$. Then, B and B^- are defined over k by 18.5.

If now $\mathfrak{t}$ is central in $\mathfrak{g}$, we let G' and $\pi : G \to G'$ be as in 17.8. The group $T' = \pi(T)$ is a maximal torus of G' and is k-split (11.14, 8.4). The groups $B' = \pi(B)$ and $B'^- = \pi(B^-)$ are the two Borel subgroups of G' (11.14) containing T'. By the previous proof, they are defined over k. Let $\sigma : G' \to G'/B'$ be the canonical projection, and $\tau = \sigma \circ \pi$. It follows from the equality $\ker d\pi = \mathfrak{t}$ that $d\pi$ maps $\mathfrak{g}_{-\alpha}$ injectively onto a supplement of $L(B')$ in $\mathfrak{g}'$. Therefore τ is separable, and (6.7) $B = \tau^{-1}(B')$ is also defined over k. Similarly, B^- is defined over k.

We now know that U_α and $U_{-\alpha}$ are defined over k. Since $d\pi$ is injective on $\mathfrak{g}_\alpha$ and $\mathfrak{g}_{-\alpha}$, the map π is a k-isomorphism of U_α (resp. $U_{-\alpha}$) onto its image. It suffices then to prove that the latter is k-isomorphic to $\mathbf{G}_a$, which reduces us to the case where $\mathfrak{t}$ is not central in $\mathfrak{g}$. If k is perfect, our assertion has already been proved (10.9, Remark). Let now k be infinite. We may then find $H \in \mathfrak{t}(k)$ such that $d\alpha(H) \neq 0$. We have then $\mathfrak{z}(H) = \mathfrak{t}$, hence (9.1) $\mathscr{Z}_G(H)^0 = T$. Using 10.6, we get then $\mathscr{Z}_B(H) = T$. In particular, the map $f: u \mapsto \operatorname{Ad} u(H) - H$ of U_α into $\mathfrak{g}$ is injective. Let X be a non-zero element of $\mathfrak{g}_\alpha(k)$. Since B/U_α is commutative, we can write

$$\operatorname{Ad} u(H) = H + c(u) \cdot X, \quad (u \in U_\alpha),$$

where $c: u \mapsto c(u)$ is clearly an injective k-morphism of U_α into $\mathbf{G}_a$. It is then bijective. By (3.9)(2), $dc(X) = -[H, X] = -\alpha(H) \cdot X \neq 0$. Hence c is separable, and yields the desired k-isomorphism of U_α onto $\mathbf{G}_a$. The argument is the same for $U_{-\alpha}$, which ends the proof.

18.8 Corollary. *Let G be connected, reductive. Then G splits over a finite separable extension of k.*

G has a maximal torus T defined over k (18.2). The latter splits over a finite separable extension k' of k (8.11). G splits over k' by the theorem.

Bibliographical Note

In characteristic zero, 18.2 is due to Chevalley: 18.2(i) is proved in [12b] and 18.2(ii) in J.M.S. Japan **6** (1954), 303–324. It has been established by Rosenlicht [26] over infinite perfect fields, and by Grothendieck ([15], Exp. XIV) in general. The proof given here is taken from [2]. The paper of Chevalley quoted above also contains a result (Proposition 3) which appears to be essentially equivalent to the rationality over k of the variety of Cartan subgroups of G, when k is of characteristic zero (see remark to 18.2 for references to the general case of this theorem).

18.7 is due to Cartier (unpublished). More general results can be found in [15, Exp. XXII]. Here, we have followed [3].

§19. Cartan Subgroups of Solvable Groups

In this section, G is a connected solvable group defined over k.

We prove here some refinements of the conjugacy theorems already established. The first lemma is a sharpening of (∗) in the proof of 10.6(4), but note that 10.6(4) is used in its proof.

19.1 Lemma. *Let T be a maximal torus of G defined over k. Then every semi-simple element of $G(k)$ is conjugate by an element of $\mathscr{C}^\infty G(k)$ to an element of $T(k)$.*

Proof. We use induction on $\dim G$. If G is nilpotent, then T contains all semi-simple elements of G (10.6(3)) and there is nothing to prove. So assume G not nilpotent, i.e. $\mathscr{C}^{\infty}G \neq \{1\}$, and let $\pi: G \to G' = G/\mathscr{C}^{\infty}G$ be the canonical projection. The torus $T' = \pi(T)$ is maximal in G', and also defined over k.

Let $g \in G(k)$ be semi-simple. Since G' is nilpotent, $\pi(g) \in T'$, as already pointed out. The map π induces a bijective map of T onto T' and $d\pi$ also induces a bijection of $L(T)$ onto $L(T')$, since $T \cap \mathscr{C}^{\infty}G = \{1\}$ and $L(T) \cap L(\mathscr{C}^{\infty}G) = \{0\}$. There exists then a unique $t \in T(k)$ such that $\pi(t) = \pi(g)$. We want to show that g is conjugate to t under $\mathscr{C}^{\infty}(G)(k)$.

There exists a unique element $u \in \mathscr{C}^{\infty}G(k)$ such that $g = u \cdot t$. On the other hand, by 10.6, we can find $v \in \mathscr{C}^{\infty}G(k)$ such that $({}^{v}g)^{-1} \in T$. We have then necessarily ${}^{v}g^{-1} = t^{-1}$, since π is bijective on T. Then $g \cdot v \cdot g^{-1} \cdot v^{-1} = u$, i.e. $u \in c_g(\mathscr{C}^{\infty}G)$. It follows from 9.3 that there exists $w \in \mathscr{C}^{\infty}G(k)$ such that $(g, w) = u$. We have then $w \cdot g^{-1} \cdot w^{-1} = g^{-1} \cdot u = t^{-1}$, whence also $w \cdot g \cdot w^{-1} = t$.

19.2 Theorem. *Any two Cartan subgroups (resp. maximal tori) defined over k of G are conjugate under an element of $\mathscr{C}^{\infty}G(k)$.*

Proof. Let T, T' be two maximal tori defined over k and C, C' their centralizers. Then C and C' are Cartan subgroups defined over k. Conversely, if D is a Cartan subgroup defined over k, then its unique maximal torus is defined over k, maximal in G, and D is its centralizer. Therefore it suffices to prove either that C and C' or that T and T' are conjugate under $\mathscr{C}^{\infty}G(k)$. We distinguish two cases:

(i) *k is infinite.* Then T contains an element t such that $\mathscr{Z}_G(t) = \mathscr{Z}_G(T) = T$ (8.18). By 19.1, there exists $g \in C^{\infty}G(k)$ such that ${}^{g}t \in T$, hence such that ${}^{g}C \supset \mathscr{Z}_G(T') = C'$. But then ${}^{g}C = C'$.

(ii) *k is finite.* Let

$$V = \{x \in G \mid x \cdot C \cdot x^{-1} = C'\}.$$

It is an algebraic set (1.7), which is k-closed, hence defined over k (since k is perfect) and non-empty, since C and C' are conjugate under $\mathscr{C}^{\infty}G$ (10.6). It is a homogeneous space under the group $H = \mathscr{N}_G(T) \cap \mathscr{C}^{\infty}G$. But $\mathscr{N}_G(T) = \mathscr{Z}_G(T) = C$ (10.6, 12.1). Moreover, H is connected by 9.3 and defined over k by the Corollary to 9.2, therefore $V(k)$ is not empty (16.5).

19.3 Corollary. *Let T be a maximal torus of G defined over k, S a torus of G defined over k and $L \subset G(k)$ a subgroup consisting of semisimple elements. Then S and L are conjugate to subgroups of T under $\mathscr{C}^{\infty}G(k)$.*

Proof. The centralizer $\mathscr{Z}_G(L)$ of L is defined over k (9.2, Cor.) and L is contained in a torus (10.6(5)), hence in a maximal one. Therefore the maximal tori of $\mathscr{Z}_G(L)$ are maximal in G and all contain L (by conjugacy of maximal tori in $\mathscr{Z}_G(L)$); one of them is defined over k (18.2). Similarly S is contained in a maximal torus defined over k. We then apply the theorem.

Bibliographical Note

Theorem 19.2 is due to M. Rosenlicht [28]. The proof given here is essentially the one of [4: §11].

§20. Isotropic Reductive Groups

In this section, G is a connected reductive k-group, and Γ the Galois group of k_s over k.

20.1 Definition. The group G is *isotropic over k* if it contains a non-trivial k-split subtorus and is *anisotropic over k* otherwise.

We shall be interested in the case where G is not commutative and $\mathscr{D}G$ is isotropic over k and shall prove analogues of the structure theorems of §14 for the k-rational points of G.

20.2 Let T be a maximal torus of G and $\Phi = \Phi(T, G)$ the set of roots of G with respect to T. It is a reduced root system (14.8). For $\alpha \in \Phi$, the eigenspace

$$\mathfrak{g}_\alpha = \{X \in \mathfrak{g} \mid \mathrm{Ad}\,t(X) = t^\alpha \cdot X (t \in T)\}$$

is one-dimensional. It is the Lie algebra of a unique one-dimensional unipotent subgroup U_α normalized by T (13.18(4)).

20.3 Lemma. *Let T be a maximal torus defined over k of G and H a closed connected subgroup normalized by T. Then the following three conditions are equivalent*: (i) *H is defined over k*; (ii) *H is k-closed*; (iii) *$(H \cap T)^o$ is k-closed and $\Phi(T, H)$ is Γ-invariant.*

Proof. The implications (i)$\Rightarrow$(ii)$\Rightarrow$(iii) are clear. There remains to prove that (iii)$\Rightarrow$(i). Assume (iii). The group T splits over k_s (8.11) and the U_α are defined over k_s (18.8); so is $T \cap H$ (8.4 Cor.). The groups $U_\alpha(k_s)$ $(\alpha \in \Phi(T, H))$ are permuted by Γ and $(T \cap H)(k_s)$ is Γ-invariant. Since the groups $(T \cap H)^o(k_s)$ and $U_\alpha(k_s)$ $(\alpha \in \Phi(T, H))$ generate a dense subgroup of H, (i) follows from AG, 14.4.

20.4 Proposition. *Let S be a k-split subtorus of G. Then $\mathscr{Z}(S)$ is the Levi subgroup of a parabolic k-subgroup of G.*

Proof. Let T be a maximal torus of G defined over k and containing S (which exists by 9.2, Cor. and 18.2) and $\Phi = \Phi(T, G)$. Fix a non-zero element $\lambda \in X_*(S)$ which is "regular" in the sense that any root which restricts non-trivially on S is not zero on λ. Since Φ is finite, λ obviously exists. Let

$$\Psi = \{\alpha \in \Phi \mid \langle \alpha, \lambda \rangle > 0\}. \tag{1}$$

Then

$$\Phi = \Phi(T, \mathscr{Z}(S)) \coprod \Psi \coprod (-\Psi). \tag{2}$$

Ψ belongs to Φ^+ for a suitable ordering; it is closed and moreover, in the notation of 14.5, we have

$$(3) \qquad (\alpha, \Psi) \subset \Psi \quad \text{if} \quad \alpha \in \Phi(T, \mathscr{Z}(S)).$$

By 13.20, $\mathscr{Z}(S)$ is generated by T and the $U_\alpha(\alpha \in \Phi(T, \mathscr{Z}(S)))$. It follows then from (3) and 14.5 that the U_β $(\beta \in \Psi)$ directly span a unipotent subgroup U_Ψ normalized by $\mathscr{Z}(S)$. Since λ is defined over k, Ψ is invariant under Γ, hence U_Ψ is defined over k (20.3). In view of (2), $Q = \mathscr{Z}(S) \cdot U_\Psi$ is a parabolic k-subgroup, with Levi subgroup $\mathscr{Z}(S)$ and unipotent radical U_Ψ, which proves the proposition. Note that

$$(4) \qquad \Phi(T, Q) = \{\alpha \in \Phi \mid \langle \alpha, \lambda \rangle \geqq 0\}.$$

20.5 Proposition. *Let P be a parabolic subgroup of G defined over k. Then $\mathscr{R}P$ and $\mathscr{R}_uP$ are defined over k. The Levi k-subgroups P are the centralizers of the maximal tori defined over k of $\mathscr{R}P$. Any two are conjugate by a unique element of $\mathscr{R}_uP(k)$. Given a Levi k-subgroup L of P, the unique parabolic subgroup P^- opposite to P and containing L (see* 14.21) *is defined over k. The natural map $G(k) \to (G/P)(k)$ is surjective.*

Proof. The maximal tori of P are maximal in G. One of them, say T, is defined over k (18.2). Since $\mathscr{R}P$ and $\mathscr{R}_uP$ are k-closed and normalized by T, they are defined over k by 20.3. By 14.19, the Levi subgroups of P are the centralizers $\mathscr{Z}_G(S)$ of the maximal tori of $\mathscr{R}P$. In view of 9.3, 6.12, 20.4 the group $\mathscr{Z}_G(S)$ is defined over k if and only if S is. The conjugacy assertion then follows from 19.2. Extending the groundfield to k_s, we write $P = \mathscr{Z}_G(T_\theta) \cdot \mathscr{R}_uP$ as in 14.1, where everything is defined over k_s. The sets of roots $\Phi(T, \mathscr{Z}_G(T_\theta))$ and $\Phi(\theta)^+ = \Phi(T, \mathscr{R}_uP)$ are defined over k, (20.3), hence so is $-\Phi(\theta)^+ = \Phi(T, \mathscr{R}_uP^-)$. Thus P^- is defined over k_s and $P^-(k_s)$ is stable under Γ, hence P^- is defined over k.

Let $\pi: G \to G/P$. If k is finite, it is surjective on rational points (16.5) (recall that P is connected). Let now k be infinite. π induces an isomorphism of $\mathscr{R}_uP^-$ onto a k-open subset U of G/P, therefore $\pi: \mathscr{R}_uP^-(k) \to U(k)$ is surjective. Then for any $g \in G(k)$ the image of $G(k)$ will contain $g \cdot U(k)$. The union of the $g \cdot U$ $(g \in G(k))$ is a dense open set which is invariant under $G(k)$, whose k-rational points are contained in the image of $G(k)$. Its (closed) complement F is also $G(k)$-invariant. But $G(k)$ is Zariski dense in G (18.3), therefore F is G-invariant. Being proper, it must be empty.

20.6 Proposition. *Let P be a proper parabolic k-subgroup of G, L a Levi k-subgroup of P and S the identity component of the center of L. Then:*

(i) $L = \mathscr{Z}(S_d)$;

(ii) *G contains a proper parabolic k-subgroup if and only if it contains a non-central k-split torus;*

(iii) *the group P is minimal if and only if S_d is a maximal k-split torus of G;*

(iv) *if P is minimal, $\mathscr{N}(S) \cap P = \mathscr{Z}(S)$.*

Proof. Fix a maximal torus T of L defined over k. With respect to T, we use the notation and conventions of 14.17. Thus

$$P = P_\theta, \quad L = \mathscr{Z}(T_\theta), \quad R_u P = U_{\Phi(\theta)^+},$$

for some proper set θ of simple roots. The Galois group Γ of k_s/k operates on $X(T_\theta/Z)$, where $Z = (\mathscr{C}G)^o$, and leaves the set of restrictions $\bar{\alpha}$ of elements of $\Phi(\theta)^+$ stable. Those are linear combinations with positive integral coefficients of elements in θ', where $\theta' = \Delta - \theta$, (cf. 14.17), which are linearly independent. Therefore it permutes the elements of $\bar{\theta}'$. There exists $\lambda \in X_*(T_\theta/Z)$ on which the $\bar{\alpha} \in \bar{\theta}'$ take the same strictly positive integral value. It is therefore fixed under Γ, hence defined over k. Therefore $(T_\theta/Z)_d \neq \{1\}$ and $T_{\theta.d}$ is not contained in Z (8.15). Moreover

$$[\theta] = \{\alpha \in \Phi(T, G) | \langle \bar{\alpha}, \lambda \rangle = 0\}.$$

Therefore there exists a one-dimensional k-split torus S' in T_θ such that $\mathscr{Z}(S') = L$. A fortiori, $L = \mathscr{Z}(S_d)$ and (i) is proved.

(ii) We have just seen that if G has a proper parabolic k-subgroup, then it contains a non-central k-split subtorus. Assume conversely S is such a torus. Let λ be a non-zero element of $X_*(S)$. Fix a maximal torus T defined over k in $\mathscr{Z}_G(S)$ containing S. This exists by (18.2) since $\mathscr{Z}_G(S)$ is defined over k (9.3). Fix an ordering on $\Phi(T, G)$ such that $\alpha \in \Phi^+$ implies $\langle \alpha, \lambda \rangle \geqq 0$ and let $\psi = \{\alpha \in \Phi | \langle \alpha, \lambda \rangle \geqq 0\}$. Then ψ is a closed set of roots containing Φ^+. There exists therefore a unique $\theta \subset \Delta$ such that $\psi = [\theta] \cup \Phi(G)^+$ [9:VI, 1.7]. The corresponding parabolic subgroup P_θ is at first defined over k_s. But ψ is invariant under Γ, hence P_θ is defined over k (20.3). It is proper since S is not central. This concludes the proof of (ii).

(iii) We have again $P = L \cdot \mathscr{R}_u P$, with $L = \mathscr{Z}(S_d)$. Assume S_d is not maximal among k-split tori and let S' a k-split torus properly containing S. Note that the projection $L \to L' = L/S_d$ maps S onto the center of L'. The latter is therefore anisotropic over k and S'/S_d is a non-central k-split torus of L'. By (ii), L' contains a proper parabolic k-subgroup Q'. Then the inverse image Q of Q' in L is a proper parabolic k-subgroup of L. But then $Q \cdot \mathscr{R}_u P$ is a proper closed k-subgroup of P containing a Borel subgroup, hence is parabolic and P is not minimal. Assume now P is not minimal and let Q be a parabolic k-subgroup of G properly contained in P. It contains necessarily $\mathscr{R}_u P$ and $Q/\mathscr{R}_u P$ is a proper parabolic k-subgroup of $P/\mathscr{R}_u P \cong L$. It follows from (ii) that L contains a non-central k-split torus S'. Then, $S' \cdot S_d$ is a k-split torus properly containing S_d, and S_d is not maximal.

(iv) Let $n \in \mathscr{N}(S) \cap P$. We can write $n = x \cdot y$ with $x \in \mathscr{Z}(S)$ and $y \in \mathscr{R}_u P$. We have then $y \in \mathscr{R}_u P \cap \mathscr{N}(S)$ and, 10.6, applied to $S \cdot \mathscr{R}_u P$, shows that $y \in \mathscr{Z}(S)$. Therefore $n \in \mathscr{Z}(S)$.

20.7 Proposition. *Let P and Q be two parabolic k-subgroups of G. Then:*

(i) *$P \cap Q$ is defined over k and contains the centralizer of a maximal k-split torus.*

(ii) *If P and Q are minimal, they are opposite if and only if they contain two opposite Borel subgroups.*

Proof. (i) The group $P \cap Q$ is k-closed. To show that it is defined over k, it suffices therefore to prove it is defined over k_s. So assume now $k = k_s$. We choose a Borel k-subgroup B of P, a maximal k-split torus $T \subset B$ and use the setup of 14.16. The group B splits over k, therefore it has a fixed point x in $(G/Q)(k)$ (15.2). There exists $g \in G(k)$ mapping onto x (20.5). Then the isotropy group of x is ${}^gQ$. It contains B. We have shown therefore the existence of $g \in G(k)$ such that ${}^gQ \supset B$. By the Bruhat decomposition over k (recall $k = k_s$), we can write $g = a^{-1} \cdot n^{-1} \cdot b$ $(a, b \in B(k), n \in \mathcal{N}(T)(k))$. Of course ${}^bP = P$. On the other hand, the relation ${}^gQ \supset B$ gives ${}^bQ \supset {}^{na}B = {}^nB$, therefore ${}^bQ \cap {}^bP \supset T$ and $Q \cap P \supset b^{-1} \cdot T \cdot b$. This shows that $P \cap Q$ contains a maximal torus of G defined over k. It is then defined over k (20.3).

We now drop the assumption $k = k_s$. To prove the second part of (i), we may assume P and Q to be minimal among parabolic k-subgroups. We have seen that $P \cap Q$ contains a maximal torus of G defined over k. The group $S = T \cap \mathcal{R}P$ is a maximal torus of $\mathcal{R}P$ which is defined over k (since it is defined over k_s, as any closed subgroup of T, and is k-closed). Therefore $\mathcal{Z}(S)$ is a Levi k-subgroup of P (20.5), and S_d is a maximal k-split torus of G (20.6); then necessarily $L = \mathcal{Z}(T_d)$. The same argument shows that $\mathcal{Z}(T_d)$ is a Levi k-subgroup of Q. This concludes the proof of (i).

(ii) Assume P and Q are minimal. Then by 20.6 and (i) $P \cap Q$ contains a Levi k-subgroup of both P and Q. Therefore P and Q are opposite if and only if $\mathcal{R}_uP \cap \mathcal{R}_uQ = \{1\}$. But this is also a necessary and sufficient condition for P and Q to contain opposite Borel subgroups (14.21(ii)).

20.8 Lemma. *Let P and Q be parabolic subgroups of G. Then the set $M(P, Q)$ of elements $g \in G$ such that ${}^gP$ and Q contain opposite parabolic subgroups is a dense open set of the form $Q \cdot x \cdot P$ for some $x \in G$. It contains an element of $G(k)$ if either k is infinite or P and Q are defined over k.*

Proof. Fix a pair of opposite Borel subgroups B, B^-. For $a, b \in G$, it is easily seen that

$$M({}^aP, {}^bQ) = b \cdot M(P, Q) \cdot a^{-1}. \tag{1}$$

Obviously, ${}^bQ \cdot x \cdot {}^aP = b \cdot Q \cdot b^{-1} \cdot x \cdot a \cdot P \cdot a^{-1}$. Therefore, in proving the first assertion, we may replace P and Q by conjugates, and in particular assume that $B \subset P$ and $B^- \subset Q$. Then (1) shows that $M(P, Q) \supset Q \cdot P$, in particular $M(P, Q) \supset B^- \cdot B$, which is open in G (14.21(iii)). Let now $x \in M(P, Q)$. The groups ${}^xP$ and Q contain opposite Borel subgroups C and C^-. Since C^- and B^- are conjugate in Q, there exists $q \in Q$ such that ${}^{qx}P$ contains a Borel subgroup ${}^qC$ opposite to B^-. But the Borel subgroups opposite to B^- are conjugate under $\mathcal{R}_uB^-$ (14.21). Therefore we may assume q chosen so that ${}^{qx}P$ contains B. Then ${}^{qx}P = P$ (11.17), $qx \in P$ (11.16) and $x \in Q \cdot P$. This proves the first assertion.

If k is infinite, then $G(k)$ is Zariski-dense (18.3) and therefore has a non-empty intersection with $M(P, Q)$. Let now k be finite. Then P and Q contain Borel subgroups defined over k and any two such subgroups are conjugate under $G(k)$ (16.6). It suffices therefore to show that G has a pair of opposite Borel k-subgroups, but this is easy: Let B be a Borel k-subgroup. It has a maximal torus T defined over k (16.6). Then $\Phi(T; B)$ is a set of positive roots Φ^+ for a suitable ordering on $\Phi = \Phi(T, G)$ and is stable under the Galois group of k_s/k. Then so is $-\Phi^+$ and therefore the Borel subgroup opposite to B and containing T is defined over k (20.3).

20.9 Theorem.

(i) *The minimal parabolic k-subgroups of G are conjugate under $G(k)$.*

(ii) *The maximal k-split tori of G are conjugate under $G(k)$.*

(iii) *If P and P' are parabolic k-subgroups conjugate under $G(K)$, then they are conjugate under $G(k)$.*

Proof. (i) If k is finite, then the minimal parabolic k-subgroups are Borel subgroups defined over k and any two of those are conjugate under $G(k)$ (16.6). So assume k to be infinite. Let P and Q be minimal parabolic k-subgroups. The sets $M(P, P)$ and $M(Q, P)$ (notation of 20.7) are open dense, hence so is their intersection and the latter contains an element $g \in G(k)$ (18.3). Then ${}^gP$ and P on the one hand, ${}^gQ$ and P on the other, contain opposite Borel subgroups, their unipotent radicals intersect only at the identity (14.21) and 20.6 implies that ${}^gP$ and ${}^gQ$ are opposite to P. They are then conjugate by an element of $\mathscr{R}_uP(k)$ (20.5).

(ii) Let S, S' be maximal k-split tori. By 20.4, $\mathscr{Z}(S)$ and $\mathscr{Z}(S')$ are Levi k-subgroups of parabolic k-subgroups P, P'. By 20.6, P and P' are minimal, and therefore, by (i), conjugate over k. We may therefore assume $P = P'$. Then $\mathscr{Z}(S)$ and $\mathscr{Z}(S')$ are conjugate by some element $g \in \mathscr{R}_uP(k)$ (20.5). But S (resp. S') is the greatest k-split subtorus of the center of $\mathscr{Z}(S)$ (resp. $\mathscr{Z}(S')$), hence ${}^gS = S'$.

(iii) Since P is equal to its normalizer, the group P has a unique fixed point x on G/Q, which is therefore k-closed. If $y \in G/Q$ is fixed under $\mathscr{R}_uP$ and $g \in G$ maps onto y, then ${}^gQ \supset \mathscr{R}_uP$, hence ${}^gQ = P$ by 14.22(iii). Therefore x is the unique fixed point of $\mathscr{R}_uP$. But $\mathscr{R}_uP$ splits over k_s and has therefore a fixed point in $(G/Q)(k_s)$ (15.2). Thus x is also rational over k_s, whence $x \in (G/Q)(k)$. By 20.5, we can find $h \in G(k)$ projecting on x. Then ${}^hQ = P$.

Bibliographical Note

Most results of this section are taken from [4].

We have limited ourselves to reductive groups. However, the conjugacy under $G(k)$ of maximal k-split tori, or of maximal connected k-split solvable groups, is true in any connected k-group, as was announced in [7]. This is very easily deduced from 20.9 if $\mathscr{R}_u G$ is defined over k, but requires new arguments otherwise.

§21. Relative Root System and Bruhat Decomposition for Isotropic Reductive Groups

In this section, G is a connected reductive group defined over k, S a maximal k-split torus of G and Z_d the greatest central k-split torus of G.

21.1. The maximal k-split tori of G are conjugate under $G(k)$ (20.9) and in particular have the same dimension. The latter is called the *k-rank* or *rank relative to k* of G and will be denoted $r_k(G)$. Conjugacy, (together with 8.4, Cor.) also implies that $S' = (S \cap \mathscr{D}G)^o$ is a maximal k-split torus of $\mathscr{D}G$, and $S = S' \cdot Z_d$, with $S' \cap Z_d$ finite. Therefore

$$(1) \qquad r_k(G) = r_k(\mathscr{D}G) + r_k((\mathscr{C}G)^o).$$

The k-rank of $\mathscr{D}G$ is also called the *semisimple k-rank of G*. This section has content only if it is > 0.

Let S be a maximal k-split torus of G. We denote by ${}_k\Phi$ or ${}_k\Phi(G)$ the set $\Phi(S, G)$ of roots of G with respect to S. Its elements are called *k-roots* or *roots relative to k* of G (with respect to S). This set is empty if and only if S is central, i.e. $\mathscr{D}G$ is anisotropic over k. The k-roots are element of $X(S)$, but they are trivial on Z_d, and therefore may also be viewed as elements of $X(S')$ or $X(S/Z_d)$. In fact, the restriction to S' is an isomorphism from $\Phi(S, G)$ onto $\Phi(S', \mathscr{D}G)$, as is clear from the definitions.

Similarly, we introduce a Weyl group relative to k, namely the quotient $\mathscr{N}(S)/\mathscr{Z}(S)$, to be denoted ${}_kW(S, G)$, or ${}_kW(G)$ or ${}_kW$, as the context requires it. This group operates faithfully on S or S', or their groups of characters or cocharacters, leaving Z_d pointwise fixed. Again the restriction to S' provides an identification of ${}_kW(S, G)$ with ${}_kW(S', \mathscr{D}G)$.

Given a field E of characteristic zero and a finitely generated free $\mathbf{Z}$-module Y, we let $Y_E = Y \bigotimes_{\mathbf{Z}} E$. The group ${}_kW$ operates in particular on $X(S)_{\mathbb{R}}$, $X(S')_{\mathbb{R}}$, $X_*(S)_{\mathbb{R}}$ or $X_*(S')_{\mathbb{R}}$. Often, we assume these vector spaces to be endowed with an admissible scalar product (14.7), i.e. a positive definite scalar product invariant under ${}_kW$.

21.2 Theorem. *The rank of ${}_k\Phi(G)$ is equal to $r_k(\mathscr{D}G)$. The group ${}_kW(G)$, viewed as a group of automorphisms of $X_*(S)_{\mathbb{Q}}$ endowed with an admissible scalar product, is generated by the reflections to the hyperplanes which annihilate a k-root. Every connected component of $\mathscr{N}(S)$ meets $G(k)$.*

Proof. For $\alpha \in {}_k\Phi$, let S_α be the identity component of $\ker \alpha$. The rank of ${}_k\Phi$ is the codimension of the intersection of the S_α. Since the identity component of the latter is Z_d, our first assertion follows from 21.1(1).

We show now that if $r_k(\mathscr{D}G) \neq 0$, then $\mathscr{N}(S)(k) \neq \mathscr{Z}(S)(k)$. By 20.5 and 20.6, there exist two opposite parabolic k-subgroups P and Q having $\mathscr{Z}(S)$ as a common Levi subgroup. By 20.9, we can find $x \in G(k)$ such that ${}^xP = Q$. The group ${}^x\mathscr{Z}(S)$ is a Levi k-subgroup of Q hence (20.5) there exists $q \in Q(k)$ such that ${}^{qx}\mathscr{Z}(S) = \mathscr{Z}(S)$, and therefore ${}^{qx}S = S$. Then $qx \in \mathscr{N}(S)(k)$. But Int qx transforms $\mathscr{R}_uP$ onto $\mathscr{R}_uQ$, therefore $qx \notin \mathscr{Z}(S)$.

Next we show that ${}_kW$ contains the reflections to the hyperplanes $X_*(S_\alpha)_{\mathbb{R}}$ of $X_*(S)_{\mathbb{R}}$ $(\alpha \in {}_k\Phi)$. Let $\alpha \in {}_k\Phi$ and $M = \mathscr{Z}(S_\alpha)$. It is connected (11.12), defined over k and its Lie algebra is the set of fixed point of S in $L(G)$ (9.2, Cor.), hence $L(M) \neq L(\mathscr{Z}(S))$ and therefore $M \neq \mathscr{Z}(S)$. The identity component of the maximal central k-split torus of M, which contains S_α by definition and belongs to S, is $\neq S$, hence equal to S_α. As a consequence of 21.1, we see that $r_k(\mathscr{D}M) = 1$. Therefore, as was already proved, $\mathscr{N}_M(S)(k)$ contains an element x not belonging to $\mathscr{Z}(S)(k)$. The element x induces an orthogonal transformation of $X_*(S)_{\mathbb{Q}}$ which is not the identity but fixes pointwise $X_*(S_\alpha)_{\mathbb{Q}}$. It is therefore the reflection to that hyperplane. It will be denoted by r_α.

Let W' be the subgroup of ${}_kW$ generated by the r_α $(\alpha \in {}_k\Phi)$. The previous argument shows that any connected component of $\mathscr{N}(S)$ in the inverse image of W' contains an element of $G(k)$. To conclude the proof, there remains to show that $W' = {}_kW$.

Since W' is simply transitive on the chambers defined by the hyperplanes $X_*(S_\alpha)_{\mathbb{R}}$ (see V, §2, Theorem 2 in [9]), it suffices to show that if $w \in {}_kW$ leaves a Weyl chamber C stable, then it is the identity. There exists a point $x \in C$ fixed under w. Let $n \in \mathscr{N}(S)$ be a representative of w. Let T be a maximal torus of G defined over k and containing S. Then ${}^wT$ and T are maximal tori in $\mathscr{Z}(S)$. We may therefore, after multiplying w by an element of $\mathscr{Z}(S)$, assume that w leaves T stable. By 20.4 and its proof, there exists a parabolic k-subgroup P with Levi subgroup $\mathscr{Z}(S)$, whose roots with respect to T are the $\alpha \in \Phi(T, G)$ such that $\langle \alpha, x \rangle \geqq 0$. By 20.6, P is minimal. Since x is fixed under w and w leaves $\Phi(T; G)$ stable, n leaves $\Phi(T, P)$ stable and therefore P. But P is equal to its normalizer, hence $n \in P$, i.e. $n \in \mathscr{N}(S) \cap P$ and, by 20.6(iv), $n \in \mathscr{Z}(S)$.

21.3 Corollary. *The group ${}_kW$, operating via inner automorphisms, is simply transitive on the set $\mathscr{P}_S$ of minimal parabolic k-subgroups containing $\mathscr{Z}(S)$.*

Let $w \in {}_kW$, $w \neq 1$. It is represented by an element $n \in \mathscr{N}(S)(k)$ not contained in $\mathscr{Z}(S)$. Therefore ${}^nP \neq P$, in view of 20.6(iv) and the fact that P is its own normalizer. Thus ${}_kW$ operates freely on $\mathscr{P}_S$. Let now $Q \in \mathscr{P}_S$. By 20.9, there exists $x \in G(k)$ such that ${}^xQ = P$. Then ${}^xL$ and L are two Levi k-subgroups of P and there exists $p \in P(k)$ such that ${}^{p \cdot x}L = L$ (20.5). Thus $p \cdot x \in \mathscr{N}(S)(k)$ and ${}^{px}Q = P$, showing that ${}_kW$ is transitive on $\mathscr{P}_S$. Here $L = \mathscr{Z}(S)$.

21.4 Corollary. *Let l be an extension of k, and T a maximal l-split torus of G containing S. Let $\mathscr{N}(S, T) = \mathscr{N}(S) \cap \mathscr{N}(T)$ and ${}_lW_k = \mathscr{N}(S; T)/\mathscr{Z}(T)$. Then $\mathscr{N}(S) = \mathscr{N}(S; T) \cdot \mathscr{Z}(S)$ and ${}_kW$ is the restriction of ${}_lW_k$ to S.*

Let $w \in {}_kW$ and $n \in \mathcal{N}(S)(k)$ a representative. The tori T and ${}^nT$ are maximal l-split in $\mathcal{Z}(S)$ hence conjugate under an element of $Z(S)(l)$ (20.9). Therefore $\mathcal{N}(S)(k) \subset \mathcal{N}(S, T) \cdot \mathcal{Z}(S)$. Since $\mathcal{N}(S)(k)$ meets every connected component of $\mathcal{N}(S)$ by the theorem, this proves the first assertion. The second one is just a reformulation.

21.5 Lemma. *Let Ψ be a subset of ${}_k\Phi(G)$ which contains all the rational multiples of its elements belonging to ${}_k\Phi(G)$. Let $w \in {}_kW(G)$ be a product of reflections r_α ($\alpha \in \Psi$) and $c \in X(S)$. Then $w(c) - c$ is a linear combination with integral coefficients of elements in Ψ.*

Proof. Let T be a maximal torus containing S and $j: X(T) \to X(S)$ the restriction homomorphism. Let $\eta = j^{-1}(\Psi) \cap \Phi(T, G)$. By 21.4, $W = W(T, G)$ contains at least one element w_α whose restriction to S is r_α ($\alpha \in {}_k\Phi$). Such an element acts trivially on the hyperplane $V_\alpha = X_*(S_\alpha)_{\mathbb{R}}$ of $X_*(S)_{\mathbb{R}}$. Therefore it is a product of reflections r_β in W, where β belongs to the set of roots which are zero on V_α (see V.3.3, Prop. in [9]). For any $\alpha \in \Psi$, these roots all belong to η, in view of our assumption on Ψ. On the other hand j is surjective, since S is a direct factor of T (8.5, Cor.). We are now reduced to the case where $S = T$, ${}_kW = W$ and ${}_k\Phi = \Phi(T, G)$ is a root system (14.8). In this case, we can use Prop. 27 in VI, 1.10 of [9].

21.6 Theorem. *Let $S' = (S \cap \mathcal{D}G)^o$ and assume $S' \neq \{1\}$. Let $(\ ,\ )$ be an admissible scalar product on $X(S')_{\mathbb{R}}$; let $\alpha \in \Phi(S, G)$ and $\gamma \in X(S)$. Then $2(\alpha, \gamma)/(\alpha, \alpha) \in \mathbb{Z}$. In particular ${}_k\Phi(S', \mathcal{D}G)$ is a root system in $X(S')_{\mathbb{R}}$, whose Weyl group is ${}_kW$.*

Proof. The restriction $X(S) \to X(S')$ is an isomorphism of ${}_k\Phi(S, G)$ onto ${}_k\Phi(S', \mathcal{D}G)$ and the rank of ${}_k\Phi$ is equal to the k-rank of $\mathcal{D}G$ (21.2). Therefore ${}_k\Phi$ generates $x(S')_{\mathbb{R}}$ over $\mathbb{R}$. On the other hand,

$$r_\alpha(\gamma) = \gamma - 2(\alpha, \gamma) \cdot (\alpha, \alpha)^{-1} \alpha,$$

hence $2(\alpha, \gamma)/(\alpha, \alpha) \in \mathbb{Z}$ by 21.5. This and 21.2 show that all the conditions for a root system with Weyl group ${}_kW$ (see VI, 1.1 in [9]) are fulfilled by the restrictions to S' of the elements of ${}_k\Phi$.

21.7 Remark. The root system ${}_k\Phi$ may be reducible and we shall describe in §22 a decomposition of G as an almost direct product reflecting the reducibility of ${}_k\Phi$. If k is not a splitting field for G, it may also happen that ${}_k\Phi$ is not reduced (if irreducible, it is then of type $\mathbf{BC}_n$ in the notation of [9]). Given $\alpha \in {}_k\Phi$ we let (α) be the set of roots which are positive multiples of α. Thus (α) consists either of α or of α and 2α. Let also Φ_{nd} be the set of non-divisible roots, i.e. of roots α such that $\alpha/2$ is not a root. Then Φ is the disjoint union of the subsets (α), where α runs through Φ_{nd}. For $\alpha \in {}_k\Phi$ we denote as usual by $\mathfrak{g}_\alpha$ the corresponding eigenspace in $\mathfrak{g}$. We let also

$$\mathfrak{g}_{(\alpha)} = \bigoplus_{\beta \in (\alpha)} \mathfrak{g}_\beta.$$

Thus either $\mathfrak{g}_{(\alpha)} = \mathfrak{g}_\alpha$ or $\mathfrak{g}_{(\alpha)} = \mathfrak{g}_\alpha \oplus \mathfrak{g}_{2\alpha}$. We have

$$\mathfrak{g} = L(\mathscr{Z}(S)) \oplus \bigoplus_{\alpha \in {}_k\Phi} \mathfrak{g}_\alpha = L(\mathscr{Z}(S)) \oplus \bigoplus_{\alpha \in \Phi_{nd}} \mathfrak{g}_{(\alpha)}.$$

If 2α is not a root then $\mathfrak{g}_\alpha$ is commutative. If 2α is a root, it it usually not a subalgebra, but $\mathfrak{g}_{(\alpha)}$ is a (metabelian) one. This follows from the standard rule $[\mathfrak{g}_\alpha, \mathfrak{g}_\beta] \subset \mathfrak{g}_{\alpha+\beta}$. As usual, a subset ψ of ${}_k\Phi$ is said to be *closed* if $\alpha, \beta \in \psi$, $\alpha + \beta \in {}_k\Phi$ imply $\alpha + \beta \in \psi$. Note that now we have also to allow $\alpha = \beta$ in checking this condition. We let ψ_{nd} be the set of $\alpha \in \psi$ which are not divisible in ψ, i.e. such that $\alpha/2 \notin \psi$. Thus

$$\psi = \bigcup_{\alpha \in \psi_{nd}} (\alpha).$$

21.8. *Relations between absolute and relative roots.* Let T be a maximal torus of G defined over k and containing S and $j: X(T) \to X(S)$ the restriction homomorphism. By definition, we have

$$(1) \qquad {}_k\Phi \subset j(\Phi) \subset {}_k\Phi \cup \{0\}.$$

An ordering on Φ and one on ${}_k\Phi$ are said to be compatible if

$$(2) \qquad {}_k\Phi^+ \subset j(\Phi^+) \subset {}_k\Phi^+ \cup \{0\}.$$

Compatible orderings always exist. More precisely, given an ordering on ${}_k\Phi$, there is always an ordering on Φ compatible with it. Let us give one such construction. Let T' be a subtorus of T such that $T = T' \cdot S$ and $T' \cap S$ is finite (8.5 Cor.). If $\alpha \in \Phi$ is zero on S, then it is not zero on T' (since it is not zero on T), hence it belongs to $\Phi(T', \mathscr{Z}(S))$ and every such root is obtained in this way. Fix $c \in X_*(S)_{\mathbb{R}}$ in the positive Weyl chamber and choose $c' \in X_*(T')_{\mathbb{R}}$ on which no element of $\Phi(T', \mathscr{Z}(S))$ is zero. Then define $\alpha \in \Phi$ to be positive if either $\langle \alpha, c \rangle > 0$ or $\langle \alpha, c \rangle = 0$ and $\langle \alpha, c' \rangle > 0$.

Let Δ and ${}_k\Delta$ the simple roots for these orderings. Since every element of a root system is a linear combination of simple roots with integral coefficients all of the same sign, we see that

$$(3) \qquad {}_k\Delta \subset j(\Delta) \subset {}_k\Delta \cup \{0\}.$$

Let $\Delta^o = \{\beta \in \Delta \mid j(\beta) = 0\}$. Then $\Phi(T, \mathscr{Z}(S)) = [\Delta^o]$. We have

$$(4) \qquad \Delta = \Delta^o \coprod \coprod_{\alpha \in {}_k\Delta} (\eta(\alpha) \cap \Delta),$$

where, for $\alpha \in {}_k\Phi$, $\eta(\alpha)$ is the set of roots mapping onto α under j.

21.9 Proposition. (i) *Let $\alpha \in {}_k\Phi$. There exists a unique closed connected unipotent k-subgroup $U_{(\alpha)}$ normalized by $\mathscr{Z}(S)$ with Lie algebra $\mathfrak{g}_{(\alpha)}$.*

(ii) *Let ψ be a closed subset of ${}_k\Phi^+$. There is a unique closed connected* *unipotent k-subgroup U_ψ normalized by $\mathscr{Z}(S)$ with Lie algebra the sum of the $\mathfrak{g}_\alpha$ $(\alpha\in\psi)$. It is directly spanned by the $U_{(\alpha)}$ $(\alpha\in\psi_{nd})$ taken in any order.*

Proof. (i) We use the setup and notation of 21.7 and assume Φ endowed with an ordering compatible with the given one on ${}_k\Phi$. For $\alpha\in{}_k\Phi$ the set $\eta(\alpha)$ is closed, and all its elements have the same sign as α (in particular, it is special in the sense of 14.5). Clearly

$$\mathfrak{g}_{(\alpha)} = \bigoplus_{\beta\in\eta(\alpha)} \mathfrak{g}_\beta. \tag{5}$$

By 13.20, 14.5, the U_β $(\beta\in\eta(\alpha))$ directly span a T-invariant closed connected unipotent subgroup $U_{\eta(\alpha)}$ with Lie algebra $\mathfrak{g}_{(\alpha)}$ and which is uniquely determined by these properties. Moreover, $\eta(\alpha)$ is invariant under the Galois group Γ of k_s/k, therefore $U_{\eta(\alpha)}$ is defined over k. This group will be our $U_{(\alpha)}$. By 13.20, $\mathscr{Z}(S)$ is generated by T and the U_β $(\beta\in\Phi(T,\mathscr{Z}(S))$. It is clear that if $\beta\in\Phi(T,\mathscr{Z}(S))$, $\gamma\in\eta(\alpha)$ and $i\beta+j\gamma\in\Phi$ for some strictly positive integers i, j then $i\beta+j\gamma\in\eta(\alpha)$. It follows then again from 14.5 that U_β normalizes $U_{\eta(\alpha)}$. Therefore $\mathscr{Z}(S)$ normalizes $U_{(\alpha)}$ and (i) is proved.

(ii) Let $\eta(\psi)$ be the union of the $\eta(\alpha)$ $(\alpha\in\psi_{nd})$. Then, again by 13.20, 14.5 the U_β $(\beta\in\eta(\psi))$ directly span a closed connected unipotent group $U_{\eta(\psi)}$ normalized by T, with Lie algebra the sum of the $\mathfrak{g}_{(\alpha)}$ $(\alpha\in\psi_{nd})$. The U_β may be taken in any order, hence $U_{\eta(\psi)}$ is directly spanned by the $U_{(\alpha)}$, $(\alpha\in\psi_{nd})$. It is therefore defined over k and normalized by $\mathscr{Z}(S)$.

21.10 Remarks. (1) The proof shows more precisely that

$$U_{(\alpha)} = U_{(\eta(\alpha))} \qquad U_\psi = U_{(\eta(\psi))},$$

where on the right-hand side we have the groups defined in 14.5 with respect to absolute roots. This allows one to carry some other results of 14.5 to the present situation. In particular, assume that $(\alpha,\psi)\subset\psi$ for some $\alpha\in{}_k\Phi$, in the notation of 14.5. Then, from the definitions, it is clear that $(\eta(\alpha),\eta(\psi))\subset\eta(\psi)$. Therefore 14.5 implies that

$$(U_{(\alpha)}, U_\psi)\subset U_\psi.$$

(2) If $(\alpha)=\{\alpha\}$, then the sum of two elements in $(\eta(\alpha))$ is not a root, therefore $U_{(\alpha)}$ is commutative. On the other hand, if 2α is a root, then $U_{(\alpha)}$ is not commutative. It is metabelian, though, and it can be shown that its center is $U_{2\alpha}$ (see [6:4.10]).

(3) We shall see later that U_ψ is k-isomorphic to an affine space.

21.11. *Standard parabolic k-subgroups.* We fix a minimal parabolic k-subgroup P containing $\mathscr{Z}(S)$, let P^- be the opposite parabolic k-subgroup containing $\mathscr{Z}(S)$, and U (resp. U^-) the unipotent radical of P (resp. P^-). We assume ${}_k\Phi$ given the ordering such that P is associated to the positive k-roots. U (resp. U^-) is directly spanned by the $U_{(\alpha)}(\alpha\in{}_k\Phi^+_{nd})$ (resp. $U_{(\alpha)}(\alpha\in-{}_k\Phi^+_{nd})$).

We extend to the present case some definitions of 14.17. For $I \subset {}_k\Delta$, we let $[I]$ be the set of k-roots which are linear combinations of elements in I and set

$$\Psi(I) = {}_k\Phi(I)^+ = {}_k\Phi^+ - [I]. \tag{1}$$

In the notation of 14.5, we have again

$$(\alpha, {}_k\Phi(I)^+) \subset {}_k\Phi(I)^+ \quad \text{if} \quad \alpha \in [I]. \tag{2}$$

Let $S_I = \left(\bigcap_{\alpha \in I} \ker \alpha\right)^o$. Then

$$[I] = \Phi(S, \mathscr{Z}(S_I)).$$

It follows from 21.9, 21.10 that $\mathscr{Z}(S_I)$ is generated by $\mathscr{Z}(S)$ and the $U_{(\alpha)}$ $(\alpha \in [I])$. It then follows from (2) and 21.10(1) that $\mathscr{Z}(S_I)$ normalizes $U_{\Psi(I)}$. Their intersection is reduced to $\{1\}$ and that of their Lie algebras to zero. The semi-direct product ${}_kP_I = \mathscr{Z}(S_I) \cdot U_{\Psi(I)}$ is therefore a parabolic k-subgroup with Levi k-subgroup $\mathscr{Z}(S_I)$ and unipotent radical $U_{\Psi(I)}$. If I is the empty set, we get P back. It is also denoted P_I.

Given P, we call *standard* a parabolic k-subgroup containing P.

21.12 Proposition. *The parabolic k-subgroups* ${}_kP_I$ $(I \subset {}_k\Delta)$ *are distinct and are all the standard parabolic k-subgroups. Any parabolic k-subgroup of G is conjugate to one and only one* ${}_kP_I$, *by an element of G which may be chosen in G(k).*

Proof. Let Q be a parabolic k-subgroup, L a Levi k-subgroup of Q and C the greatest k-split central torus of L. Then $L = \mathscr{Z}(C)$ (see 20.5, 20.6). Let D be a maximal k-split torus containing C and T a maximal torus of G defined over k and containing D. Choose $\lambda \in X_*(C)$ as in 20.4, i.e. such that (with Φ standing for $\Phi(T, G)$),

$$\begin{aligned} \Phi(T,Q) &= \{\alpha \in \Phi \mid \langle \alpha, \lambda \rangle \geqq 0\} \\ \Phi(T,L) &= \{\alpha \in \Phi \mid \langle \alpha, \lambda \rangle = 0\} \\ \Phi(T,\mathscr{R}_uQ) &= \{\alpha \in \Phi \mid \langle \alpha, \lambda \rangle > 0\}. \end{aligned} \tag{1}$$

Since ${}_k\Phi = \Phi(D, G)$ may be viewed as the set of non-zero restrictions of elements of Φ, we can also write

$$\begin{aligned} \Phi(D,Q) &= \{\alpha \in {}_k\Phi \mid \langle \alpha, \lambda \rangle \geqq 0\}, \\ \Phi(D,L) &= \{\alpha \in {}_k\Phi \mid \langle \alpha, \lambda \rangle = 0\}, \\ \Phi(D,\mathscr{R}_uQ) &= \{\alpha \in {}_k\Phi \mid \langle \alpha, \lambda \rangle > 0\}. \end{aligned} \tag{2}$$

By 20.9, there exists $g \in G(k)$ such that ${}^gD = S$. In view of 21.2, 21.3 and the transitivity of ${}_kW$ on the Weyl chambers, we may also insure that Int g brings λ into the positive Weyl chamber defined by ${}_k\Phi^+$. Replacing Q by ${}^gQ$, we may assume that Q satisfies those conditions and that $D = S$. Then

${}_k\Phi(D, Q) \supset {}_k\Phi^+$, hence $Q \supset P$. Moreover, a linear combination with non-zero coefficients of the same sign of elements of ${}_k\Delta$ vanishes on λ if and only if each of those simple k-roots vanishes on λ. Therefore ${}_k\Phi(S, L) = [I]$, where I is the set of simple k-roots vanishing on λ. As a result $L = \mathscr{Z}(S_I)$ and $Q = {}_kP_I$. The uniqueness of I follows from 11.17.

21.13 Proposition. *Let P be a parabolic k-subgroup of G and L a Levi k-subgroup of P.*

(i) *If Q is a parabolic k-subgroup of G, then $(Q \cap P)\cdot\mathscr{R}_uP$ (resp. $L \cap Q$) is a parabolic k-subgroup of P (resp. L), which is minimal if Q is.*

(ii) *The parabolic k-subgroups of G contained in P are the semi-direct products of $\mathscr{R}_uP$ by the parabolic k-subgroups of L. For two parabolic k-subgroups Q, Q' of P the following conditions are equivalent:* (a) *Q and Q' are conjugate under P;* (b) *Q and Q' are conjugate under $P(k)$;* (c) *$Q \cap L$ and $Q' \cap L$ are conjugate under L;* (d) *$Q \cap L$ and $Q' \cap L$ are conjugate under $L(k)$.*

Proof. (i) By 14.22, we know that $P \cap Q$ is connected, $(P \cap Q)\cdot\mathscr{R}_uP$ is parabolic in P (hence in G) and $Q \cap L$ is parabolic in L. By 20.7, the group $P \cap Q$ contains the centralizer of a maximal k-split torus S' and is defined over k. Its semi-direct product with $\mathscr{R}_uP$ is then also defined over k. After conjugation by an element of $P(k)$ we may assume that $S' = S$ (20.9). Since $\mathscr{Z}_G(S)$ contains a maximal torus of G defined over k (18.2), the group $P \cap L$ is defined over k by 20.3. This proves the first part of (i). If Q is minimal then only one of $U_{(\alpha)}, U_{(-\alpha)}$ $(\alpha \in [I])$ can belong to $Q \cap L$, whence the minimality assertion.

(ii) The group $(Q \cap P)\cdot\mathscr{R}_uP$ is the semidirect product of $(Q \cap L)$ and $\mathscr{R}_uP$. If Q is contained in P, then it contains $\mathscr{R}_uP$ by 14.22. Therefore two parabolic k-subgroups of P are conjugate in P if and only if their intersections with L are conjugate in L. The remaining part of (ii) then follows from 20.5 and 20.9.

21.14. We keep the previous notation. For $w \in {}_kW$, let

(1) $${}_k\Phi_w = \{\alpha \in {}_k\Phi^+_{nd} \mid w^{-1}(\alpha) > 0\} \quad {}_k\Phi'_w = \{\alpha \in {}_k\Phi^+_{nd} \mid w^{-1}(\alpha) < 0\}.$$

These are closed sets of positive roots, whose disjoint union is ${}_k\Phi^+_{nd}$, to which we may apply 21.9. Therefore

(2) $$U = U'_w \cdot U_w \quad \text{where} \quad U'_w = U_{{}_k\Phi'_w} \quad U_w = U_{{}_k\Phi_w}.$$

Note that

(3) $$w^{-1}\cdot U_w \cdot w \subset U, \quad w^{-1}\cdot U'_w \cdot w \subset U^-.$$

More precisely

(4) $$U_w = U \cap {}^wU \quad U'_w = U \cap {}^wU^-.$$

as can be seen using 21.9 and 13.20.

Lemma. *Let $n, n' \in \mathcal{N}(S)$ and w the image of n in ${}_kW$. The double coset $U \cdot n \cdot U$ is locally closed in G and the product map defines an isomorphism of varieties of $U'_w \times \{n\} \times U$ onto $U \cdot n \cdot U$, which is defined over k if $n \in \mathcal{N}(S)(k)$. The double cosets $U \cdot u \cdot U$ and $U \cdot n' \cdot U$ are equal if and only if $n = n'$.*

Proof. The set $U \cdot n \cdot U$ is an orbit of $U \times U$ acting on G by left and right translations, hence is open in its closure (1.8), i.e. is locally closed. By (2), (3) we have

$$U \cdot n \cdot U = U'_w \cdot U_w \cdot n \cdot U = U'_w \cdot n \cdot U$$

therefore the product map $\phi: U'_w \times \{n\} \times U \to U \cdot n \cdot U$ is surjective. Composed with the left translation by n^{-1}, it is the restriction to a closed subvariety of the product map $U^- \times U \to U^- \cdot U$. But the latter is an isomorphism since $U \cap U^- = \{1\}$ and $L(U) \cap L(U^-) = \{0\}$. The group U being defined over k, the map is defined over k if $n \in \mathcal{N}(S)(k)$. This proves the first assertion.

Assume now that $U \cdot n \cdot U = U \cdot n' \cdot U$. We want to prove that $n = n'$, i.e. that $N \cap U \cdot n \cdot U = \{n\}$ or, equivalently by the above, $N \cap U'_w \cdot n \cdot U = \{n\}$ where $N = \mathcal{N}_G(S)$. In view of (3) it suffices to show that $N \cap U^- \cdot U = \{1\}$. Let $n = v \cdot u (u \in U,\ v \in U^-)$. For any $s \in S$, we have clearly

$$(v^{-1} \cdot {}^n s \cdot v \cdot {}^n s^{-1}) \cdot ({}^n s \cdot s^{-1}) \cdot (s \cdot u \cdot s^{-1} \cdot u^{-1}) = 1.$$

The three factors defined by the brackets belong respectively to U^-, S and U respectively, hence are equal to one (see 14.21). Therefore $u, v \in \mathcal{Z}(S)$, whence $u = v = 1$ and $n = 1$.

21.15 Theorem. *Let $N = \mathcal{N}(S)$ and P, U be as above. Then $G(k) = U(k) \cdot N(k) \cdot U(k)$ and is the disjoint union of the classes $P(k) \cdot w \cdot P(k)$ $(w \in {}_kW)$. The system ${}_k\mathcal{T} = (G(k), P(k), N(k), R)$, where R is the set of reflections r_α to the simple k-roots, is a Tits system.*

Proof. Let $g \in G(k)$. By 20.7, the group $P \cap {}^gP$ contains the centralizer of a maximal k-split torus S'. The groups $\mathcal{Z}(S)$ and $\mathcal{Z}(S')$ are both Levi k-subgroups of P, hence are conjugate by an element of $U(k)$ (20.5). There exists therefore $x \in P(k)$ such that ${}^{x \cdot g}P \supset \mathcal{Z}(S)$. Then, by 21.3, we can find $n \in N(k)$ such that ${}^{n \cdot x \cdot g}P = P$. Since P is its own normalizer (11.16), we get $n \cdot x \cdot g \in P(k)$ and therefore $g \in P(k) \cdot N(k) U(k)$. But $P(k) = U(k) \cdot \mathcal{Z}(S)(k)$ and $\mathcal{Z}(S)(k) \subset N(k)$, so that $g \in U(k) \cdot N(k) \cdot U(k)$, as claimed. This also implies that $G(k)$ is the union of the classes $P(k) \cdot w \cdot P(k)$ $(w \in {}_kW)$. We have to show that they are distinct. Assume $w' \in P(k) \cdot w \cdot P(k)$ and let n, n' be representatives in $N(k)$ of w and w'. There exist then $u, v \in U(k)$ and $a, b \in \mathcal{Z}(S)(k)$ such that $n' = u \cdot a \cdot n \cdot b \cdot v$, whence $n' = a \cdot n \cdot b$ by 21.14, and $w' = w$ by definition.

There remains to prove the last assertion. We have $N(k) \cap P(k) = \mathcal{Z}(S)(k)$ by 20.6, hence $T = N(k) \cap P(k)$ is normal in $N(k)$ and $N(k)/T = {}_kW$, therefore $({}_kW, R)$ is a Coxeter system.

In view of the definition of a Tits system (see (14.15)), where our $G(k)$, $P(k)$, $N(k)$ and R are G, B, N, S there, it remains to check:

For $r \in R$ and $w \in {}_kW$, we have

$$r \cdot P(k)\{w, r \cdot w\} \cdot P(k) = P(k)\{w, r \cdot w\} \cdot P(k), \tag{1}$$

$$r \cdot P(k) \cdot r \neq P(k). \tag{2}$$

Let $\alpha \in {}_k\Delta$ be the k-root such that $r = r_\alpha$. Replacing w by $r \cdot w$ if necessary, we may assume that $w^{-1}(\alpha) > 0$. We have $\mathscr{Z}(S_\alpha) \cap U = U_{(\alpha)}$ and the roots of S in the unipotent radical of ${}_kP_{\{\alpha\}}$, to be denoted V in this proof, are all positive k-roots but those in (α). Therefore $U = U_{(\alpha)} \cdot V$ and we have

$$\begin{aligned} r \cdot P(k) \cdot \{w, r \cdot w\} P(k) &= \mathscr{Z}(S)(k) \cdot r \cdot U(k) \cdot \{w, r \cdot w\} \cdot P(k) \\ &= \mathscr{Z}(S)(k) \cdot V(k) r \cdot U_{(\alpha)}(k)\{w, r \cdot w\} \cdot P(k) \\ r \cdot P(k) \cdot \{w, r \cdot w\} \cdot P(k) &\subset P(k) \mathscr{Z}(S_\alpha)(k) \cdot w \cdot P(k). \end{aligned}$$

By our first assertion, applied to $\mathscr{Z}(S_\alpha)$:

$$\mathscr{Z}(S_\alpha)(k) = \mathscr{Z}(S)(k) \cdot U_{(\alpha)}\{1, r\} \cdot U_{(\alpha)}(k),$$

hence

$$r \cdot P(k) \cdot \{w, r \cdot w\} \cdot P(k) \subset P(k) \cdot \{1, r\} U_{(\alpha)}(k) \cdot w \cdot P(k).$$

Since $w^{-1}(\alpha) > 0$, we have $w^{-1} \cdot U_{(\alpha)}(k) w \subset P(k)$, and we get

$$r \cdot P(k) \cdot \{w, r \cdot w\} \cdot P(k) \subset P(k)\{w, r \cdot w\} P(k).$$

If we multiply both sides by r, we get the opposite inclusion, and (1) is proved.

The group $r \cdot U \cdot r$ contains $U_{(-\alpha)}$, hence is $\neq U$ and (2) follows.

21.16. By the general theory, the standard parabolic subgroups of the Tits system ${}_k\mathscr{T}$ are the groups $P(k) \cdot {}_kW_I \cdot P(k)$. $I \subset R$. If we identify ${}_k\Delta$ and R in the obvious way $(\alpha \mapsto r_\alpha)$, and use the theorem, we see that $P(k) \cdot {}_kW_I \cdot P(k) = {}_kP_I(k)$ in the notation of 21.11. Therefore, $Q \mapsto Q(k)$ is a bijective correspondence between the parabolic k-subgroups of G and the parabolic subgroups of ${}_k\mathscr{T}$. In particular, a parabolic k-subgroup is completely determined by its k-rational points.

We shall denote $P \cdot w \cdot P$ by $C(w)$ or ${}_kC(w)$ if it appears necessary to emphasize k. If $n \in \mathscr{N}(S)(k)$ represents w, then $C(w)$ is k-isomorphic to $U'_w \cdot n \cdot P$ hence

$$C(w)(k) = U'_w(k) \cdot n \cdot P(k) = P(k) \cdot n \cdot P(k). \tag{1}$$

$C(w)(k)$ is the $C(w) = BwB$ of the general theory of Tits systems. An easy induction starting from 21.15 (see Prop. 2 in IV, §2 of [9]) yields

$${}_kP_I(k) \cdot w \cdot {}_kP_J(k) = P(k) \cdot {}_kW_I \cdot w \cdot {}_kW_J \cdot P(k) \quad (w \in {}_kW; I, J \subset {}_k\Delta) \tag{2}$$

As a consequence. $w \mapsto {}_kP_I(k) \cdot w \cdot {}_kP_J(k)$ yields a bijection

$${}_kW_I \backslash {}_kW / {}_kW_J \xrightarrow{\sim} {}_kP_I(k) \backslash G(k) / {}_kP_J(k). \tag{3}$$

Also

$$G(k)/_kP_J(k) = \bigcup_{w \in {}_kW/_kW_J} U(k)\cdot w\cdot {}_kP_J(k)/_kP_J(k)$$

We shall describe the right-hand side more precisely later.

Our next goal is to prove that U_ψ is k-isomorphic to an affine space. That follows from the much more general results quoted in 15.13. Rather than invoking them, we shall use the features of the present situation to find a self-contained way to this statement. The lemmas below are in fact some of the (elementary) ingredients involved in the proofs of the statements in 15.13.

21.17 Lemma. *Let U be a connected unipotent k-group, on which a k-torus T operates k-morphically. Assume there is a non-trivial character c of T and a structure of vector space over k_s on U so that any $t\in T(\bar{k})$ operates by the dilation $u\mapsto t^c\cdot u$. Then this is already so over k and c is defined over k.*

The set of weights of T in U reduces to $\{c\}$ and is on the other hand invariant under $\Gamma = \mathrm{Gal}(k_s/k)$, hence c is defined over k.

In view of our assumptions, we have $\gamma(r\cdot u) = {}^\gamma c\cdot {}^\gamma u$ for all $\gamma\in\Gamma$, $r\in k_s$ and $u\in U(k_s)$. It suffices then to show that $U(k_s)$ has a vector space basis consisting of elements in $U(k)$. There exists clearly a finite Galois extension k' of k such that $U(k')$ contains a vector space basis of $U(k_s)$. Let $\Gamma' = \mathrm{Gal}(k'/k)$ and $(u_1,\ldots,u_m)$ a maximal set of elements in $U(k)$ which are linearly independent over k_s. For every $u\in U(k')$ and $a\in k'$, we have that $\sum_{\gamma\in\Gamma'} {}^\gamma a\cdot {}^\gamma u$ is rational over k, hence a linear combination of the u_i's. By the theorem of the normal basis, each ${}^\gamma u$ is also a linear combination of the u_i's, hence those form a basis of $U(k_s)$ over k_s.

21.18 Lemma. *Let V, W, Z be irreducible k-varieties, $\alpha: V\to W$ a dominant k-morphism, $\beta: W\to Z$ a morphism and assume that $\beta\circ\alpha$ is defined over k. Then β is defined over k.*

Proof. We may replace Z by an open dense affine subset and have to prove that $\beta^o(k[Z])\subset k(W)$. We know that $(\beta\circ\alpha)^o(k[Z])\subset k[V]$ and that α^o is injective. Therefore it suffices to show that

$$\alpha^o(K(W))\cap k(V)\subset \alpha^o(k(W)). \tag{1}$$

Let f be in the left-hand side. There exist u_i, $v_i\in\alpha^o(k(W))$ and $a_i\in K$, linearly independent over k, $(i=1,\ldots,m)$ such that $f\left(\sum_i a_i\cdot u_i\right) = \sum_i a_i\cdot v_i$, and therefore such that

$$\sum_i a_i(f\cdot u_i - v_i) = 0. \tag{2}$$

But K and $k(V)$ are linearly disjoint over k (AG.12.1), therefore (2) implies $f\cdot u_i = v_i$ $(1 \leqq i \leqq m)$, which proves the lemma.

21.19 Lemma. *Let L be a connected k-group and H a a connected unipotent k-subgroup of L which is k-isomorphic to the additive group of some vector space E over k. Assume that $\pi: L \to L/H$ has a global morphic cross-section σ. Then it has one defined over k.*

Proof. We identify H with E and write the group law on H additively. We fix a basis (e_i) $(i = 1, \ldots, m)$ of E over k and denote by h_i the ith coordinate of $h \in H$ with respect to that basis. Let

$$F = \{g, gh\} \quad (g \in L, h \in H)$$

be the graph of the operation of H on L defined by right translations. Then $\theta:(x, y) \mapsto x^{-1}\cdot y$ is a k-morphism of L onto H. Let $\alpha_i: L \to K$ be defined by $g \mapsto (g^{-1}\cdot\sigma(\pi(g)))_i$. It belongs to $K[L]$. If $g \in L$, $h \in H$, then $\pi(g\cdot h) = \pi(g)$ and

$$(g\cdot h)^{-1}\cdot\sigma(\pi(gh) = h^{-1}\cdot(g^{-1}\cdot\sigma(\pi(g))).$$

This can be written

$$\alpha_i(g\cdot h) = \alpha_i(g) - h_i, \quad (g \in L, h \in H, i = 1, \ldots, m). \tag{1}$$

For each i, there exist elements $c_{i,j} \in K$ $(1 \leqq j \leqq J(i))$, linearly independent over k, with $c_{i1} = 1$ and $\varphi_{ij} \in k[L]$ such that

$$\alpha_i(g) = \sum_j c_{i,j}\cdot\varphi_{i,j}(g), \quad (g \in L, i = 1, \ldots, m), \tag{2}$$

and (1) yields

$$\sum_j c_{i,j}\cdot\varphi_{i,j}(g\cdot h) = \sum_j c_{i,j}\cdot\varphi_{i,j}(g) - h_i \quad (g \in L, h \in H, i = 1, \ldots, m). \tag{3}$$

By the linear disjointness of K and $k(L \times H)$ over k, this gives

$$\varphi_{i,1}(g\cdot h) = \varphi_{i,1}(g) - h_i, \quad (g \in G, h \in H, i = 1, \ldots, m). \tag{4}$$

$$\varphi_{i,j}(g\cdot h) = \varphi_{i,j}(g) \quad (j = 2, \ldots, J(i)). \tag{5}$$

From (4) we see that there exist $\psi_{i,j} \in k[L/H]$ $(j = 2, \ldots, J(i))$ such that $\varphi_{i,j} = \pi^o_{i,j}(\psi_{i,j})$. We definé a morphism of varieties $\mu: L/H \to H$ by

$$\mu(z) = -\sum_i e_i\cdot\left(\sum_{j>1} c_{i,j}\cdot\psi_{i,j}(z)\right) \quad (z \in L/H), \tag{6}$$

and a new cross section $\sigma': L/H \to L$ by

$$\sigma'(z) = \sigma(z)\cdot\mu(z) \quad (z \in L/H).$$

We claim that σ' is defined over k. We have

$$g^{-1}\cdot\sigma'(\pi(g)) = g^{-1}\sigma(\pi(g))\cdot\mu(\pi(g)).$$

Using (3), (6), we get

$$(g^{-1}\cdot\sigma'(\pi(g)))_i = \alpha_i(g) - \sum_{j>1} c_{i,j}\cdot\varphi_{ij}(g)$$

and therefore, in view of (2):

$$(g^{-1}\cdot\sigma'(\pi(g)))_i = \varphi_{i1}(g), \quad (i=1,\ldots,m).$$

Let $\varphi_1 : L \to H$ be defined by

$$\varphi_1(g) = \sum_i e_i\cdot\varphi_{i1}(g) \quad (g\in L).$$

It is a k-morphism of varieties. Then

$$\sigma'\circ\pi : g \mapsto g\cdot\varphi_1(g) \quad (g\in L)$$

is a k-morphism of L into L. By 21.18, applied to $\alpha=\pi$ and $\beta=\sigma'$, the morphism σ' is defined over k.

21.20 Theorem. (i) *Fix an ordering on* ${}_k\Phi = \Phi(S,G)$ *and let* ψ *be a closed subset of* ${}_k\Phi^+$. *Then* U_ψ *is* k*-isomorphic, as a variety, to an affine space.*

(ii) *Let* P *be a parabolic* k*-subgroup of* G. *Then* G/P *is rational over* k. *If* k *is infinite* $P(k)$ *is Zariski-dense in* P.

Proof. (i) By 21.9, U_ψ is directly spanned, over k, by the $U_{(\alpha)}(\alpha\in\psi_{nd})$. This reduces us to the case where $\psi=(\alpha)$ for some $\alpha\in{}_k\Phi$. Assume first that 2α is not a root. Then $U_{(\alpha)}$ is a commutative group, which is directly spanned over k_s by the groups U_β, $(\beta\in\eta(\alpha))$ in the notation of 21.9; those are isomorphic to $\mathbb{G}_a$ over k_s. Then our assertion follows from 21.17.

Assume now that 2α is a root. The previous argument applies to $U_{(2\alpha)}$ and $U_{(\alpha)}/U_{(2\alpha)}$. It follows from 21.9 that the groups U_β $(\beta\in\eta(\alpha))$, taken in any order, directly span a subvariety of $U_{(\alpha)}$ which defines a morphic cross section for the projection $U_{(\alpha)}\to U_{(\alpha)}/U_{(2\alpha)}$. It is only defined over k_s, but 21.19 then provides one which is defined over k. Therefore (see 6.14), $U_{(\alpha)}$ is k-isomorphic, as a variety, to $(U_{(\alpha)}/U_{(2\alpha)})\times U_{(2\alpha)}$, hence to an affine space over k.

(ii) The unipotent radical of any parabolic k-subgroup of G is conjugate over k to a group U_ψ, with ψ as before (21.11), therefore is k-isomorphic, as a variety, to an affine space. This holds in particular for the unipotent radical $\mathscr{R}_uP^-$ of a parabolic k-subgroup P^- opposite to P. The projection $G\to G/P$ defines a k-isomorphism of $\mathscr{R}_uP^-$ onto a k-open subset of G/P, as follows from 14.21, therefore G/P is rational over k.

P is the semi-direct product over k of $\mathscr{R}_uP$ and a Levi k-subgroup L (20.5). Assume k to be infinite. Then $\mathscr{R}_uP(k)$ is dense in $\mathscr{R}_uP$ by (i) and $L(k)$ is dense in L by 18.3, whence the second assertion of (ii).

21.21. Our next goal in this section is the description of the relative closure in $G(k)$ of a Bruhat cell. We first recall or state some facts about Weyl groups and root systems to be used below. We keep the notation of 21.14. The length $l(w)$ of $w \in {}_kW$ is defined with respect to the set R of simple reflections r_α $(\alpha \in {}_k\Delta)$.

(i) *Given $J \subset {}_k\Delta$ and $w \in {}_kW$ there exists a unique element, to be denoted w^J or $w(w, J)$, of smallest length in the coset $w \cdot {}_kW_J$.*

See IV, §1, Exercise 3 in [9]. In fact, a proof will also be contained in 21.24. We denote by ${}_kW^J$ the set of w^J $(w \in {}_kW)$. It is a set of representatives for the left cosets $w_k \cdot W_J$. Each $w \in {}_kW$ has a unique expression $w = w^J \cdot w_J$, with $w^J \in {}_kW^J$ and $w_J \in {}_kW_J$.

(ii) *Let $J_i \subset {}_k\Delta$, write W_i for ${}_kW_{J_i}$ $(i = 1, \dots, n)$ and let $w = w_1 \cdots w_n$ with $w_i \in W_i$. Then there exists $w_i' \in W_i$ such that $w = w_1' \cdots w_n'$, $l(w_i') \leqq l(w_i)$ and $l(w) = \sum_i l(w_i')$.*

This follows from the exchange condition for Coxeter systems [9:IV, 1.5]. As a direct consequence of (i) and (ii) we get

(iii) *Let $w \in W_1 \cdots W_n$. Then $w^{J_n} \in W_1 \cdots W_{n-1}$.*

Finally, by Corollary 2, p. 158 of [9], we have

(iv) $l(w) = \mathrm{Card}_k\, \Phi'_w$,

where (see 21.14)

$$ {}_k\Phi'_w = w({}_k\Phi^-) \cap {}_k\Phi^+_{nd}. $$

21.22 Proposition. *Let $w, w' \in {}_kW$ be such that $l(w) + l(w') = l(w \cdot w')$. Then*

$$ (1) \qquad C(w) \cdot C(w') = C(w \cdot w'), $$

$$ (2) \qquad C(w)(k) \cdot C(w')(k) = C(w \cdot w')(k). $$

Proof. We first assume that $w = r_\alpha (\alpha \in {}_k\Delta)$ and set $\psi = {}_k\Phi'_{w'}$. Then

$$ (3) \qquad {}_k\Phi'_w = \{\alpha\}, \quad {}_k\Phi'_{ww'} = \{\alpha, r_\alpha(\psi)\}. $$

Let n and n' be representatives in $\mathcal{N}(S)_k$ of w and w'. Then

$$ (4) \quad C(w) = U_{(\alpha)} \cdot n \cdot P \quad C(w') = U_{(\psi)} \cdot n' \cdot P, \quad C(w \cdot w') = U_{(\alpha)} U_{r_\alpha(\psi)} \cdot n \cdot n' \cdot P. $$

All these decompositions are k-isomorphisms of varieties so that

$$ (5) \qquad c(w)(k) = U_{(\alpha)}(k) \cdot n \cdot P(k), \quad C(w')(k) = U_{(\psi)}(k) \cdot n' \cdot P(k), $$

$$ (6) \qquad C(w \cdot w')_k = U_{(\alpha)}(k) \cdot U_{(r_\alpha(\psi))}(k) \cdot n \cdot n' \cdot P(k). $$

We have, taking the above into account

$$ C(w) \cdot C(w') = U_{(\alpha)} \cdot n \cdot U_{(\psi)} \cdot n' \cdot P = U_{(\alpha)} \cdot U_{(r_\alpha(\psi))} \cdot n \cdot n' \cdot P = C(w \cdot w'), $$

and this is also valid for k-rational points. This proves (1) and (2) in the case where w is a simple reflection. They follow for fixed w' by induction on $l(w)$.

21.23. If $n \in \mathcal{N}(S)(k)$ represents $w \in {}_kW$, we also denote by $w(P)$ the conjugate $n \cdot P \cdot n^{-1}$ of P. Fix $J \subset {}_k\Delta$. By applying 21.3 to a Levi k-subgroup of ${}_kP_J$ containing $\mathscr{Z}(S)$, we see that $v \mapsto v(P)$ is a bijection of ${}_kW_J$ onto the set of minimal parabolic k-subgroups of ${}_kP_J$ containing $\mathscr{Z}(S)$.

For any $w \in {}_kW$, we know that $Q_w = (w^{-1}(P) \cap {}_kP_J) \cdot \mathscr{R}_{u\,k}P_J$ is a minimal parabolic k-subgroup of ${}_kP_J$ (21.13), evidently containing $\mathscr{Z}(S)$. Any $v \in {}_kW_J$ normalizes $\mathscr{R}_{u\,k}P_J$ and ${}_kP_J$, therefore

$$(1) \qquad Q_{w \cdot v} = v^{-1}(Q_w) \quad (w \in {}_kW, v \in {}_kW_J).$$

In view of our initial remark, there exists a unique $w_o \in w.\, {}_kW_J$ such that $P = Q_{w_o}$. We claim

$$(2) \qquad w_o = w^J$$

By definition of W^J, this is (i) in the next lemma.

We let $\Psi(J)^- = -\Psi(J)$, where, as in 21.11, $\Psi(J)$ is the set of non-divisible k-roots occurring in the Lie algebra of $\mathscr{R}_{u\,k}P_J$.

Lemma. *Let $v \in {}_kW_J, v \neq 1$. Then*

(i) $l(w_o) < l(w_o \cdot v)$.
(ii) ${}_k\Phi'_{w_o} = w_o \cdot v(\Psi(J)^-) \cap {}_k\Phi^+$.
(iii) ${}_k\Phi'_{v^{-1}} = w^{-1}({}_k\Phi^-_{nd}) \cap [J]^+_{nd}, \quad ([J]^+ = [J] \cap {}_k\Phi^+)$.

Proof. As recalled in 21.21:

$$(1) \qquad l(w) = \operatorname{Card}(w^{-1}({}_k\Phi^+_{nd}) \cap {}_k\Phi^-) = \operatorname{Card}(w({}_k\Phi^-_{nd}) \cap {}_k\Phi^+).$$

The condition imposed on w_o is equivalent to

$$(2) \qquad w_o^{-1}({}_k\Phi^+_{nd}) \cap [J] = [J]^+_{nd},$$

therefore

$$(3) \qquad l(w_o) = \operatorname{Card} w_o^{-1}({}_k\Phi^+) \cap \Psi(J)^-.$$

$$(4) \qquad {}_k\Phi'_{w_o} = w_o^{-1}({}_k\Phi^+) \cap \Psi(J)^- = w_o(\Psi(J)^-) \cap {}_k\Phi^+.$$

The element v leaves $\Psi(J)$, $\Psi(J)^-$ and $[J]$ stable, but not $[J]^+$. We have then

$$(5) \qquad \operatorname{Card}(v^{-1} \cdot w_o^{-1}({}_k\Phi^+_{nd}) \cap \Psi(J)^-) = \operatorname{Card}(w_o^{-1}({}_k\Phi^+) \cap \Psi(J)^-)$$

$$(6) \qquad \operatorname{Card} v^{-1} w_0^{-1}({}_k\Phi^+_{nd}) \cap [J]^- \neq 0.$$

Since $l(w_o \cdot v)$ is the sum of the left-hand sides of (5) and (6), these relations, together with (3), prove (i). Since v leaves $\Psi(J)^-$ invariant, (4) also implies (ii). Moreover

$$(7) \qquad {}_k\Phi'_{v^{-1}} = v^{-1}({}_k\Phi^-_{nd}) \cap {}_k\Phi^+ = v^{-1}([J]^-) \cap [J]^+_{nd}.$$

(since $v^{-1}(\Psi(J)^-) \cap \Psi(J) = \varnothing$). On the other hand

$$w_o^{-1}({}_k\Phi^-_{nd}) \cap [J] = [J]^-_{nd}$$

and therefore

$$(8) \qquad v^{-1} w_o^{-1}({}_k\Phi^-_{nd}) \cap [J]^+ = v^{-1}([J]^-) \cap [J]^+_{nd}.$$

The assertion (iii) now follows from (7) and (8).

21.24 Lemma. *Let $J \subset {}_k\Delta$, and $w \in {}_kW$. Let $g \in P(k)\cdot w \cdot P(k)$ and $h \in G(k)$ and assume that $h^{-1}\cdot P \cdot h = (g^{-1}\cdot P \cdot g \cap P_J) \cdot \mathcal{R}_u P_J$. Then*

$$g \cdot h^{-1} \in P(k)\cdot w^J \cdot P(k).$$

Proof. Set $g' = g \cdot h^{-1}$ and let $u \in {}_kW$ be such that $g' \in C(u)(k)$. Since $h^{-1}\cdot P \cdot h$ is in P_J, we have $h \in {}_kP_J(k)$ (see 11.17) therefore, in view of 21.16,

$$g = g' \cdot h \in P(k)\cdot u \cdot P_J(k) \subset P(k)\cdot u \cdot {}_kW_J \cdot P(k)$$

whence $w \in u \cdot {}_kW_J$ and $w(w, J) = w(u, J)$. But we also have

$$P = (g'^{-1}\cdot P \cdot g' \cap P_J)\cdot \mathcal{R}_u P_J$$

therefore 21.23 shows that $u = w(u, J) = w(w, J)$, which is our claim.

21.25 Lemma. *Let $P_1, \ldots, P_n$ be parabolic k-subgroups containing P. Then $P_1 \cdots P_n$ is closed and $(P_1 \cdots P_n)(k) = P_1(k) \cdots P_n(k)$.*

Proof. There is nothing to prove if $n = 1$, so we argue by induction on n.

Assume $P_1 \cdots P_{n-1}$ is closed. Of course, $P_1 \cdots P_{n-1}$ is right invariant under P, hence $P_1 \cdots P_{n-1}$ is the full inverse image of its image in G/P, therefore the latter is closed. Since G/P is complete, the canonical projection $G/P \to G/P_{\dot{n}}$ is closed, hence the image of $P_1 \cdots P_{n-1}$ in $G/P_{\dot{n}}$ is also closed. But then, so is its full inverse image $P_1 \cdots P_{\dot{n}}$ in G.

Let $J_i \subset_k \Delta$ be such that $P_i = P_{J_i} (i = 1, \ldots, n)$. Note that in view of 21.15, it follows by induction on n from Prop. 2 in §2 of [9:IV] that

$$P_1(k) \cdots P_n(k) = P(k)\cdot W_{J_1} \cdots W_{J_n} \cdot P(k).$$

Let $g \in (P_1 \cdots P_n)(k)$. Then there exists $w \in W_{J_1} \cdots W_{J_n}$ such that $g \in P(k)\cdot w \cdot P(k)$. By 21.21 (iii), we have $w(w, J_n) \in W_{J_1} \cdots W_{J_{n-1}}$. By 21.13, the group

$$P' = (g^{-1}\cdot P \cdot g \cap P_n)\cdot \mathcal{R}_u P_n$$

is a minimal parabolic k-subgroup. There exists therefore $h \in P_n(k)$ such that $h^{-1}\cdot P \cdot h = P'$. But then $g \cdot h^{-1} \in P_1 \cdots P_{n-1}$ by 21.21 (iii). Since

$$(P_1 \cdots P_{n-1})(k) = P_1(k) \cdots P_{n-1}(k)$$

by induction assumption, our second claim is proved.

21.26 Theorem. *Assume k to be infinite. Let $w \in {}_kW$ and $w = s_1 \cdots s_n$ be a reduced decomposition of w. Then the set $A_w = \{s_{i_1} \cdots s_{i_m} | m \in \mathbf{N}, 1 \leqq i_1 < \cdots < i_{i_m} \leqq q\}$ depends only on w, not on the reduced decomposition, and we have*

$$\overline{C(w)}(k) = \overline{C(w)(k)} \cap G(k) = \bigcup_{v \in A_w} C(v)(k). \tag{1}$$

Proof. The relation (1) and 21.15 imply that A_w depends only on w. It suffices to prove (1). It follows from 21.20 that $U_w(k)$ and $P(k)$ are dense in U_w and P respectively. Since $C(w)(k) = U'_w(k)\cdot n \cdot P(k)$, where $n \in \mathcal{N}(S)(k)$ represents w, it also follows that $C(w)(k)$ is dense in $C(w)$. This proves the first equality in (1).

For $X \subset G(k)$, let us write $A(X)$ for the relative closure $\bar{X} \cap G(k)$ of X in

$G(k)$. If $X_1, \ldots, X_m \in G(k)$, then we see from (AG, 6.6) that

$$A(X_1 \cdots X_m) = A(A(X_1) \cdots A(X_q)). \tag{2}$$

Write P_i for P_{s_i} $(i = 1, \ldots, q)$. By 21.25, $P_1 \cdots P_q$ is closed and we have

$$A(P_1(k) \cdots P_q(k)) = P_1(k) \cdots P_q(k). \tag{3}$$

$C(s_i)(k)$ is dense in $C(s_i)$ (21.21), hence in P_i and the first equality of (1), together with the Bruhat decomposition (21.15), gives

$$A(C(s_i)(k)) = P_i(k) = C(s_i)(k) \cup P(k). \tag{4}$$

From (2) and 21.22, we get

$$A(C(w)(k)) = A(C(s_1)(k) \cdots C(s_q)(k)) = A(A(C(S_1)(k)) \cdots A(C(s_q)(k)))$$
$$A(C(w)(k)) = P_1(k) \cdots P_q(k) = (C(s_1)(k) \cup P(k)) \cdots (C(s_q)(k) \cup P(k)).$$

By repeated application of 21.22, we see that the last term is equal to the last term of (1), and (1) is proved.

21.27 Proposition. *Assume $G(k)$ to be endowed with a topology $\mathcal{T}$ having the following properties:*

(a) *$\mathcal{T}$ is finer than the topology induced by the Zariski topology.*
(b) *The product map $G(k) \times G(k) \to G(k)$ is continuous, the left-hand side being endowed with the product topology.*
(c) *For every $\alpha \in {}_k\Delta$, the group $P(k)$ is not open in $P_{\{\alpha\}}(k)$.*

Then the closure of $C(w)(k)$ with respect to $\mathcal{T}$ is equal to its relative closure $\overline{C(w)(k)} \cap G(k)$ in the Zariski topology, described in 21.26.

Proof. In the previous proof, the Zariski topology was used only through the relations (2), (3), (4). It suffices therefore to see they are satisfied by the relative closure $A_{\mathcal{T}}$ with respect to $\mathcal{T}$. Obviously (3) follows from (a) and (2) from (b). Let $\alpha \in {}_k\Delta$ and $r = r_\alpha$. If $P(k)$ is not open in $P_{\{\alpha\}}(k)$ then it meets the relative $\mathcal{T}$-closure of $C(r)(k)$. Therefore $A_{\mathcal{T}}(C(r)(k) \cdot P(k)) \supset P(k)$. But $C(r)(k) = (C(r)(k)) \cdot P(k)$ whence $A_{\mathcal{T}}(C(r)(k)) \supset P(k)$ and 21.26(4) is satisfied in $\mathcal{T}$.

21.28 Corollary. *Assume k endowed with a non-discrete topology $\mathcal{E}$, satisfying the separation axiom T_1 and with respect to which it is a topological ring. Then the closure of $C(w)(k)$ with respect to the topology $\mathcal{T}$ of $G(k)$ associated to $\mathcal{E}$ coincides with the relative Zariski closure.*

By definition, $\mathcal{T}$ is the coarsest topology such that the restrictions to $G(k)$ of any element in $k[G]$ is continuous. It obviously satisfies (a) and (b) above. By 21.21, $U_{-(\alpha)}$ is k-isomorphic to an affine space. Since k is not discrete, $U_{-(\alpha)}(k) \cap P(k) = \{1\}$ is not open in $U_{-(\alpha)}(k)$, hence $P(k)$ is not open in $P_{\{\alpha\}}(k)$ and 21.27(c) is satisfied, so that 21.27 applies to the present situation.

21.29 Proposition. *Let $J \in {}_k\Delta$, $w \in {}_kW$ and π (resp. π_J, resp. σ) the canonical projection $G \to G/P$ (resp. $G \to G/{}_kP_J$, resp. $G/P \to G/{}_kP_J$).*

(i) *We have* $P\cdot w\cdot {}_kP_J = P\cdot w^J\cdot {}_kP_J$ *and* $\pi_J(C(w)) = \pi_J(C(w^J))$.

(ii) *The canonical projections* $\sigma: \pi(C(w)) \to \pi_J(C(w^J))$ *and* $\pi_J(k): \pi(C(w)(k)) \to \pi_J(C(w^J)(k))$ *are surjective. They are injective if and only if* $w \in {}_kW^J$, *in which case* $\alpha: \pi(C(w)) \to \pi_J(C(w))$ *is a k-isomorphism of varieties.*

(iii) *We have* $(C(w)\cdot {}_kP_J)(k) = C(w)(k)\cdot {}_kP_J(k)$.

(iv) *If* $w, w' \in {}_kW^J$ *are distinct, then* $\pi_J(C(w)) \cap \pi_J(C(w')) = \emptyset$. *The sets* $\pi_J(C(w))$, $(w \in {}_kW^J)$ *form a partition of* $G(k)/{}_kP_J(k)$.

Proof. In the notation of 21.21, we have by 21.22

$$(1) \qquad C(w) = C(w^J)\cdot C(w_J), \quad C(w)(k) = (C(w^J)(k))\cdot(C(w_J)(k)),$$

therefore

$$(2) \qquad C(w)\cdot {}_kP_J = C(w^J)\cdot {}_kP_J, \quad (C(w)(k))\cdot {}_kP_J(k) = (C(w^J)(k))\cdot {}_kP_J(k).$$

This implies (i) and the first assertion of (ii). Let n, n^J and n_J be representatives in $\mathcal{N}(S)(k)$ of w, w^J and w_J respectively. The subset $X = n^J\cdot U'_{w_J}\cdot n_J$ belongs to $C(w)(k)$, is mapped injectively by π in $\pi(C(w)(k))$ and we have $\pi_J(X) = \{n^J\cdot {}_kP_J\}$. Therefore if π_J is injective, then $U'_{w_J} = \{1\}$, and $w_J = 1$, $w = w^J$.

We have the obvious relation $U'_{w^J}\cdot n \subset C(w)$ and, by (2),

$$U_{w^J}\cdot n\cdot {}_kP_J = C(w)\cdot {}_kP_J.$$

Moreover, 21.23(ii) shows that $n^{-1}\cdot U'_{w_J}\cdot n \subset \mathcal{R}_{u\cdot k}P_J$, therefore the product map $\{U'_{w^J}\cdot n_k\cdot P_J \to C(w)\cdot {}_kP_J$ is a k-isomorphism of varieties. From this, (ii) and (iii) follow immediately.

Let $u, v \in {}_kW^J$. If $\pi_J(C(u))$ and $\pi_J(C(v))$ have a non-empty intersection, then the double cosets

$$C(u)\cdot {}_kP_J = P\cdot u\cdot {}_kP_J \quad \text{and} \quad C(v)\cdot {}_kP_J = P\cdot v\cdot {}_kP_J$$

also have a non-empty intersection, hence are equal and therefore $u \in v\cdot {}_kW_J$ (21.16). Since $u, v \in {}_kW^J$, they are equal. This proves the first part of (iv); the second one then follows from (ii), 21.16 and the surjectivity of $\pi_J(k)$.

Bibliographical Note

Up to 21.21, the results of this section are contained in [4: §5]. The proofs are similar, with one apparent exception: To establish 21.21 there, use was made of results of Rosenlicht quoted in 15.13, but not proved here. We have extracted from his arguments the minimum needed here to handle $U_{(\alpha)}$. More precisely, 21.17 is lemma 3.16 of [4], 21.18 lemma 1 of [28] and the proof of 21.19 follows closely that of a similar statement on p. 100 of [28].

The remaining part of this section is taken from [5]. For k algebraically closed, 21.26 was proved by C. Chevalley (unpublished). For $k = \mathbf{C}$, with respect to the analytic topology, 21.28 is proved in R. Steinberg's Yale Notes, p. 107.

§22. Central Isogenies

In this section, G is a connected k-group and T a maximal torus of G defined over k. Until 22.10, G is assumed to be reductive.

The relative notions introduced in the previous sections, such as k-rank or relative root systems, are not invariant under arbitrary isogenies, in contrast with the corresponding absolute analogues. They are quite obviously so under separable isogenies, but this class is too narrow. The appropriate concept here is that of a central isogeny, to which this section is devoted. In order to arrive as directly as possible to those conservation theorems, we first confine ourselves to reductive groups, using a notion of central isogeny adapted to that case. From 22.11 on, we shall discuss central isogenies in the general case.

22.1 Lemma. *Let N be a closed normal subgroup of G. Then the following conditions are equivalent*: (i) *N is central*; (ii) $N \subset T$; (iii) *N is contained in the intersection Z of the maximal tori of G*; (iv) *N consists of semisimple elements*; (v) *For every $\alpha \in \Phi(T, G)$, the group U_α is not contained in N*; (vi) *N^o is central.*

If N is k-closed and satisfies these conditions, then it is defined over k.

Proof. We first recall that the maximal tori of G are conjugate (11.3) and that their intersection is the center $\mathscr{C}G$ of G (13.17, Cor. 2). In particular, every normal subgroup of G contained in T is central. With that taken into account, the implications

$$(\text{i}) \Leftrightarrow (\text{ii}) \Leftrightarrow (\text{iii}) \Rightarrow (\text{iv}) \Rightarrow (\text{v})$$

are obvious. The group N^o, being normalized by T, is generated by $(N^o \cap T)^o$ and the groups U_α it contains (13.20), therefore (v) implies $N^o \subset T$, and (vi). Let us now prove that (vi)$\Rightarrow$(ii). Assume (vi). Then $N^o \subset T$, hence $N^0 \subset \mathscr{C}(G)$ as pointed out above. In order to show that $N \subset T$, it suffices to show that N/N^o is contained in T/N^o in G/N^o, but this is clear since, N/N^o being finite and normal, is central. The equivalence of the six conditions is now proved. If N satisfies them, it belongs to T, hence is defined over k_s (8.2, 8.11) and therefore over k if it is moreover k-closed.

22.2 Lemma. *Let $\mathfrak{m}$ be an ideal of $\mathfrak{g}$ stable under* $\operatorname{Ad} G$. *Then the following conditions are equivalent*: (i) $\mathfrak{m} \subset \mathscr{Z}_{\mathfrak{g}}(G)$; (ii) $\mathfrak{m} \subset \mathfrak{t}$; (iii) *$\mathfrak{m}$ is contained in the intersection $\mathfrak{z}$ of the Lie algebras of the maximal tori of G*; (iv) *$\mathfrak{m}$ consists of semi-simple elements*; (v) *$\mathfrak{m}$ does not contain any algebra $\mathfrak{u}_\alpha$* ($\alpha \in \Phi(T, G)$). *We have the equalities*

$$\mathfrak{z}(\mathfrak{g}) = \mathscr{Z}_{\mathfrak{g}}(G) = \mathfrak{z}. \tag{1}$$

Proof. By the Corollary to 9.2,

$$\mathscr{Z}_{\mathfrak{g}}(T) = \mathfrak{t} = L(\mathscr{Z}_G(T)), \tag{2}$$

therefore (i)$\Rightarrow$(ii). We have (ii)$\Rightarrow$(iii) by the conjugacy of maximal tori (11.3) and (iii)$\Rightarrow$(i) because every torus centralizes its Lie algebra and G is generated by its maximal torii (12.1 and 13.16, Cor. 2). The implications (iii)$\Rightarrow$(iv)$\Rightarrow$(v) are obvious. The ideal $\mathfrak{m}$, being stable under T is the direct sum of the $\mathfrak{u}_\alpha$ it contains and of its intersection with $\mathfrak{t}$, hence (v)$\Rightarrow$(ii). This shows the equivalence of (i) to (v).

The center $\mathfrak{z}(\mathfrak{g})$ of $\mathfrak{g}$ is stable under every automorphism of $\mathfrak{g}$, hence under $\mathrm{Ad}\, G$. It does not contain any $\mathfrak{u}_\alpha$, since $\mathfrak{u}_\alpha$ is not central in the Lie algebra of the subgroup of G locally isomorphic to $\mathbf{SL}_2$ generated by U_α and $U_{-\alpha}$. Therefore $\mathfrak{z}(\mathfrak{g}) \subset \mathscr{Z}_{\mathfrak{g}}(G)$, and (1) follows in view of the equivalence of (i) and (iii).

22.3. A surjective k-morphism $f: G \to G'$ is *quasi-central* if its kernel N is central in G. It is *central* if it is quasi-central and $\ker df_e$ is central in $\mathfrak{g}$.

If f is quasi-central, then $\ker f$, which is necessarily k-closed, is defined over k by 22.1. Since any maximal torus of G' is the image of a maximal torus of G (11.14), 22.1 shows that f is quasi-central if and only if the inverse image of a maximal torus of G' is a maximal torus of G.

If f is an isogeny, it is automatically quasi-central (22.1). If moreover it is separable, then, df being injective, it is central. Thus a separable isogeny of a reductive group is central.

As examples, in characteristic two, the standard isogeny of $\mathbf{SL}_2$ onto $\mathbf{PGL}_2$ (17.6) is central, but the isogeny of $\mathbf{PGL}_2$ onto $\mathbf{SL}_2$ is not. Similarly, in characteristic p, the isogeny of $\mathbf{SL}_p$ onto $\mathbf{PGL}_p$ is central, but not separable.

The definition of quasi-central goes over *verbatim* to an arbitrary affine algebraic group, but the usual notion of central isogeny is different in the general case (see 22.11); however both notions are equivalent for reductive groups (22.15).

22.4 Proposition. *Let $f: G \to G'$ be a surjective morphism. The following conditions are equivalent*:

(i) *f is central.*
(ii) *The restriction of f to any closed connected unipotent subgroup U is an isomorphism onto $f(U)$.*
(iii) *The induced map $f^*: X(f(T)) \to X(T)$ maps $\Phi(f(T), G')$ onto $\Phi(T, G)$.*

Assume (i). By 22.1 and 22.2, the kernels of f and df consist of semi-simple elements, therefore f is injective on U and df on $L(U)$, whence (ii). If (ii) holds, then f maps U_α ($\alpha \in \Phi(T, G)$) isomorphically onto a one-dimensional subgroup normalized by $f(T)$, and (iii) follows. Note that $f(U_\alpha)$ is necessarily of the form U'_β, ($\beta \in \Phi(f(T), G')$). Assume (iii). Let $N = \ker f$. It is invariant, hence N^o is generated by $(N \cap T)^o$ and the U_α it contains (13.20). By (iii), df is injective on each $\mathfrak{u}_\alpha$ hence $N^o \subset T$, and by 22.1 N is central, contained in T. Thus f is quasi-central. Ker df is also stable under T, hence direct sum of its intersection with $\mathfrak{t}$ and of some $\mathfrak{u}_\alpha$. Some df is injective on $\mathfrak{u}_\alpha$, it follows that $\mathfrak{m} \subset \mathfrak{t}$, hence $\mathfrak{m} \subset z(\mathfrak{g})$ by 22.2. Therefore f is central.

22.5 Corollary. *Assume f to be central and surjective. Then* $\operatorname{Im} df$ *contains all the nilpotent elements of $\mathfrak{g}'$. Let T' be a maximal torus of G' and H' a k-subgroup of G'. Then $\mathfrak{g}' = df(\mathfrak{g}) + \mathfrak{t}'$. If $H' \subset T'$ or $H' \supset T'$ then $f^{-1}(H')$ is defined over k.*

Proof. Every nilpotent element X of $\mathfrak{g}'$ is tangent to a closed connected unipotent subgroup, say U' (14.26). Any maximal unipotent subgroup of $f^{-1}(U')^o$ maps onto U' under f (11.14). Then 22.4(ii) shows that $X \subset \operatorname{Im} df$. On the other hand, if U' and U'^- are two opposite maximal unipotent subgroups normalized by T', then $\mathfrak{g}'$ is the direct sum of the Lie algebras of U', U'^- and T'. Since $L(U')$ and $L(U'^-)$ belong to the image of df, the first assertion follows.

If $H' \supset T'$, then $\mathfrak{g}' = df(\mathfrak{g}) + \mathfrak{h}'$, and H is defined over k by 6.13. Assume now that $H' \subset T'$. Its centralizer is defined over k (9.2, Cor.), contains a maximal torus of G', hence also a maximal torus T'_o defined over k of G' (18.2). Then $T_o = f^{-1}(T'_o)$ is defined over k, as we just saw, and is a maximal torus (22.3). Being a k-closed subgroup of a torus defined over k, the group H is defined over k (8.2, 8.11).

22.6 Theorem. *Let G' be a k-group and $f: G \to G'$ a surjective k-morphism. Assume either f central or k perfect:*

(i) *The parabolic k-subgroups of G' (resp. G) are the images (resp. the inverse images) by f of the parabolic k-subgroups of G (resp. G').*

(ii) *The maximal k-split tori of G' (resp. G) are the images (resp. the maximal k-split tori of the inverse images) of the maximal k-split tori of G (resp. G').*
 Assume f central.

(iii) *Let S be a maximal k-split torus of G and $S' = f(S)$. Then f maps $\mathcal{N}_G S$ (resp. $\mathcal{Z}_G(S)$) onto $\mathcal{N}'_G(S')$ (resp. $\mathcal{Z}_{G'}(S')$) and induces an isomorphism of ${}_kW(S, G)$ onto ${}_kW(S', G')$. The homomorphism $(f|_S)^*: X(S') \to X(S)$ maps $\Phi(S', G')$ isomorphically onto $\Phi(S, G)$. If $\alpha \in \Phi(S', G')$ and $\beta = (f|_S)^*(\alpha)$ then f induces an isomorphism of $U_{(\beta)}$ onto $U'_{(\alpha)}$.*

Proof. (i) Let P (resp. P') be a parabolic k-subgroup of G (resp. G'). It follows from 11.14 that $f(P)$ (resp. $f^{-1}(P')$) is a parabolic subgroup. $f(P)$ is clearly defined over k, and $f^{-1}(P')$ is defined over k by 22.5 if f is central, by general principles (AG, 12.2) if k is perfect.

(ii) If S is a k-split torus in G, then $f(S)$ is a k-split torus in G' (8.15). It suffices therefore to show that if S' is a maximal k-split torus of G', then $f^{-1}(S') = H$ contains a k-split torus S mapping onto S'. The group S' is contained in a maximal torus of G' hence H is defined over k again by 22.5 if f is central, by general principles if k is perfect. Let T be a maximal torus of H defined over k (18.2). Then $f(T)$ is a maximal torus of $f(H)$ (11.14), hence is equal to S'. But then T_d maps onto $S'_d = S'$ (8.15).

(iii) By Corollary 2 to 11.14, $f(\mathcal{Z}_G(S)) = \mathcal{Z}_{G'}(S')$, hence also $\mathcal{Z}_G(S) = f^{-1}(\mathcal{Z}_{G'}(S'))$ (since $\ker f \subset \mathcal{Z}_G(S)$ in view of 22.1). By (ii) we know that if P (resp. P') is a minimal parabolic k-subgroup of G (resp. G') then $f(P)$ (resp.

$f^{-1}(P'))$ is a minimal parabolic k-subgroup of G' (resp. G). The previous remark shows that P (resp. P') contains $\mathscr{Z}_G(S)$ (resp. $\mathscr{Z}_G(S')$) if and only if $f(P)$ (resp. $f^{-1}(P')$) contains $\mathscr{Z}_{G'}(S')$ (resp. $\mathscr{Z}_G(S)$). By 21.3, $\mathscr{N}_G(S)$ permutes transitively the minimal parabolic k-subgroups containing $\mathscr{Z}_G(S)$ and ${}_kW(S,G)$ acts simply transitively on them. It follows that $f(\mathscr{N}_G(S))$ acts transitively on the set of minimal parabolic k-subgroups of G' containing $\mathscr{Z}_{G'}(S')$ and that $f({}_kW(S,G))$ is simply transitive on this set. Again by 21.3, this proves the first part of (iii).

Let P, P^- be two opposite minimal parabolic k-subgroups of G with intersection $\mathscr{Z}_G(S)$ and U, U^- their unipotent radicals. Then $f(P)$ and $f(P^-)$ are two opposite minimal parabolic k-subgroups of G', with intersection $\mathscr{Z}_{G'}(S')$ and unipotent radicals $f(U)$ and $f(U^-)$. Moreover, f induces a k-isomorphism of U and U^- onto their images. Let T_o be a maximal torus of G defined over k in $\mathscr{Z}_G(S)$, hence containing S. Then $T'_o = f(T_o)$ is a maximal torus of G' defined over k containing S'. We already know (22.4), that if $\alpha \in \Phi(T,G)$, then $f(U_\alpha) = U'_{\alpha'}$, where α' is the root of G' with respect to T'_o mapped onto α by $f^*_{|T_o}$. This implies first that $f^*_{|S}$ is a bijection of $\Phi(S',G')$ onto $\Phi(S,G)$. Let $\beta \in \Phi(S,G)$. Then $U_{(\beta)}$ is directly spanned by the one-parameter groups U_α where α runs through the set $\psi(\beta)$ of elements of $\Phi(T,G)$ whose restriction to S is a positive integral multiple of β (21.9). This, combined with the similar fact for G' and the previous remark, implies the last assertion.

22.7 Corollary. *Under the assumptions of* (i), (ii), *the k-rank of G is the sum of the k-ranks of G' and of* $(\ker f)^o$. *In particular, if f is moreover a k-isogeny, then* $r_k(G) = r_k(G')$.

Recall that $N = \ker f$ is contained in any maximal torus and is defined over k. If S is a maximal k-split torus, then $(S \cap N)^o$ is the maximal k-split torus of N. Since $f(S)$ is a maximal k-split torus of G', the first assertion follows. If f is moreover an isogency, then $N^o = \{1\}$, whence the second assertion.

22.8. Let H be a connected algebraic k-group, H_i $(i = 1, \ldots, m)$ a finite set of closed connected normal k-subgroups. Recall (see p. xi) that H is the almost direct product of the H_i's if the product mapping

$$\mu: H_1 \times \cdots \times H_m \to H$$

of the inclusions $H_i \to H$ is surjective, with finite kernel. In that case, μ is an isogeny, which is defined over k if the H_i's are so.

H is *almost k-simple* if it is semi-simple and has no closed connected normal k-subgroup of strictly positive dimension.

22.9 Proposition. *Assume G is an almost direct product over k of connected normal k-subgroups* G_i $(i = 1, \ldots, m)$. *Then the canonical isogency* $\mu: \tilde{G} =$

$G_1 \times \cdots \times G_m \to G$ is central. If S is a maximal k-split torus of G, then $S_i = (S \cap G_i)^o$ is a maximal k-split torus of G_i and S is the almost direct product of the S_i's. In particular,

$$rk_k(G) = \sum_i rk_k(G_i).$$

Proof. Ker f is finite, hence central. Let $X \in \ker df$. We can write $X = \sum X_i$, with $X_i \in \mathfrak{g}_i$ and must show that $X_i \in \mathfrak{z}(\mathfrak{g}_i)$. Let $G(i)$ be the subgroup generated by the G_j's $(i \neq j)$, and $X(i)$ the sum of the X_j $(j \neq i)$. Since $df(X) = 0$, we have $df(X_i) = -df(X(i))$ and $df(X_i)$ belongs to the intersection of $\mathfrak{g}_i$ and $L(G(i))$. The group G_i centralizes $G(i)$, and therefore also its Lie algebra. Hence $X_i \in \mathfrak{z}(\mathfrak{g}_i)$.

Let S_i be a maximal k-split torus in G_i $(i = 1, \ldots, m)$, $\tilde{S}$ the product of the S_i's. It is a maximal k-split torus of $\tilde{G}$ and $S = \mu(\tilde{S})$ is a maximal k-split torus of G by 22.6. By construction, it is the almost direct product of the $\mu(S_i)$ and clearly $\mu(S_i) = (\mu(\tilde{S}) \cap G_i)^o$. Thus S satisfies our conclusion. By conjugation (20.9), this is then true for any maximal k-split torus of G. The last assertion is now obvious.

22.10 Theorem. *Let G be semi-simple. Let G_i $(i \in I)$ be the minimal elements among the closed connected normal k-subgroups of G of strictly positive dimension.*

(i) *G is the almost direct product of the G_i, which are almost k-simple.*

(ii) *Let J be the set of $i \in I$ for which $rk_k G_i > 0$. Then ${}_kW(G) = \Pi_{j \in J} {}_kW(G_j)$ and ${}_k\Phi(G)$ is the direct sum of the k-root systems ${}_k\Phi(G_j)$ $(j \in J)$, each of which is irreducible.*

Proof. (i) By 14.10, we have a unique decomposition of G as an almost direct product of its almost simple closed connected normal subgroups N_a $(a \in A)$. Each of those is generated by subgroups U_α invariant under some maximal torus T. We may choose T defined over k (18.2). Since T and G split over a separable extension (18.8), the U_α are defined over k_s, hence so are the N_a. They are then permuted by $\Gamma = \mathrm{Gal}(k_s/k)$. Let now H be a connected normal k-subgroup of G. By 14.10, it is the almost direct product of some of the N_a, which are then permuted by Γ. But, then the other N_a's are also permuted by Γ, hence generate a normal k-subgroup H', such that G is the almost direct product of H and H'. From this (i) follows by an easy induction on the dimension. More precisely, we see that if $\{A_i\}$ $(i = 1, \ldots, m)$ are the orbits of Γ in A, then we may take for G_i the subgroup generated by the N_j $(j \in A_i)$.

(ii) The first assertion is clear for the product $\tilde{G}$ of the G_i's. By 22.6, it also holds for G. There remains to see that if $rk_k(G) > 0$ and G is almost k-simple, then ${}_k\Phi(G)$ is irreducible. The proof is essentially the same as in 14.10: Assume ${}_k\Phi(G) = \Phi_1 \oplus \Phi_2$. Then, by 14.5 and 21.9, $U_{(\alpha)}$ and $U_{(\beta)}$ commute if

$\alpha \in \Phi_1, \beta \in \Phi_2$. Therefore the closed k-groups G_1 and G_2 generated by the $U_{(\alpha)}$ for $\alpha \in \Phi_1$ and $\alpha \in \Phi_2$ respectively commute with one another. They are both normalized by $\mathcal{N}_G(S)$, where S is the maximal k-split torus underlying the definition of the $U_{(\alpha)}$. But G is generated by the $U_{(\alpha)}$ and $\mathcal{N}_G(S)$. Therefore G_1 and G_2 are normal in G and G is not almost k-simple.

22.11. For the sake of completeness, we discuss now the notion of central isogeny in the general case.

Let H, H' be k-groups and $f: H \to H'$ be a morphism. As before, it is said to be quasi-central if its kernel belongs to $\mathscr{C}H$. In that case, the commutator map $H \times H \to H$, which sends (x, y) onto $x \cdot y \cdot x^{-1} \cdot y^{-1}$, factors through $f(H) \times f(H)$, i.e. there exists a map $\kappa: f(H) \times f(H) \to H$ defined by

$$\kappa(f(x), f(y)) = x \cdot y \cdot x^{-1} \cdot y^{-1} \quad (x, y \in H). \tag{1}$$

κ is obviously unique. The morphism f is *central* if it is quasi-central and κ is a morphism of varieties.

In the context of the functorial definition of an affine algebraic group (1.5), this condition can also be expressed by saying that for any commutative K-algebra C, the kernel of $f(C): H(C) \to H'(C)$ *is central in* $H(C)$. In the case of a connected reductive group, the definition in 22.3 amounts to require that last condition for the algebra of dual numbers, as can be seen from 3.20.

22.12 Lemma. *Let X, X', Y be algebraic varieties, Z an affine variety, $f: X \to X'$ a surjective morphism, $\psi: X \times Y \to Z$ a morphism and $\psi': X' \times Y \to Z$ a map such that $\psi = \psi' \circ (f \times \mathrm{Id.})$. Then ψ' is a morphism if and only if the restriction of ψ' to $X' \times \{y\}$ is a morphism for every $y \in Y$.*

The latter condition is evidently necessary. Assume it is fulfilled. Our assertion is local in X' and Y, so we may assume them to be affine. Let $\{U_i\}$ be a finite open affine cover of X. Replacing X by the product U of the U_i and f by the composition of f with the canonical map $\prod_i U_i \to X$, we are reduced to the case where X is affine, too.

We have to show then that if $a \in K[Z]$, then $\psi'^{o}(a) \in K[X'] \otimes K[Y]$. Since f is surjective, f^o is injective and this amounts to proving that $\psi^o(a) \in f^o(K[X']) \otimes K[Y]$. We can write

$$\psi^o(a) = \sum_{1 \leqq i \geqq r} b_i \otimes c_i \quad (b_i \in K[X], c_i \in K[Y], i = 1, \ldots, r),$$

where we may assume the c_i's to be linearly independent over K. We may therefore find points $y_1, \ldots, y_r$ in Y such that the determinant of the $c_i(y_j)$ is $\neq 0$. Since the restriction of ψ' to $X' \times \{y_j\}$ is a morphism, we have $\sum_i c_i(y_j) \cdot b_i \in f^o(K[X'])$ for all j's whence $b_i \in f^o K[X']$, $(i = 1, \ldots, r)$.

22.13 Proposition. *Let $f: H \to H'$ be a morphism of algebraic groups. The following conditions are equivalent*:

(i) *f is central;*

(ii) *for every $h \in H$, there exists a morphism of varieties $\varphi_h: f(H) \to H$ such that $\varphi_h(f(x)) = x \cdot h \cdot x^{-1}$ for all $x \in H$;*

(iii) *there exists a morphism of varieties $\iota: f(H) \times H \to H$ such that $\iota(f(x), y) = x \cdot y \cdot x^{-1}$.*

Proof. Assume (i). With κ as in 22.11(1), we have,

$$\varphi_h(x') = \kappa(x', f(h)) \cdot h, \quad (x' \in f(H)),$$

which proves that (ii) holds. The implication (ii)$\Rightarrow$(iii) follows from the previous lemma where $X = Z = H, X' = f(H), Y = H$, and $\psi(x, y) = x \cdot y \cdot x^{-1}$. Similarly, the lemma with $X = Z = H, X' = f(H)$, $\psi'(x, y') = x \cdot \iota(y', x^{-1})$, (whence $\psi' = \kappa$), yields (iii)$\Rightarrow$(i).

22.14 Lemma. *Let G' be an algebraic group, $f: G \to G'$ a quasi-central surjective morphism, U^+ and U^- two opposite maximal unipotent subgroups of G normalized by T and X a variety on which G acts morphically by an action $\psi: G \times X \to X$. Then a morphism $\psi': X \to X$ such that $\psi = \psi' \circ (f \times Id.)$ exists if and only if this is true for the restrictions of ψ to $U^\pm$ and T.*

This latter condition is evidently necessary. Assume it. Let

(1) $$\nu^\pm: f(U^\pm) \times X \to X \quad \text{and} \quad \tau: f(T) \times X \to X$$

be the morphisms defined by

(2) $$\psi|_{U^\pm \times X} = \nu^\pm \circ (f \times Id.), \quad \psi|_{T \times X} = \tau \circ (f \times Id.).$$

The existence of τ implies that ker f operates trivially on X, hence that there exists a set-theoretic action

$$\psi': G' \times X \to X \quad \text{such that} \quad \psi = \psi' \circ (f \circ Id.).$$

Clearly, $\psi^\pm$ and τ are the restrictions of ψ' to $f(U^\pm) \times X$ and $f(T) \times X$ respectively. Let $u_+ \in f(U^+)$, $u_- \in f(U^-)$ and $t \in f(T)$. Then

(3) $$\psi'(U_+ t U_-, x) = \psi'(U_+, \psi'(t, \psi'(U_-, X))) = \nu^+(U_+, \tau(t, \nu^-(U_-, x))).$$

Let $V = f(U^+) \cdot f(T) \cdot f(U^-)$. The relation (3) and the assumption show that the restriction of ψ' to $V \times X$ is a morphism of varieties. On the other hand, for $g \in G$, the map $x \mapsto \psi(g, x) = \psi'(f(g), x)$ is an automorphism of X, therefore the restriction of ψ' to $(f(g) \cdot V) \times X$ is also morphic. Since the open sets $(f(g) \cdot V) \times X$ $(g \in G)$ cover $G' \times X$, the lemma is proved.

22.15 Proposition. *Let $f: G \to G'$ be a morphism of algebraic groups. Then f is central in the sense of 22.3 if and only if it is central in the sense of 22.11.*

Proof. Assume f to be central in the sense of 22.11. Given $g \in G$, consider the commutator map $\gamma: x \mapsto g \cdot x \cdot g^{-1} \cdot x^{-1}$. By assumption, there exists a morphism $\gamma': f(G) \to G$ such that $\gamma = \gamma' \circ f$, whence (see 3.16)

$$(d\gamma)_1 = \operatorname{Ad} g - 1 = (d\gamma')_1 \circ df;$$

therefore $\ker df \subset \ker(\operatorname{Ad} g - 1)$ for all $g \in G$, which implies $\ker df \subset \mathfrak{z}(\mathfrak{g})$ by 22.2. This shows that f is central in the sense of 22.3. Assume now the latter. Let U^+ and U^- be two opposite maximal unipotent subgroups of G normalized by T. By 22.4, the morphism $U^\pm \to f(U^\pm)$ is an isomorphism. Let $\alpha^\pm$ be its inverse. Since f is quasi-central, there exists a map $\iota: f(G) \times G \to G$ such that $\iota(f(x), y) = x \cdot y \cdot x^{-1}$ $(x \cdot y \in G)$. In order to prove that f is central in the sense of 22.11, it suffices by 22.13 to show that ι is a morphism of varieties. By 22.14, it is enough to prove that the restrictions $v^\pm$ and τ of ι to $f(U^\pm) \times G$ and $f(T) \times G$ are morphisms. The map $v^\pm$ is the composition of $\alpha^\pm \times Id.$ and of the morphism $(u, x) \mapsto u \cdot x \cdot u^{-1}$, hence is a morphism. There remains to see that τ is a morphism. By 22.12, we have only to show that for every $g \in G$, the map $\tau_g: f(T) \times G \to G$ defined by $\tau_g(t') = \tau(t', g)$ is a morphism. But we have $\tau_{g \cdot h} = \tau_g \circ \tau_h$ $(g, h \in G)$, therefore it suffices to prove this when g runs through a generating set for G, for instance for $g \in T \cup U^+ \cup U^-$. If $g \in T$, then τ_g is the constant map $f(T) \to \{g\}$ and if $g \in U^\pm$, then

$$\tau_g(t') = \alpha^\pm(t' \cdot f(g) \cdot t'^{-1}),$$

which shows that τ_g is a morphism.

Bibliographical Note

All the results and proofs in this section may be found in §2 of [5]. But there, the only notion of central isogeny is the general one (22.11) and we have slightly rearranged this material to suit our purposes.

§23. Examples

In this section, we describe the Tits system in $G(k)$ for various classical groups G.

For $n \geqq 1$, we denote by $e_{i,j}$ the $n \times n$ matrix all coefficients of which are zero except for the (i, j)-th one, which is equal to one. We let p be the characteristic exponent of k. For the classification of root systems, we refer to [9:VI].

23.1 Let D be a finite dimensional central K-algebra endowed with a k-structure such that $D(k)$ is a central division algebra over k, whose degree is denoted d. Thus, over K, the algebra D is isomorphic to $\mathbf{M}_d(K)$. We let D^* or $\mathbf{GL}_1(D)$ (resp. D_1 or $\mathbf{SL}_1(D)$) be the multiplicative group of elements of D with reduced norm Nr different from zero (resp. equal to one). The previous isomorphism maps D_1 onto $\mathbf{SL}_d(K)$. The group D^* is the almost direct product of its center S, a one-dimensional k-split torus, and of $\mathbf{SL}_1(D)$. Let us show that the latter group is anisotropic over k. Let T be a k-split torus contained in D. There is a k-basis (u_i) $(1 \leqq i \leqq d^2)$ of $D(k)$ and characters λ_i of T such that

$$t \cdot u_i \cdot t^{-1} = t^{\lambda_i} \cdot u_i \quad (t \in T; i = 1, \ldots, d^2).$$

However

$$Nr(t^{\lambda_i} \cdot u_i) = t^{d \cdot \lambda_i} \cdot Nr(u_i)$$

and, since $tu_i t^{-1}$ and u_i are conjugate, they have the same reduced norm. Moreover, $Nr(u_i) \neq 0$ since $D(k)$ is a division algebra. As a result, λ_i is trivial, hence T is central in D, therefore $T = S$ and $T \cap D_1$ is finite. In view of (22.8), this also shows that D^* has k-rank one.

23.2 *The groups* $\mathbf{GL}_n(D)$ *and* $\mathbf{SL}_n(D)$. Let D be as before and fix $n \in \mathbf{N}$, $n \geqq 2$. We view D^n as a right module over D. The group $G = \mathbf{GL}_n(D)$ can be identified to the group of invertible D-linear transformations of D^n. Let M be the group of diagonal matrices in G. It is isomorphic to $(D^*)^n$. Its center S is a k-split torus of dimension n. Let $y_i \in X^*(S)$ be the character of S such that s^{y_i} is the i-th diagonal entry of $s \in S$ $(1 \leqq i \leqq n)$. The y_i's form obviously a basis of $X^*(S)$.

It is elementary that $M = \mathscr{Z}_G(S)$. It follows then from 23.1 that S is maximal k-split in M, hence in G and therefore G is of k-rank n.

For $i \neq j$, let $U_{i,j}$ be the group of matrices $1 + t \cdot e_{i,j} (t \in D)$. It is unipotent, k-isomorphic to the additive group of D, and normalized by M. In particular

$$s(1 + te_{ij})s^{-1} = (1 + s^{y_i - y_j} \cdot te_{ij}) \quad (u \in U_{i,j}, 1 \leqq i \neq j \leqq n; s \in S).$$

The Lie algebra $\mathfrak{g}$ of G is the direct sum of the Lie algebras $L(M)$ and $L(U_{i,j})$ of M and of the $U_{i,j}$'s. The $U_{i,j}$'s for $i < j$ (resp. $i > j$) directly span a unipotent k-subgroup U^+ (resp. U^-) normalized by M. As a consequence, $M \cdot U^+$ and $M \cdot U^-$ are opposite minimal parabolic k-subgroups and $\Phi(S, G)$ is of type $\mathbf{A}_{n-1}$. Moreover, each k-root has multiplicity d^2. The normalizer of S is the group of monomial matrices and $\mathscr{N}_G(S)/M$ is the symmetric group in n letters.

The isomorphism $D \tilde{\to} \mathbf{M}_d(K)$ induces one of $\mathbf{M}_n(D)$ (resp. G) onto $\mathbf{M}_{nd}(K)$ (resp. $\mathbf{GL}_{nd}(K)$). On K^{nd} we use the "telescopic" basis

$$f_i \cdot e_a \quad (i = 1, \ldots, n, a = 1, \ldots, d).$$

The group T of diagonal matrices of $\mathbf{GL}_{nd}(K)$ is a maximal torus, the group of characters $X^*(T)$ of which has a basis $x_{i,a}$, where $x_{i,a}$ is the character which associates to $t \in T$ its (i, a)-th diagonal entry. The roots of G with respect to T are the characters

$$x_{i,a} - x_{j,b} \quad ((i, a) \neq (j, b); 1 \leqq i, j \leqq n, 1 \leqq a, b \leqq d).$$

From this the restriction $\Phi(T, G) \to \Phi(S, G)$ is easy to describe, namely: the roots $x_{i,a} - x_{i,b} (a \neq b)$ restrict to zero and $x_{i,a} - x_{j,b}$ $(1 \leqq a, b \leqq d)$ to $y_i - y_j (i \neq j)$.

Let $G' = \mathbf{SL}_n(D)$. The isomorphism $\mathbf{M}_n(D) \tilde{\to} \mathbf{M}_{nd}(K)$ maps it onto $\mathbf{SL}_{nd}(K)$. It is a k-form of $\mathbf{SL}_{nd}$. For each field k' between k and K, the group $\mathbf{SL}_n(D)(k')$ is known to be generated by the unipotent subgroups $U_{i,j}(k')$ and to be the derived group of $G(k')$. The group G is the almost direct product of its center and of G'. Hence the latter has k-rank $n-1$. The group G (resp. G') is k-split if and only if $d=1$. Then $G = \mathbf{GL}_n$ and $G' = \mathbf{SL}_n$.

23.3 *The symplectic group* $\mathbf{Sp}_{2m}$. Let V be a finite dimensional vector space over K endowed with a k-structure, n the dimension of V, and F a bilinear form on $V(k)$ which is alternating, i.e. $F(x,x)=0$ for all $x \in V(k)$, and defined over k, i.e. k-valued on $V(k)$. If $p \neq 2$, our assumption is equivalent to $F(x,y)+F(y,x)=0$ for all $x,y \in V$. In characteristic two, F is symmetric. We assume F to be non-degenerate, which forces n to be even, say $n=2m$. As is well-known, $V(k)$ has a *symplectic basis* $(e_{\pm i})$ $(i=1,\ldots,m)$, i.e. such that

$$(1) \qquad F(e_i, e_{-i}) = 1 \quad (i=1,\ldots m), \quad F(e_{\pm i}, e_{\pm j}) = 0 \quad \text{if} \quad i \neq j.$$

The group G of elements in $\mathrm{GL}(V)$ preserving F is the symplectic group $\mathbf{Sp}_{2m}$. It is defined over k. The torus S consisting of the transformations $e_i \mapsto s_i e_i, e_{-i} \mapsto s_i^{-1} e_{-i}$ is k-split and a maximal torus of G. We let (y_i) be the basis of $X^*(S)$ such that s^{y_i} is s_i in the previous notation. Let L_i be the subgroup of G leaving $[e_i, e_{-i}]$ stable and fixing the other basis vectors. It is isomorphic to $\mathbf{SL}_2$, centralizes the kernel of y_i, its upper triangular unipotent subgroup is stable under S, and corresponds to the root $2y_i$. Let $i \neq j$. The two subspaces $E_{i,j} = [e_i, e_j]$ and $E_{-i,-j} = [e_{-i}, e_{-j}]$ are in duality with respect to F and isotropic for F. The subgroup of G which leaves these two subspaces stable and fixes the other basis vectors contains $\mathbf{SL}_2$, acting by the identity representation on $E_{i,j}$ and the contragredient representation on $E_{-i,-j}$. It centralizes the kernel of $y_i - y_j$ in S. Its upper triangular unipotent subgroup is associated to the root $y_i - y_j$. Similarly, by considering $[e_i, e_{-j}]$ and $[e_{-i}, e_j]$, we produce a unipotent subgroup associated to the root $y_i + y_j$. On the other hand, it is easily checked, and well-known, that $\mathbf{Sp}_{2m}$ is of dimension $m(2m+1)$. Therefore we have produced all the roots, up to sign, and get the root system of type $\mathbf{C}_m$. We leave it to the reader to check that the parabolic k-subgroups are the stability groups of the rational isotropic flags.

We have tacitly assumed that G is connected. This can be deduced for instance from the classical fact that G is generated by transvections, and each of those is contained in a one-dimensional unipotent subgroup isomorphic to $\mathbf{G}_a$ (see e.g. [D:II, §5]). It is also an immediate consequence of the previous discussion, which we temporarily view as applying to G^o. First it is easily seen that S is its own centralizer in G. By the conjugacy of maximal tori, any connected component of G contains an element n normalizing S. Int n induces an automorphism of the root system $\mathbf{C}_m$. But the latter has no automorphism except those given by its Weyl group, whence the existence of $n' \in G^o$ normalizing S such that $n' \cdot n^{-1}$ centralizes S, whence $n \in G^o$. We also see that $G \subset SL(V)$ since $S \subset SL(V)$, obviously.

23.4 *Orthogonal groups* ($p \neq 2$). We assume here $p \neq 2$. Let V be a finite dimensional vector space over K endowed with a k-structure, Q a quadratic form on V defined over k and F the associated symmetric bilinear form. We assume F to be non-degenerate. Let $G = \mathrm{SO}(Q)$ be the special orthogonal group of Q, i.e. the group of linear transformations of determinant one leaving Q, (or, equivalently, F) invariant.

Recall that Q is said to be isotropic over k if it is vanishes on some non-zero element of $V(k)$ and anisotropic over k otherwise. We claim first that Q is isotropic over k if and only if G is isotropic over k (20.1).

Assume first G to be isotropic over k and let S be a non-trivial k-split subtorus of G. There exists a non-trivial character λ of S and $v \in V(k) - \{0\}$ such that $s \cdot v = s^{\lambda} \cdot v$ for all $s \in S$. Then $Q(v) = 0$ since, on the one hand, $Q(s \cdot v) = Q(v)$ and on the other $s \cdot v = s^{\lambda} \cdot v$ implies $Q(s \cdot v) = s^{2\lambda} \cdot v$ for all $s \in S$. Therefore Q is isotropic over k. Now assume the latter. Then V is the direct sum of a hyperbolic plane E defined over k and of the orthogonal complement E' of E. There exists a k-basis e_1, e_2 of E such that

$$F(e_1, e_1) = F(e_2, e_2) = 0 \quad \text{and} \quad F(e_1, e_2) = 1. \tag{1}$$

Then the transformations $s(x)$, $(x \in K^*)$ defined by

$$s(x) \cdot e_1 = x \cdot e_1, s(x) \cdot e_2 = x^{-1} e_2 \quad s(x) \cdot f = f \quad (f \in E') \tag{2}$$

belong to G and form a k-split torus, hence G is isotropic over k.

From now on, assume that Q is isotropic over k and let q be its Witt index over k (dimension of a maximal isotropic subspace defined over k). By Witt's theorem, $V(k)$ contains q linearly independent hyperbolic planes $H_1, \ldots, H_q$ and the restriction Q_o of Q to the orthogonal complement V_o of their direct sum is anisotropic. On H_i we choose a basis (e_i, e_{n-q+i}) satisfying the same conditions as (e_1, e_2) in (1) $(i = 1, \ldots q)$ and let $e_{q+1}, \ldots, e_{n-q}$ be a basis of $V_o(k)$. Then it is clear (see (2) above) that the diagonal torus consisting of elements with diagonal entries

$$(s_1, \ldots, s_q, 1, \ldots, 1, s_1^{-1}, \ldots, s_q^{-1}), \quad (s_1, \ldots, s_q \in K^*), \tag{3}$$

is a q-dimensional k-split subtorus of G. We claim it is maximal k-split. Indeed, if T is a torus of G containing S, it leaves the fixed point set V_o of S stable; if it is k-split, its restriction to V_o is a k-split torus contained in $\mathrm{SO}(Q_o)$, hence is reduced to the identity by our initial remark, which implies that $T = S$. Therefore $q = r_k G$. We identify $\mathrm{SO}(Q_o)$ to the subgroup of G leaving fixed e_i for $i \notin [q+1, n-q]$. Then $\mathscr{Z}_G(S) = S \times \mathrm{SO}(Q_o)$.

If $q = [n/2]$, then V_o is zero- or one-dimensional, $\mathrm{SO}(Q_o)$ is reduced to the identity, $\mathscr{Z}_G(S) = S$, hence S is a maximal torus of G. Therefore G is k-split. If $n = 2q + 2$, then $\mathrm{SO}(Q_o)$ is commutative, $\mathscr{Z}_G(S)$ is a maximal torus $\supset S$ and the minimal parabolic k-subgroups are Borel subgroups, hence G is quasi-split over k. If $n > 2q+2$, then $SO(Q_o)$ is not commutative and G is not quasi-split.

We denote by y_i the character of S such that $s^{y_i} = s_i$ is the i-th diagonal entry of S $(i = 1, \ldots, q)$. With respect to the basis $(e_{i,j})$ of $\mathfrak{gl}_n$, the weights of S for the adjoint action in $\mathfrak{gl}_n$ are given by the following table

(4)

$j \backslash i$	$[1,q]$	$[q+1,n-q]$	$[n-q+1,n]$
$[1,q]$	$y_i - y_j$	y_i	$y_i + y_{j-n+q}$
$[q+1,n-q]$	$-y_i$	0	y_{j-n+q}
$[n-q+1,n]$	$-y_{i-n+q} - y_j$	$-y_{i-n+q}$	$-y_{i-n+q} + y_{j-n+q}$.

We may assume the e_i's $(q < i \leqq n-q)$ chosen so that Q_o is diagonal. Then so is the associated form F_o to Q_o. We write the elements of $\mathbf{M}_n(K)$ in block form, as 3×3 matrices corresponding to the above partition of $[1,n]$. Then F has entries (I_q, F_o, I_q) on the second diagonal, where I_q is the $q \times q$ identity matrix. The Lie algebra $\mathfrak{g}$ of G is the subalgebra of $\mathfrak{gl}_n$ given by

$$\mathfrak{g} = \{A = \mathfrak{gl}_n \mid A \cdot F + F \cdot {}^tA = 0\}. \tag{5}$$

Writing $A = (A_{a,b})$ $(1 \leqq a, b \leqq 3)$, we get from (5)

$$\begin{aligned} A_{1,3} + {}^tA_{1,3} = A_{1,2} \cdot F_o + {}^tA_{2,3} = A_{1,1} + {}^tA_{3,3} &= 0 \\ A_{2,2} \cdot F_o + F_o \cdot {}^tA_{2,2} &= 0 \\ A_{3,2} \cdot F_o + {}^tA_{2,1} = A_{3,1} + {}^tA_{3,1} &= 0. \end{aligned} \tag{6}$$

From this and (4) we see that the k-roots of G with respect to S are

$$\begin{aligned} &y_i - y_j (i \neq j), && \text{with multiplicity one,} \\ &\pm (y_i + y_j)(i \neq j) && \text{with multiplicity one,} \\ &\pm y_i && \text{with multiplicity } n - 2q. \end{aligned} \tag{7}$$

Therefore $\Phi(S, G)$ is of type $\mathbf{D}_q$ if $n = 2q$, of type $\mathbf{B}_q$ otherwise.

If we add to (6) the conditions

$$A_{2,1} = A_{3,1} = 0, \ A_{1,1} \quad \text{upper triangular,} \tag{8}$$

then we get the Lie algebra of a minimal parabolic k-subgroup. It is the isotropy subgroup of the full isotropic flag over k

$$[e_1] \subset [e_1, e_2] \subset \cdots \subset [e_1, e_2, \ldots, e_q].$$

The standard parabolic k-subgroups are the stabilizers of the rational isotropic flags $V_1 \subset \cdots \subset V_a$, where V_i is of dimension $d(i)$ and spanned by

$$e_1, \ldots, e_{d(i)} \quad (1 \leqq d(1) < d(2) < \cdots < d(a) \leqq q).$$

It follows from Witt's theorem that the parabolic k-subgroups are the stability groups of the rational isotropic flags in V.

Remarks. As in 23.3, we have implicitely assumed that G is connected. Again, this follows easily from the previous discussion, viewed as applying to G^0. To see this we may assume that $k = K$, hence $q = [n/2]$. We have already remarked that S is its own centralizer. If n is odd, the argument is the same as for the symplectic group. Let now $n = 2q$ be even. If $q = 1$, then using the basis e_1, e_2 of (1), we see that $G = \mathbf{GL}_1$. Let now $q > 1$. A connected component of G contains an element n normalizing S. Moreover, after multiplying n by an element of $\mathcal{N}_{G^o}(S)$, we may assume that it leaves stable the set of positive roots for a given ordering, hence that it permutes the elements of a set Δ of simple roots. In the previous notation, we may take for Δ the set consisting of the roots $y_i - y_{i+1}$ $(i = 1, \ldots, q-1)$ and $y_{q-1} + y_q$. Then Int n leaves pointwise fixed the one-dimensional torus S_o on which these simple roots take equal values. It follows that S_o is given by the relations

$$s_i = s_{q-1}^{q-i} \quad (i = 1, \ldots, q-1), s_q = 1.$$

In particular, the characters s_i, s_i^{-1} are distinct on S_o. Therefore Int n leaves invariant the lines $k \cdot e_i$ $(i \neq q, q+1)$ and the plane spanned by e_q, e_{q+1}. This implies $n \in S$, in particular $n \in G^o$.

23.5 *Quadratic forms in characteristic two.* Our next goal is the discussion of orthogonal groups in characteristic two. As a preliminary, we collect here a few notions and facts on quadratic forms in characteristic two (see [D] for instance).

Let again V be a finite dimensional vector space over K, of dimension n, endowed with a k-structure. A *quadratic form* on $V(k)$ is a k-valued function satisfying the identity

$$(1) \qquad Q(a \cdot x + b \cdot y) = a^2 Q(x) + b^2 Q(y) + a \cdot b \cdot F(x, y), \quad (a, b \in k, x, y \in V(k)),$$

where F is a bilinear form on $V(k)$, which is determined by Q since (1) implies

$$(2) \qquad F(x, y) = Q(x + y) + Q(x) + Q(y) \; (x, y \in V(k)).$$

We also see from (1) that

$$(3) \qquad Q(a \cdot x) = a^2 \cdot Q(x) \quad (a \in k, \; x \in V(k))$$

$$(4) \qquad F(x, x) = 0 \quad (x \in V(k))$$

hence F is alternating. Its rank is even and will be denoted $2m$. We let V^o be the radical of its extension to V. It is defined over k.

The form Q is said to be *non-degenerate* if $Q(x) \neq 0$ for every $x \in V^o(k) - \{0\}$. We always assume this to be the case. Then $d = n - 2m$ is called the *defect of* Q. We also assume $d \leqq 1$. Therefore $d = 0$ if $n = 2m$ is even, $d = 1$ if $n = 2m + 1$ is odd. We also view Q as a quadratic form on V by extension of scalars. Clearly, as a form on V, it is still non-degenerate, of defect $\leqq 1$.

If $(e_1, \ldots, e_n)$ is a basis of V, then a simple induction from (1) shows that

$$Q(x_1 \cdot e_1 + \cdots + x_n \cdot e_n) = \sum_{i=1}^{i=n} x_1^2 \cdot Q(e_i) + \sum_{1 \leqq i < j \leqq n} x_i \cdot x_j \cdot F(e_i, e_j). \tag{5}$$

If n is odd, we assume e_n to span V^o. Then, given F, it is easily seen that (5) defines a non-degenerate quadratic form of defect $\leqq 1$ for any set of values $Q(e_i)$ such that $Q(e_n) \neq 0$ if n is odd.

A subspace $E \subset V$ is *singular* for Q if Q is zero on E. We see again from (1) that if A and B are two singular subspaces orthogonal with respect to F, then their sum is singular. In particular the sum of two singular vectors orthogonal with respect to F is singular. Let us say that a two-dimensional subspace E of V is hyperbolic if it has a basis e, f such that

$$Q(e) = Q(f) = 0 \quad F(e, f) = 1. \tag{6}$$

In the associated coordinates x_1, x_2 on E we have

$$\begin{gathered} Q(x, x) = x_1 \cdot x_2, \quad F(x, y) = x_1 \cdot y_2 + x_2 \cdot y_1, \\ (x = x_1 e + x_2 f, y = y_1 e + y_2 f). \end{gathered} \tag{7}$$

It is again true (see [D]) that a non-singular plane in $V(k)$ containing a non-zero singular element of $V(k)$ is hyperbolic. From this we see that if q is the maximal dimension of a singular subspace in $V(k)$, then there exist q hyperbolic planes $H_i (1 \leqq i \leqq q)$ defined over k, and orthogonal with respect to F. The integer q is called the index of Q, and Q is anisotropic over k if $q = 0$.

23.6 *Orthogonal groups in characteristic two.* We go on with the setup of 23.5 and denote by G the orthogonal group $O(Q)$ of Q, i.e. the subgroup of $GL(V)$ preserving Q. It automatically belongs to $SL(V)$, as will be clear below. It preserves F, as follows from 23.5 (2), hence also V^o, and is the identity on V^o.

Let $n = 2m > 3$. Fix a symplectic basis (e_i) of $V(k)$ for F. We assume that

$$F(e_i, e_{m+i}) = 1 \quad \text{for} \quad 1 \leqq i \leqq m, \quad F(e_i, e_j) = 0 \quad \text{if} \quad j - i \neq \pm m. \tag{1}$$

i.e., written as a 2×2 matrix of m-blocks, F is given by

$$F = \begin{pmatrix} 0 & I_m \\ I_m & 0 \end{pmatrix}, \quad (I_m : m \times m \text{ identity matrix}). \tag{2}$$

G is a subgroup of $\mathbf{Sp}_{2m} = \mathbf{Sp}(F)$. It follows from 23.5 that $g \in \mathbf{Sp}(F)$ belongs to G if and only $Q(g(e_j)) = Q(e_j)$ for $j = 1, \ldots, 2m$. i.e. if $g = (g_{i,j})$ satisfies the conditions

$$\sum_{i=1}^{i=n} g_{i,j}^2 Q(e_i) + \sum_{1 \leqq i \leqq m} g_{i,j} \cdot g_{m+i,j} = Q(e_j) \quad (j = 1, \ldots, 2m). \tag{3}$$

We claim that G is defined over k. It is the intersection of $\mathbf{Sp}_{2m}$, which is defined over k, and of the $2m$ irreducible quadrics Q_j, defined by (3), also defined over k; it suffices therefore to show that these varieties are smooth at 1 and intersect properly there. If we write a $2n \times 2n$ matrix in $n \times n$ blocks, then the Lie algebra of $\mathbf{Sp}_{2m}$ consists of the matrices $\begin{pmatrix} X_1 & X_2 \\ X_3 & X_4 \end{pmatrix}$ satisfying the conditions

$$(4) \qquad {}^tX_1 + X_4 = {}^tX_2 + X_2 = {}^tX_3 + X_3 = 0,$$

(which translate the relation ${}^tX \cdot F + F \cdot X = 0$). On the other hand, the tangent space at 1 to Q_j is the coordinate hyperplane $X_{m+j,j} = 0$ if $j \leqq m$ and $X_{j-m,j} = 0$ if $j > m$. Therefore the intersection is proper. This also shows that the Lie algebra $\mathfrak{g}$ of G consists of the matrices satisfying (4), in which moreover X_2 and X_3 have zeroes on the diagonal. This is independent of the values of Q on the basis vectors. Therefore, the Lie algebras of the groups $\mathrm{O}(Q)$, where Q runs through the quadratic forms with associated form F, coincide. If Q has maximal index m, then $Q(e_i) = 0$ for all i and

$$(5) \qquad Q(x) = \sum_{1 \leqq i \leqq m} x_i \cdot x_{m+i}.$$

We let $\mathbf{O}_{2m}$ denote its orthogonal group.

Let now $n = 2m + 1$, and π the canonical projection $V \to V/V^o = V_1$. The form F induces a non-degenerate alternating form on V_1, which we also denote by F. We choose a basis (e_i) of $V(k)$ such that e_n spans V^o and the e_i's ($i \leqq 2m$) project onto a basis of $V_1(k)$ satisfying (1). Then the e_i's also satisfy (1) and moreover $F(e_i, e_n) = 0$ for $i \leqq 2m$. Consequently $g \in \mathrm{GL}(V)$ belongs to G if and only if it fixes e_n, preserves F and satisfies the same conditions (3) as before. Let us write an element of $\mathrm{GL}(V)$ fixing e_n in the form

$$(6) \qquad \begin{pmatrix} A & 0 \\ a & 1 \end{pmatrix}$$

where a is a $1 \times 2m$ matrix, 0 the $2m$ zero column vector and A a $2m \times 2m$ matrix. Then g preserves F if and only if A belongs to $\mathbf{Sp}_{2m}$. Let G_1 be the group of such matrices. It is defined over k, and its Lie algebra consists of matrices

$$\begin{pmatrix} X_1 & X_2 & 0_m \\ X_3 & X_4 & 0_m \\ a_1 & a_2 & 0 \end{pmatrix}$$

where a_1, a_2 are $1 \times m$ matrices, 0_m the $m \times 1$ zero matrix, and the X_i's are $m \times m$ matrices satisfying (4). A computation of tangent spaces identical to the previous one shows again that the intersection of G_1 and of the quadrics Q_j is proper and that the Lie algebra of G is the set of matrices (6) where the X_i form the Lie algebra of $\mathbf{O}_{2m}$. Here again, the orthogonal groups of the non-degenerate forms of defect one with associated form F have all the same Lie algebra. If $Q(e_i) = 0$ for $i \leqq 2m$, then $q = m$ and Q is given by

$$Q(x) = x_n^2 Q(e_n) + \sum_{1 \leqq i \leqq m} x_i \cdot x_{m+i}.$$

We let $\mathbf{O}_n$ be the corresponding orthogonal group, when $Q(e_n) = 1$.

The projection π induces a k-morphism $G \to \mathbf{Sp}_{2m}$. We claim it is a purely inseparable isogeny. Incidentally, this will show that G is connected, since $\mathbf{Sp}_{2m}$ is so (by 23.3). Assume $\pi(g)$ is the identity. Then $g \cdot e_i = e_i + e_i \cdot e_n$ for $i \leqq 2m$ and, since $F(e_i, e_n) = 0$, we have

$$Q(e_i) = Q(g \cdot e_i) = Q(e_i) + c_i^2 \cdot Q(e_o), \quad (i = 1, \ldots, 2m),$$

hence $c_i = 0$. On the other hand, $g \cdot e_n = e_n$, hence g is the identity. Therefore π is injective. It is surjective already for dimensional reasons. A bit more strongly, it can be checked directly that if k is perfect, then $\pi(k): G(k) \to \mathbf{Sp}_{2m}(k)$ is surjective: In fact, any element $h \in \mathbf{Sp}_{2m}(k)$ defines $h_1 \in G_1(k)$ in the obvious manner. We still dispose of the $2m$ first entries of the last row to modify h_1 to an element $g \in G$ having the same image as h_1 under π. The conditions for this are

$$Q(h_1 \cdot e_i + a_{n,i} e_n) = Q(e_i), \quad (i = 1, \ldots, 2m),$$

which can be written

$$a_{n,i}^2 \cdot Q(e_n) + Q(h_1 \cdot e_i) = Q(e_i), \quad (i = 1, \ldots, 2m),$$

an equation which can be solved uniquely since k is assumed perfect and $Q(e_n) \neq 0$

The previous considerations show that $d\pi$ maps $\mathfrak{g}$ onto the Lie algebra of $\mathbf{O}_{2m}$, therefore the kernel of $d\pi$ is $2m$-dimensional. It consists of the matrices for which $X_i = 0$ $(i = 1, \ldots, 4)$ in the above notation, therefore it is commutative, made up of nilpotent elements.

In view of the existence of hyperbolic planes, the proof that the index q of Q is equal to the k-rank of G is exactly the same as in 23.3. We assume now $q \geqq 1$ and want to show that the k-root system is the same as when $p \neq 2$, and also determine explicitely the kernel of $d\pi$ when n is odd.

Choose q hyperbolic planes $H_1, \ldots, H_q$ defined over k. Let V_o be the orthogonal complement to their direct sum, with respect to F. It contains V^o and is defined over k. Let (e_{2i-1}, e_{2i}) be a standard k-basis of H_i (see 23.3 (6)). We complete (e_i) $(i = 1, \ldots, 2q)$ by a k-basis of V_o to get a basis which is a symplectic basis if $n = 2m$, such that e_n spans V^o and the e_i $(i \leqq 2m)$ project to a symplectic basis of $V_1 = V/V^o$ if n is odd. Then, in the corresponding coordinates x_i, Q has the form

$$Q(x) = \sum_{i=2q+1}^{i=n} x_i^2 \cdot Q(e_i) + \sum_{1 \leqq i \leqq m} x_{2i-1} \cdot x_{2i}. \tag{7}$$

Since q is the index, the restriction of Q to V_o is anisotropic over k. Let S be the k-split torus consisting of diagonal matrices with entries $(s_1, s_1^{-1}, \ldots, s_q, s_q^{-1}, 1, \ldots, 1)$ and let, as before $y_i \in X^*(S)$ be the character assigning s_i to s.

The Lie algebra of G is that of $\mathbf{O}_n$, described explicitely earlier, from which it is clear that if $q \geqq 2$, then $\pm y_i \pm y_j$ $(1 \leqq i < j \leqq q)$ is a k-root. If $n = 2m = 2q$, then, for dimensional reasons, these are all the k-roots and $\Phi(S, G)$ is of type $\mathbf{D}_m$. Let now $n > 2q$. Fix $i \leqq q$ and $j > 2q$, and let $E_{i,j}$ be the space spanned by e_{2i-1}, e_{2i} and e_j. The restriction of Q to $E_{i,j}$ is the standard 3-form

$$Q(x) = x_{2i-1} \cdot x_{2i} + x_j^2 \cdot Q(e_j), \tag{8}$$

and $Q(e_j) \neq 0$ since Q is anisotropic over k on V_o. The subgroup $H_{i,j}$ of G which fixes the e_a's for $a \neq 2i-1$, $2i$, j is isomorphic to $\mathbf{O}_3$. The group $H_{i,j}$ is normalized by S and centralizes the kernel of y_i. Let $U(a)$ be the 3×3 matrix given by

$$U(a) = U_{i,j}(a) = \begin{pmatrix} 1 & 0 & 0 \\ a^2 \cdot Q(e_j) & 1 & 0 \\ a & 0 & 1 \end{pmatrix} \quad (a \in K). \tag{9}$$

The $U(a)$ form a one-dimensional unipotent k-subgroup $U_{i,j}$, k-isomorphic to $\mathbf{G}_a$. It is easily checked that it belongs to $H_{i,j}$. Indeed, this amounts to show that

$$Q(e_{2i-1} + a^2 \cdot Q(e_j) \cdot e_{2i} + a \cdot e_j) = Q(e_{2i-1}) = 0$$
$${}^t U(a) \cdot F_3 \cdot U(a) = F_3,$$

where F_3 is the matrix of the restriction of F to $E_{i,j}$, which is immediate. Let $s(t)$ be the diagonal 3×3 matrix with entries t, t^{-1}, 1. Then we have

$$s(t) \cdot U(a) \cdot s(t^{-1}) = U(t^{-1} \cdot a) (a \in K, t \in K^*)$$

which shows that G has the root $-y_i$ with respect to S. Since j is arbitrary in $(2q+1, n]$, this proves that $-y_i$ has at least the multiplicity $n - 2q$. On the other hand, the centralizer of S is $S \times O(Q_o)$ where Q_o is the restriction of Q to V_o. Adding up its dimension to those of the root spaces already found yields the dimension $n(n-1)/2$ of $O(Q)$. Therefore, if $n > 2q$, the root system $\Phi(S, G)$ is of type $\mathbf{B}_q$, where $\pm y_i$ has multiplicity $n - 2q$ $(1 \leqq i \leqq q)$ and $\pm y_i \pm y_j$ has multiplicity one $(1 \leqq i < j \leqq q)$.

To describe the kernel of $d\pi$, for n odd, we may assume k algebraically closed, hence $m = q$ and $Q(e_n) = 1$. Now $\pm y_i$ has multiplicity one and occurs in $H_{i,n}$, in the previous notation. The isogeny π maps $U_{i,n}(a)$ onto the matrix

$$\begin{pmatrix} 1 & 0 \\ a^2 & 1 \end{pmatrix}$$

and $H_{i,n}$ onto $\mathbf{SL}_2$. Therefore $\pi(U_{i,n})$ is the root group corresponding to the root $-2y_i$ in the symplectic group, and the restriction of π to $U_{i,n}$ is the map $a \mapsto a^2$. Hence the Lie algebra of $U_{i,n}$ belongs to the kernel of $d\pi$, and $\ker d\pi$ is the sum of the Lie algebras of the root groups corresponding to the roots $\pm y_i$ $(i = 1, \ldots, m)$.

23.7 *Division algebras with involutions.* We collect here a few known notions and facts on division algebras with involution, to be used in the next section. We again assume $p \neq 2$.

Recall that an *involution* $\sigma: a \mapsto a^\sigma$ of an algebra A over a field E is an antiautomorphism of order ≤ 2, i.e. a E-linear bijection of A onto itself satisfying the conditions

$$(a \cdot b)^\sigma = b^\sigma \cdot a^\sigma, \sigma^2 = 1 \; (a, b \in A).$$

It defines an isomorphism of A onto its opposite algebra A^o. It is said to be of the first (resp. second) kind if it is (resp. is not) the identity on the center of A. The algebra A is always assumed to have an identity so that E is identified to a subfield of the center fixed under σ.

Let now D be a finite dimensional K-algebra defined over k, such that $D(k)$ is a division algebra over its center k', whose degree over k' is denoted d, endowed with an involution σ. We shall assume to be in one of the following cases:

(*i*) $k' = k$. Then σ is necessarily of the first kind. Under the isomorphism

$$D := D(k) \bigotimes_k K = \mathbf{M}_d(K)$$

the involution σ extends to one of D. Since $X \mapsto {}^tX$ is already one, there exists, by the Skolem–Noether theorem, an element $J \in D^*$ such that

$$X^\sigma = J \cdot {}^tX \cdot J^{-1} \quad (X \in D):$$

The relation $X^{\sigma\sigma} = I_d$ implies that $J \cdot {}^tJ^{-1}$ is central in D, hence of the form $\delta \cdot I_d$ $(\delta \in K^*)$. But then it follows from $J = \delta \cdot {}^tJ$ that $\delta^2 = 1$, hence

$${}^tJ = \delta J \quad (\delta = \pm 1)$$

which implies

$$J^\sigma = \delta J.$$

We shall say that σ is of type δ. Let $D^\pm$ be the eigenspace of σ in D for the eigenvalue ± 1. It is defined over k and $D = D^+ \oplus D^-$. The dimension of D^- is that of the subgroup of elements in $D^* \doteq \mathbf{GL}_d(K)$ which are fixed under the automorphism $g \mapsto (g^\sigma)^{-1}$, i.e. such that $gJ{}^tg = J$. It is an orthogonal (resp. symplectic) group if $\delta = 1$ (resp. $\delta = -1$), hence of dimension $d(d-1)/2$ (resp. $d(d+1)/2$)). It follows that

$$g \cdot J \cdot {}^tg = J, \dim D^\pm = d(d \pm \delta)/2. \tag{1}$$

In particular, the two types of involution of the first kind are distinguished by the dimension of D^+.

Let $U \in D^*$. It is readily checked that $\tau: X \mapsto UX^\sigma U^{-1}$ is an involution if and only if $U^\sigma = \varepsilon U$, where $\varepsilon = \pm 1$ and that, in that case, τ is of type $\varepsilon\delta$. Now σ is the identity if and only if $d = 1$. If $d > 1$, both D^+ and D^- are $\neq 0$, hence if $D(k)$ possesses an involution, then it has involutions of both types. In particular d is always even (in fact, A. Albert has shown that it is a power of two).

(ii) σ is of the second kind and its fixed point set on k' is k. Then k' is a separable quadratic extension of k and σ induces on it the non-trivial automorphism of k' over k. Under the natural extension of σ to $k' \bigotimes_k K$, trivial on the second factor, the latter may be identified to the direct sum of two copies of K exchanged by σ. Similarly, σ extends to $D(k) \bigotimes_k K$ and the latter can be identified to the direct sum of two copies of $\mathbf{M}_d(K)$ on which σ acts by $(x, y) \mapsto ({}^t y, {}^t x)$.

23.8 ε-σ-hermitian forms. We keep the notation and assumptions of the previous section. Let $V(k)$ be a finite dimensional right vector space over $D(k)$ and n its dimension. We view it as the space of k-points of the K-vector space $V = V(k) \bigotimes_k K$.

Let $\varepsilon = \pm 1$. An ε-σ-hermitian form F on $V(k)$ is a map $F: V(k) \times V(k) \to D(k)$ which is additive in each argument and satisfies moreover the conditions:

(1) $F(x \cdot a, y \cdot b) = a^\sigma \cdot F(x, y) \cdot b, \quad F(y, x) = \varepsilon F(x, y)^\sigma \quad (x, y \in V(k); a, b \in D(k))$.

Let (e_i) be a basis of $V(k)$ over $D(k)$. To F is associated the matrix (F_{ij}) where $F_{ij} = F(e_i, e_j)$, which we also denote by F, and the value of F on x, y, viewed as column vectors, can be written ${}^t x^\sigma \cdot F \cdot y$. The second condition of (1) is equivalent to F being ε-σ-hermitian, i.e. to

(2) $$F_{ij} = \varepsilon F_{ji}^\sigma \quad (1 \leqq i, j \leqq n).$$

We always assume F to be non-degenerate, i.e. $F(x, V(k)) = 0$ implies $x = 0$ $(x \in V(k))$.

The form is hermitian if $\varepsilon = 1$, antihermitian if $\varepsilon = -1$. If σ is of the second kind, there is no essential difference between the two because if $c \in k'$ is such that $c^\sigma = -c$, and F is ε-σ-hermitian, then $c \cdot F$ is $-\varepsilon$-σ-hermitian.

Let now σ be of the first kind. If $d = 1$, an antihermitian (resp. hermitian) form is a symplectic (resp. symmetric) form. Let $d > 1$. Then σ is of one of two types $\delta = 1, -1$, and ε has two values. A priori, this gives four possibilities but the two for which $\varepsilon\delta$ has a given value are equivalent. More precisely, let $c \in D(k)$ be such that $c^\sigma = -c$ and assume σ to be of type δ. Then, as remarked above $x \mapsto x^\tau = c \cdot x^\sigma \cdot c^{-1}$ is an involution of type $-\delta$. It is readily checked that if F is ε-σ-hermitian, then $c \cdot F$ is $-\varepsilon$-τ-hermitian.

An element $x \in V(k)$ is isotropic if $F(x, x) = 0$. The form F is isotropic over k if there exists such an $x \neq 0$, anisotropic otherwise. A subspace of $V(k)$ is isotropic if the restriction of F to it is identically zero. In particular, it consists of isotropic vectors. A hyperbolic plane in $V(k)$ is a two-dimensional subspace spanned by two isotropic vectors x, y such that $F(x, y) = 1$. With respect to the associated coordinates, F has the form

(3) $$F(x, y) = x_1^\sigma \cdot y_2 + \varepsilon x_2^\sigma \cdot y_1.$$

It is well-known that any non-degenerate, non-isotropic two-plane containing a non-zero isotropic vector is hyperbolic (cf. [D]).

The index q of F (over k) is the maximum of the dimensions of isotropic subspaces of $V(k)$. Assume $q \geqq 1$. Using the existence of hyperbolic planes recalled above, we see that V can be written as the direct sum of q mutually orthogonal hyperbolic planes $H_1, \ldots, H_q$ and of their orthogonal complement V_o. The restriction F_o of F to V_o is non-degenerate and anisotropic. In H_i we choose isotropic vectors e_i, e_{q+i} such that $F(e_i, e_{q+i}) = 1$. We complete (e_i) $(i \leqq 2q)$ to a basis of V by adding one of V_o, with respect to which F_o is diagonal, which can obviously be done. Then F has the form

$$F(x, y) = \sum_{i=1}^{i-q} x_i^{\sigma} \cdot y_{q+i} + \varepsilon x_{q+i}^{\sigma} \cdot y_i + \sum_{i > 2q} x_i^{\sigma} \cdot c_i \cdot y_i, \tag{4}$$

where

$$c_i = F(e_i, e_i), \text{ hence } c_i \neq 0, \; c_i^{\sigma} = \varepsilon c_i, \; (i = 2q + 1, \ldots, n). \tag{5}$$

Conversely, the right-hand side of (4), under the condition (5), defines a non-degenerate ε-σ-hermitian form of index at least q. In these coordinates F is given by the matrix

$$F = \begin{pmatrix} 0 & I_q & 0 \\ \varepsilon I_q & 0 & 0 \\ 0 & 0 & C \end{pmatrix}, \quad \text{where} \quad C = \begin{pmatrix} c_{2q+1} & & 0 \\ & \ddots & \\ 0 & & c_n \end{pmatrix}. \tag{6}$$

Of course, if σ is of the first kind, $d = 1$ and $\varepsilon = -1$, the conditions of (5) are incompatible and $n = 2q$, as is well-known. But, in all other cases, (5) can be fulfilled and there is no limitation on n.

23.9. *Unitary groups.* The unitary group of F in $V(k)$ is the group of $D(k)$-linear transformations of $V(k)$ which preserve F. Denote it $U(F, k)$. Its intersection with $\mathbf{SL}_n(D)(k)$ is the special unitary group $SU(F, k)$. In matrix form, these groups are defined by

$$U(F, k) = \{X \in \mathbf{GL}_n(D(k)) \mid {}^t X^{\sigma} \cdot F \cdot X = F\}, \quad SU(F, k) = U(F, k) \cap \mathbf{SL}_n(D). \tag{1}$$

If σ is of the second kind, and $c \in k'$ satisfies $c^{\sigma} = -c$, then $U(F, k) = U(c \cdot F, k)$. If σ is of the first kind, and $d > 1$, let again $c \in D(k)$ be such that $c^{\sigma} = -c$ and $\tau = (\operatorname{Int} c) o \sigma$. We noticed earlier that if σ of type δ, then τ is of type $-\delta$ and if F is ε-σ-hermitian, then cF is $-\varepsilon$-τ-hermitian. Moreover, it is immediately checked that $U(F, k) = U(cF, k)$, so we could limit ourselves to one value of ε if σ is of the second kind or σ is of the first kind and $d > 1$, but this would not bring any simplification in the later discussion.

We want to view these groups as groups of k-points of algebraic k-groups. We extend F to V, viewed as a right D-module, in the obvious way and let $U(F)$ (resp. $SU(F)$) be the group of invertible transformations of V preserving F. In matrix form, its elements are given by (1), where $D(k)$ is replaced by D.

Recall that D is isomorphic to $\mathbf{M}_d(K)$ (resp. $\mathbf{M}_d(K) \oplus \mathbf{M}_d(K)$) if σ is of the first (resp. second) kind. From this we get a natural embedding of $U(F)$ in $\mathbf{GL}_{nd}(K)$ (resp. $\mathbf{GL}_{2nd}(K)$) so that the condition (1) translates to a set of quadratic equations on the coefficients of the image with coefficients in k. Therefore $U(F)$ and $SU(F)$ are k-closed.

The condition $F(x,x)=0$ is equivalent to a set of d^2 (resp. $2d^2$) quadratic equations with coefficients in k in nd^2 (resp. $2nd^2$) variables. Therefore if $n \geqq 2$, there are always non-trivial isotropic vectors with coefficients in K. But these equations define a k-variety, as is readily seen by putting F in diagonal form, so that there are isotropic vectors already with coefficients in k_s. In other words, we can put F in the form 23.8(4), with $n-2q \leqq 1$, and $c_n=1$, if $n=2q+1$, $\varepsilon=1$, over k_s. Then it is clear that $U(F)$ and $SU(F)$ are defined over k_s. Being k-closed, they are already defined over k.

From now on, we let G stand for $SU(F)$. As in the previous cases, we see that if $x \in V(k)$ spans a line Kx stable under a k-split torus, on which the latter acts non-trivially, then x is isotropic. On the other hand, if F has index $q \geqq 1$, then, in the coordinates underlying 23.8 (4), the diagonal matrices with entries

$$(s_1,\ldots,s_q,s_1^{-1},\ldots,s_q^{-1},1,\ldots,1), \quad (s_i \in K^*,\ i=1,\ldots,q)$$

from a q-dimensional k-split subtorus S to G. From this we see that S is maximal k-split, hence that q is equal to the k-rank of G. As before, let y_i be the character of S which assigns s_i to s $(l \leqq i \leqq q)$. To describe the k-roots and their multiplicities we go over to the Lie algebra. The condition for $X \in \mathfrak{sl}_n(D)$ to belong to the Lie algebra $\mathfrak{g}$ of G is

$$(2) \qquad {}^tX^\sigma \cdot F + F \cdot X = 0.$$

We write X as a 3×3 block matrix- corresponding to the partition

$$[1,q], \quad [q+1,2q], \quad [2q+1,n]$$

of n. An easy computation shows that (2) is equivalent to the set of conditions (3), (4):

$$(3) \qquad \varepsilon^t X_{21}^\sigma + X_{21} = {}^tX_{11}^\sigma + X_{22} = {}^tX_{12}^\sigma + \varepsilon X_{12} = 0$$

$$(4) \qquad {}^tX_{31}^\sigma \cdot C + X_{23} = {}^tX_{32}^\sigma \cdot C + \varepsilon \cdot X_{13} = {}^tX_{33}^\sigma \cdot C + C \cdot X_{33} = 0$$

From (3), we see that the entries of X_{12} and X_{21} above the diagonal are arbitrary, and determine those below the diagonal. On the other hand, the diagonal entries of X_{12} and X_{21} satisfy the relation

$$(5) \qquad a^\sigma + \varepsilon \cdot a = 0.$$

This shows first that if $1 \leqq i < j \leqq q$, then $\pm y_i \pm y_j$ is a root of multiplicity d^2 (resp. $2d^2$) if σ is of the first (resp. second) kind.

The space corresponding to $2y_i$ (resp. $-2y_i$) is the i-th diagonal entry of X_{12} (resp. X_{21}). In view of (5), it is the eigenspace $D^{-\varepsilon}$ of σ with eigenvalue $-\varepsilon$ on D. In the notation of 23.7, it is of dimension $d(d-1)/2$ if $\varepsilon\delta=1$, of dimension $d(d+1)/2$ if $\varepsilon\delta=-1$, if σ is of the first kind. In particular, this root does not occur if $d=1$ and $\varepsilon\delta=1$, but it does in all other cases. If σ is of the second kind, then the elements of D are pairs (x, y) of elements of $\mathbf{M}_d(K)$ and (5) amounts to $y=-\varepsilon\cdot{}^t x$. This defines a space of dimension d^2, hence $2y_i$ has multiplicity d^2 in that case.

Assume now $n>2q$. In $\mathfrak{gl}_n(D)$, the matrices with coefficients zero outside the i-th row of X_{13} and the i-th column of X_{13} form the eigenspace for S with character s_i. Taking (4) into account, we see that the root s_i occurs in $\mathfrak{g}$ with multiplicity $(n-2q)d^2$ (resp. $2(n-2q)d^2$) if σ is of the first (resp. second) kind. In short we have shown:

(a) If σ is of the first kind of type δ, $\varepsilon\delta=1$ and $d=1$, then $\Phi(S, G)$ is of type $\mathbf{D}_q$ (resp. $\mathbf{B}_q$) if $n=2q$ (resp. $n>2q$).

(b) In the other cases, $\Phi(S, G)$ is of type $\mathbf{C}_q$ (resp. $\mathbf{BC}_q$) if $n=2q$ (resp. $n>2q$).

Remarks. (1) Let σ be of the first kind. If $d=1$, then, as already remarked, we get back over K the special orthogonal group if $\varepsilon=1$, the symplectic group if $\varepsilon=-1$. Let now $d>1$. Let us denote by $J_{n,\varepsilon}$ the matrix F in 23.8(6) for $n-2q\leqq 1$ (and $c_n=1$ if $n=2q+1$). We identify $\mathbf{M}_n(D)$ to $\mathbf{M}_n(K)\otimes\mathbf{M}_d(K)$. The form F can be written in $d\times d$ blocks as $J_{n,\varepsilon}\otimes I_d$. Furthermore, since the J in the definition of σ can be replaced by $A\cdot J\cdot{}^t A$ ($A\in\mathbf{GL}_d(K)$), we may assume σ to be $X\mapsto{}^t X$ if σ is of type 1 and $X\mapsto J_{d,-1}\cdot{}^t X\cdot J_{d,-1}^{-1}$ if σ is of type -1. Our computations show that $\mathfrak{g}$ is the Lie algebra of the subgroup of $\mathbf{SL}_n(D)$ which preserves $J_{n,\varepsilon}\otimes I_d$ if $\delta=1$, and $J_{n,\varepsilon}\otimes J_{d,-1}$ if $\delta=-1$. Therefore G is the symplectic group $\mathbf{Sp}_{nd}$ if $\varepsilon\delta=-1$, the orthogonal group $\mathbf{SO}_{nd}$ if $\varepsilon\delta=1$.

Let σ be of the second kind. Then $\mathbf{M}_n(D)$ is the direct sum of two copies of $\mathbf{M}_{nd}(K)$. It is clear that there exists a bijection v of $\mathbf{M}_{nd}(K)$, whose square is the identity, such that $X\mapsto X^\sigma$ is of the form $(x, y)\mapsto(v(y), v(x))$. The Lie algebra $\mathfrak{g}$ is the subalgebra of the direct sum of two copies of $\mathfrak{sl}_{nd}(K)$ consisting of the matrices $(x, v(x))$. Therefore G is K-isomorphic to $\mathbf{SL}_{nd}(K)$.

(2) In 23.7–23.9 we have left out the case $p=2$. If δ is of the second kind, there is little change if we assume F to be a hermitian "trace form", i.e. $F(x, x)$ is of the form $a+a^\sigma (a\in D(k))$ for all $x\in V(k)$ (cf [D]). If σ is of the first kind, then, as in 23.5, one has to adopt a different framework. Once this is done the results are similar. In fact, [T2] offers a treatment valid in all characteristics.

(3) The above provides many k-forms of the classical groups. To see how they fit in the classification, we refer to [Ti 1]: 23.2 gives the forms ${}^1\mathbf{A}_n$ there, and 23.8, for σ of the second kind, the groups of type ${}^2\mathbf{A}_{n,q}^{(d)}$. If σ is of the first kind, then 23.8 gives the groups $\mathbf{C}_{n,q}^{(d)}$ if $\varepsilon\delta=1$, the groups ${}^2\mathbf{D}_{n,q}^{(d)}$ if $\varepsilon\delta=1$. If $d=1$, $\varepsilon=1$, we get the type $\mathbf{B}_{nq}$, ${}^1\mathbf{D}_{n,q}^{(1)}$ in 23.3, 23.5. We refer to [W] for a systematic description of the classical groups in terms of algebras with involutions, from which much is borrowed in 23.7–23.9. The discussion in [W] is made assuming $p=1$, but is valid for $p\neq 2$. It is reformulated in [Ti 2] so as to hold also for $p=2$.

§24. Survey of Some Other Topics

This second edition is an "enlargement" of the first one, but in a limited way, the only one the author could contemplate at this time. A more comprehensive exposition would have to include many other topics. To orient the reader, we survey briefly here two of them, without proofs, but with references. We have also included a discussion of real reductive groups with some proofs, mainly to relate the "restricted" root system and Weyl group of Lie theory to the relative ones introduced in §21.

A. Classification

24.1 *Classification over* K. Let G be a connected semi-simple group, T a maximal torus, and $\Phi = \Phi(T, G)$ the set of roots of G with respect to T. We have seen in §14 that Φ is a reduced root system in the rational vector space $X(T)_{\mathbf{Q}} = X(T) \bigotimes_{\mathbf{Z}} \mathbf{Q}$. Let $Q = Q(\Phi)$ be the sublattice of $X(T)_{\mathbf{Q}}$ generated by the roots and $P = P(\Phi)$ the sublattice of weights, i.e. of elements $\lambda \in X(T)_{\mathbf{Q}}$ for which $2(\lambda, \alpha)(\alpha, \alpha)^{-1} \in \mathbf{Z}$ for all $\alpha \in \Phi$, where (,) is any positive non-degenerate scalar product invariant under the Weyl group. Then $Q \subset X(T) \subset P$.

Let us call "*diagram*" the datum $D = (\Phi, \Gamma)$ consisting of a reduced root system in a rational vector space V and of a lattice intermediary between $P(\Phi)$ and $Q(\Phi)$. An isomorphism of the diagram D' onto the diagram D is the obvious notion: An isomorphism of the ambiant vector spaces $V' \tilde{\to} V$ mapping Φ' onto Φ and Γ' onto Γ. We shall consider more generally a class of maps to be called here isogenies, too (Chevalley speaks of special isomorphisms). An isogeny or p-isogeny (p prime or equal to 1) from D' to D is an isomorphism $\lambda: V' \tilde{\to} V$ of the ambiant vector spaces mapping Γ' into Γ, such that there exists a bijection $v: \Phi \tilde{\to} \Phi'$ and for each $\alpha \in \Phi$ a natural integer m_α satisfying the conditions $\lambda(v(\alpha)) = p^{m_\alpha}.\alpha$. Under those conditions, the map $w' \mapsto \lambda w' \lambda^{-1}$ defines an isomorphism of the Weyl group W' of Φ' onto the Weyl group W of Φ, and m_α is W-invariant.

With this terminology, we can say that §14 provides a map from groups to diagrams, call it δ, which assigns the diagram $D(G, T) = (\Phi(T, G), X(T))$ to (G, T). More generally, to an isogeny $\mu:(G, T) \to (G', T')$ is naturally associated a p-isogeny $D(G', T') \to D(G, T)$, where p is the characteristic exponent of K. It is defined by the comorphism $X(T') \to X(T)$ associated to the restriction of μ to T. It is (contravariantly) functorial with respect to p-isogenies. If μ is central, which is always the case if $p = 1$, then $m_\alpha = 0$ for all α. If $p \neq 1$ and μ is the m-th power of the Frobenius isogeny, then $m_\alpha = m$ for all α. Assume G to be simple. If all the roots have the same length, m_α must be constant (since W-invariant) and we are in one of the previous cases. But there are a few cases, where $p = 2, 3$, in which m_α takes two values. This corresponds to the so-called exceptional isogenies, of which one example was described in 23.6, relating in characteristic two the split orthogonal group

$\mathbf{SO}(2n+1)$ and the symplectic group $\mathbf{Sp}_{2n}$. (In this case $m_\alpha = 0$ if α is a short root $x_i \pm x_j (i \neq j)$ and $m_\alpha = 1$ if α is a long root $2x_i$.)

The classification theorem asserts that, over K, δ defines a equivalence between the category of isomorphism classes of semisimple groups and isogenies with that of isomorphism classes of diagrams and p-isogenies.

Over $\mathbf{C}$, the fact that δ is a bijection is basically equivalent to the Killing–Cartan classification of complex semi-simple Lie algebras, completed by the description of the complex Lie groups having a given semi-simple Lie algebra, which goes back to E. Cartan and H. Weyl. In general, the result is due to C. Chevalley ([13], [C]).

Fix Φ and denote by G_Γ the group with diagram (Φ, Γ). Write G_{sc} if $\Gamma = P(\Phi)$ and G_{ad} if $\Gamma = Q(\Phi)$. The classification theorem implies in particular that G_Γ is a quotient of G_{sc} and G_{ad} a quotient of G_Γ with respect to a central isogeny. In fact, G_{ad} is the image of G_Γ under the adjoint representation, and G_{sc} is the universal covering of G_Γ in the sense that any projective rational representation of G_Γ lifts to a linear one of G_{sc}. If $K = \mathbf{C}$, then $P(\Phi)/\Gamma$ may be identified to the fundamental group of G_Γ and $\Gamma/Q(\Phi)$ to its center. We shall also say in the general case that G_{sc} is simply connected.

For the discussion, I shall divide the classification theorem into three assertions:

(*i*) Two groups with isomorphic diagrams are isomorphic,
(*ii*) any diagram is the diagram of some group,
(*iii*) (which generalizes (*i*)) a p-isogeny of diagrams is associated to an essentially unique isogeny of the groups.

Assertions (*i*) and (*iii*) are proved in the last lectures of [13]. (*i*) is also established in [17] and [32], and more generally for schemes in [15]. All these proofs proceed by a reduction to groups of rank 2, where it is then done case by case, using the classification. An *a priori* proof, which applies also to (*iii*), without use of classification, has been given by M. Takeuchi [Ta] and is also presented in [J].

A proof of (*ii*), i.e. the existence of a group with a given diagram, is already sketched in [13], but Chevalley comes back to it in [C], where he outlines the principle of a general argument based on the construction of schemes over $\mathbf{Z}$. Given a diagram D, Chevalley starts from the pair of complex groups (G, T) giving rise to it, and then constructs a suitable $\mathbf{Z}$-form $\mathbf{Z}[G]$ of the coordinate ring $\mathbf{C}[G]$ of G. The group G over K with the given diagram is then obtained by a suitable reduction mod p. A survey of this proof is given in [B1], (with one gap, corrected in the comments to that paper in [B3] Vol. 2, p. 703). [32] provides a proof operating solely over K. Again, [15] supplies a more general treatment over schemes.

24.2 *Classification over k.* The problem here is, given an almost simple K-group G, to describe the "k-forms" of G, that is, to classify, up to k-isomorphism, the k-groups G' which are isomorphic to G over K. There is no thorough treatment in the literature yet, the most complete one so far being [Ti 1]. By known principles (*loc. cit.*) it essentially suffices to consider, over the given field k, the almost k-simple groups which remain almost simple over K, i.e. which are absolutely almost simple. It also suffices to consider those which are either simply connected or of adjoint type. In fact, assume G' is a k-form. We can form its diagram starting from a maximal k-torus. Then the Galois group Γ of k_s/k operates on T, $X(T)$, on $X(T)_{\mathbf{Q}}$ and leaves Φ, therefore also $P(\Phi)$ and $Q(\Phi)$, stable. Then the groups in the central isogeny class over k of G' which have a k-form correspond to the lattices between $P(\Phi)$ and $Q(\Phi)$ which are stable under Γ.

The first step towards the classification is the consideration of the index of a k-form, which in some way reduces the isotropic case to the anisotropic one.

Let G' be an isotropic k-form of G. We use the notation of 21.8: S is a maximal k-split torus of G', T a maximal k-torus containing S and $j: X(T) \to X(S)$ the restriction homomorphism. We assume compatible orderings on $\Phi = \Phi(T, G')$ and ${}_k\Phi = \Phi(S, G')$ to have been chosen and let Δ, ${}_k\Delta$ be the corresponding sets of simple roots. The groups T and G' split over k_s (8.11, 18.8). There is an operation of Γ on Δ, to be referred to as the Δ-action, (called the $*$-action in [Ti 1]), defined as follows ([Ti 1:2.3], [4:6.2]): To $\alpha \in \Delta$ we associate first the set of parabolic subgroups of G' conjugate to $P_{\Delta - \{\alpha\}}$ (notation of 14.17). It is a conjugacy class of proper maximal parabolic subgroups. Let $\gamma \in \Gamma$. Then $\gamma(P_{\Delta - \{\alpha\}}(k_s))$ is the group of k_s-points of a proper maximal parabolic k_s-subgroup Q. There exists therefore a unique $\beta = \beta(\alpha) \in \Delta$ such that Q is conjugate under $G'(k_s)$ to $P_{\Delta - \{\beta\}}$. Then, by definition ${}_\Delta\gamma(\alpha) = \beta$. This transformation defines in fact an automorphism of the Dynkin diagram, i.e. of the root system. The k-index of G' consists of the Δ-action of Γ, together with the set Δ^o of simple roots which are zero on S.

It is not difficult to describe ${}_k\Phi$ and ${}_kW$ in terms of Φ, W and the k-index (see [Ti 1:2.5] or [4:6.13]). The quotient $\mathscr{Z}(S)/S$ is anisotropic over k, as follows from 22.7, in particular $\mathscr{D}\mathscr{Z}(S)$ is anisotropic over k. It is called the semi-simple anisotropic kernel. By Theorem 2 in 2.7.1 of [Ti 1], G' is determined up to k-isomorphism by its index and its semi-simple anisotropic kernel. As a result, the determination of the k-forms of a given K-group G is reduced to finding the possible indexes and, for a given index, the possible associated semi-simple anisotropic kernels. We refer to [Ti 1] for a discussion of these problems and an extensive table of k-forms described in this way.

B. Linear Representations

24.3. *Representations of complex semi-simple Lie algebras.* We keep the notation of 24.1. The Lie algebra $\mathfrak{t}$ of T may be identified to $X_*(T)_K = X_*(T) \otimes_{\mathbf{Z}} K$ and then $X(T) \otimes_{\mathbf{Z}} K = X(T)_K$ becomes identified to the dual $\mathfrak{t}^*$ of $\mathfrak{t}$. It is enough to check this for $\mathbf{GL}_1$, where it is clear. If $\lambda \in X(T)$, the image of $\lambda \otimes 1$ in $\mathfrak{t}^*$ is the differential $d\lambda$ of λ at s. More generally, $X(T) \bigotimes_{\mathbf{Z}} K$ is naturally contained in $K[T]$, since the latter is the group algebra of $X(T)$ (cf §8), and the identification of $X(T)_K$ with $\mathfrak{t}^*$ associates to those regular functions their differentials at the identity.

Let $\sigma: T \to \mathrm{GL}(E)$ be a finite dimensional rational representation. The eigenspace E_λ corresponding to $\lambda \in \mathfrak{t}^*$ is

$$E_\lambda = \{x \in E \mid d\sigma(t) \cdot x = \lambda(t) \cdot x \quad (t \in \mathfrak{t})\}.$$

λ is a weight of $\mathfrak{t}$ in E if $E_\lambda \neq 0$. The weights are the differentials of the weights of T in E in the sense of §8. Assume $p = 1$, then it is clear that $E_{d\lambda} = E_\lambda$ $(\lambda \in X(T))$. This is not so otherwise in general since two distinct weights of T may have the same differential.

Assume now $p = 1$. Then $X(T)_{\mathbf{Q}}$ defines a $\mathbf{Q}$-form $\mathfrak{t}^*_{\mathbf{Q}}$ of $\mathfrak{t}^*$, which is spanned by the differential of the roots. The previous equality shows that roots may be defined "infinitesimally" i.e. in Lie algebra terms, the Weyl group may be viewed as a group of automorphisms of $\mathfrak{t}$, and the roots as elements of $\mathfrak{t}^*_{\mathbf{Q}}$, forming a reduced root system in $\mathfrak{t}^*_{\mathbf{Q}}$. More generally, the differentials of the weights $\lambda \in P(\Phi)$ also identify to the weights of the root system in $\mathfrak{t}^*_{\mathbf{Q}}$. The whole theory may be developed infinitesimally, purely in Lie algebras terms, as was done over $\mathbf{C}$ by W. Killing, E. Cartan and then by many others ([10], [18], e.g.). Fix a Borel subgroup B of G containing T and let $\Phi^+ = \Phi(T, B)$ or also the set $\Phi^+(\mathfrak{t}, \mathfrak{b})$ of their differentials. A weight λ in $\mathfrak{t}^*_{\mathbf{Q}}$ is dominant if $(\lambda, \alpha) \geqq 0$ for all $\alpha \in \Phi^+$. Here again (,) is a positive non-degenerate scalar product on $\mathfrak{t}^*_{\mathbf{Q}}$ invariant under the Weyl group.

Let $\pi: \mathfrak{g} \to \mathfrak{gl}(E)$ be an irreducible finite dimensional representation of $\mathfrak{g}$. The space E contains a unique line D invariant under $\mathfrak{b}$. The weight λ_π of $\mathfrak{t}$ in D is dominant and is the highest weight of π (i.e. any other weight is equal to λ_π minus a positive linear combination of simple roots). Moreover, $\pi \mapsto \lambda_\pi$ defines a bijection between isomorphism classes of finite dimensional irreducible $\mathfrak{g}$-modules and dominant weights. For this theorem, which goes back to E. Cartan and H. Weyl, see e.g. [10], [18].

These representations do not always lift to representations of a given group G with Lie algebra $\mathfrak{g}$. They do so precisely when λ_π is the differential of a character of T.

We note also that there is a simple way to describe the highest weight of the contragredient representation π^* to π. Let i be the "opposition involution" of $\mathfrak{t}^*$ (or of $X(T)$). It assigns $-w_0(\lambda)$ to λ, where w_0 is the longest element of the Weyl group (expressed as a product of the simple reflections associated to the basis of Φ^+). Then $\lambda_{\pi^*} = i(\lambda_\pi)$.

Let Δ be the basis of Φ contained in Φ^+. For $\alpha \in \Delta$, let $\omega_\alpha \in P(\Phi)$ be defined by the conditions

$$2(\omega_\alpha, \beta) \cdot (\beta, \beta)^{-1} = \delta_{\alpha,\beta} \quad (\beta \in \Delta).$$

The ω_α are the *fundamental highest weights.* They form a basis of $P(\Phi)$ and any dominant weight (with respect to Φ^+) is a positive integral combination of the ω_α's.

24.4 *Linear representations of semi-simple groups.* We keep the notation $G, T, B, \Phi^+, (\ ,\ )$ and i of the previous sections. Given $\lambda \in X(T)$, let E^λ be the space of regular functions f on G satisfying the condition

$$f(g \cdot b) = b^{i(\lambda)} \cdot f(g) \quad (g \in G, b \in B).$$

It can be shown to be finite dimensional and $\neq 0$ if and only if λ is dominant. The group G acts on it by left translations. Let first $K = \mathbf{C}$ (or, slightly more generally, $p = 1$). Then E^λ is an irreducible G-module with highest weight λ, and $\lambda \mapsto E^\lambda$ defines a bijection between dominant characters and equivalence classes of finite dimensional irreducible representations. This is a global version of the results of 24.3, the "Borel–Weil" theorem [B3:I, 392–396]. Without assumption on p, it is still true that λ is the highest weight of E^λ, (in the sense of 24.3), has multiplicity one, and that the corresponding line D_λ is the only B-invariant line in E^λ, but E^λ need not be irreducible. However, it contains a unique irreducible subspace F^λ, which in turn contains D_λ, and the assignment $\lambda \mapsto F^\lambda$ yields again a bijection between dominant characters and equivalence classes of finite dimensional irreducible representations. This theorem is due to C. Chevalley [13], too. Proofs are also given in [17] and [32].

The G-module E^λ is called a Weyl-module. Its dimension and character are the same as those of the irreducible module so denoted for the complex group $G_\mathbf{C}$ corresponding to G under the classification. When it is not irreducible, there arise the problems of finding the character of F^λ and a Jordan–Hölder series for E^λ. Though not completely solved, they have been the object of many papers. See [H], [J] for a general discussion and numerous references. Earlier surveys are given in [B1] and [B2].

To a linear representation $\pi: G \to \mathrm{GL}(E)$ is naturally associated a representation of G by projective transformations of the projective space $\mathbf{P}(E)$ of lines in E. It is obtained by composing π with the canonical projection of $\mathrm{GL}(E)$ onto the quotient $\mathrm{PGL}(E)$ of $\mathrm{GL}(E)$ by its center, which may be identified with the group $\mathrm{Aut}\,\mathbf{P}(E)$ of projective transformations of $\mathbf{P}(E)$. Given a line $D \subset E$, we denote by $[D]$ the point of $\mathbf{P}(E)$ defined by D.

Let now E be the Weyl module E^λ. Then $[D_\lambda]$ is the only fixed point of B in $\mathbf{P}(E^\lambda)$. The orbit $G[D_\lambda]$ is closed (being an image of G/B) and, in view of the fixed point theorem 10.4, is the only closed orbit in $\mathbf{P}(E^\lambda)$ (and a fortiori in $\mathbf{P}(F^\lambda)$). The stability group of D_λ in G may be bigger than B. If so it is a standard parabolic group P_λ. It is easily seen that in the notation of 14.18, P_λ is the group $P_{J(\lambda)}$, where $J(\lambda)$ is the set of simple roots which are orthogonal to λ. Changing slightly the notation, we see that, given an irreducible representation $\pi: G \to \mathrm{GL}(E)$ of G there is associated to it a unique conjugacy class $\mathscr{P}_\pi$ of parabolic subgroups, consisting of the biggest parabolic subgroups having a fixed point in $\mathbf{P}(E)$, those fixed points forming the unique closed orbit of G in $\mathbf{P}(E)$.

In characteristic zero, a rational representation $\sigma: G \to \mathrm{GL}(E)$ is irreducible if and only if its differential $d\sigma: \mathfrak{g} \to \mathfrak{gl}(E)$ is so. This is by far not so in positive characteristic. In fact, if $p > 1$, the set $\mathscr{M}_G$ of irreducible representations whose differential is also irreducible is finite. To describe it, assume for simplicity that G is simply connected. Then $\mathscr{M}_G$ consists of those representations π whose highest weight λ_π is a linear combination $\lambda_\pi = \sum c_\alpha(\pi)\omega_\alpha$ of the fundamental highest weights with integral coefficients $c_\alpha(\pi) \in [0, p)$. It consists therefore of p^l elements, where l is the rank of G. Moreover, by a result of C.W. Curtis, their differentials are, up to equivalence, all the irreducible representations of $\mathfrak{g}$, viewed as a restricted Lie algebra. If the $c_\alpha(\pi)$ are all equal to $p-1$, then the corresponding Weyl module is already irreducible and provides an irreducible representation of degree p^l, called the Steinberg representation (for all this, see the above references).

In characteristic zero, the representations of G are fully reducible, but this is not so in positive characteristic. For the purposes of invariant theory, D. Mumford conjectured that a weaker property, "geometric reductivity", would hold. An affine algebraic group H is geometrically reductive if, given a finite dimensional representation $\pi: G \to \mathrm{GL}(V)$ and a point $v \in V - \{0\}$ fixed under G, there exists a G-invariant homogeneous polynomial in V which is not zero on v (if a linear form could always be found, this would imply full reducibility). It is easily seen that if the semi-simple groups are geometrically reductive, then so are all groups whose identity component is reductive. Geometric reductivity of reductive groups was proved first by C.S. Seshadri for $\mathbf{GL}_2$ and then by W. Haboush in general. This condition implies that the invariant polynomials on V separate the closed disjoint G-invariant sets and also that the ring of invariants is finitely generated. For a discussion and references, see Appendix 1 in [22].

24.5 *Rationality questions for representations.* So far the discussion of linear representations has been carried out over K, without any concern for fields of definition. But various rationality requirements can be investigated. We summarize here some relevant notions and results, referring for more details to [4:§12] in characteristic zero and to [Ti 3] for the general case. In the latter paper, reductive groups are also considered. For simplicity, we keep here the previous framework.

We shall denote by $\mathscr{R}(G)$ (resp. $\mathscr{R}'(G)$) the set of equivalence classes of irreducible rational linear (resp. projective) representations of G. We have a natural map $\mathscr{R}(G) \to \mathscr{R}'(G)$, which assigns to a linear representations $G \to \mathrm{GL}(E)$ the associated representation in $\mathbf{P}(E)$. If G is simply connected, this map is bijective. In general it is injective, but not surjective. An element of $\mathscr{R}(G)$ (resp. $\mathscr{R}'(G)$) is said to be *rational over k* if it is represented by a k-morphism $G \to \mathrm{GL}(E)$ (resp. $G \to \mathrm{PGL}(E)$), where E is defined over k.

It follows first from the general theory that if G splits over k, and λ is dominant, the Weyl module E^λ may be endowed with a k-structure so that $G \to \mathrm{GL}(E^\lambda)$ is defined over k. By 2.5 in [Ti 3] this is also true for the irreducible representation $G \to \mathrm{GL}(F^\lambda)$. Since G splits over k_s (18.18), this implies that any element of $\mathscr{R}(G)$ or $\mathscr{R}'(G)$ is rational over k_s.

This first of all allows one to make $\Gamma = \mathrm{Gal}(k_s/k)$ operate on $\mathscr{R}(G)$ or $\mathscr{R}'(G)$. In fact, let $\pi: G \to \mathrm{GL}(E)$ be defined over k_s and G, E over k. Given $\gamma \in \Gamma$, we define $\gamma(\pi): G(k_s) \to \mathrm{GL}(E)(k_s)$ by $\gamma(\pi(g)) = \gamma(\pi(\gamma^{-1}g))$ i.e. $\gamma(\pi) = \gamma \circ \pi o \gamma^{-1}$. This operation is compatible with equivalence (over k_s) and goes over to $\mathscr{R}(G)$ and $\mathscr{R}'(G)$. Moreover it is easily seen that if π is irreducible with highest weight λ_π, then $\gamma(\pi)$ is irreducible with highest weight ${}_\Delta\gamma(\lambda_\pi)$. Therefore the class of π is stable under Γ if and only if λ_π is stable under the Δ-action.

By 12.6 of [4] in characteristic zero, and 3.3 of [Ti 3] in general this is equivalent to either of the following conditions: (a) The image of the class $[\pi]$ of π in $\mathscr{R}'(G)$ is rational over k; (b) there exists a central division algebra D over k such that $[\pi]$ is realized by a k-morphism $G \to \mathbf{GL}_m(D)$. In (b), the algebra D is unique up to k-isomorphism and the degree of π is $m.d$, where $d^2 = [D:k]$.

We have associated in 24.4 to an irreducible representation π a conjugacy class $\mathscr{P}_\pi$ of parabolic subgroups of G. Obviously, $\mathscr{P}_\pi$ does not change if π is replaced by an equivalent representation, hence a class $\mathscr{P}_\xi$ may be assigned in this way to any $\xi \in \mathscr{R}(G)$ or $\mathscr{R}'(G)$.

Assume ξ is rational over k and let $\pi: G \to \mathrm{GL}(E)$ be a k-morphism representing ξ. If a parabolic k_s-subgroup Q leaves stable a line defined over k_s in E, then so does any Γ-conjugate of Q. Therefore the class P_ξ is stable under Γ, hence defined over k. However, it does not necessarily contain an element defined over k. If ξ is rational over k, then P_ξ contains an element defined over k if and only if the highest weight line in E as above is defined over k, i.e. if and only if the highest weight itself is invariant under the ordinary action (see 8.12) of Γ. In that case, ξ is said to be *strongly rational over k*. This holds if and only if ξ is ${}_\Delta\Gamma$-stable and the highest weight $\lambda_\pi(\pi \in \xi)$ is orthogonal to Δ^o ([4:12.10], [Ti 3:3.3]). That second condition is automatic if Δ^o is empty, i.e. if G is quasi-split over k. In that case therefore ξ is ${}_\Delta\Gamma$-stable if and only if it is strongly rational over k. The highest weights of the strongly rational representations are determined by their restrictions to S. If G is simply connected those are again all the positive integral linear combinations of $l = \dim S$ fundamental ones [4:12.13].

C. Real Reductive Groups

In this section, we feel compelled by tradition to denote by K a maximal compact subgroup rather than a universal field (which would be **C** *anyway).*

24.6 *Real reductive groups and real Lie groups.* In the twenties, E. Cartan introduced for connected non-compact semi-simple Lie groups with finite center analogues of the Cartan subalgebras, roots and Weyl groups (the latter two being often called "restricted roots" and "restricted Weyl group") by transcendental methods, relying on the use of maximal compact subgroups. If H is the identity component of the group of real points of a semi-simple **R**-group, they turn out to be part of the Tits system constructed in §21. The main purpose of this section is to describe this connection. We shall do so in a slightly more general context, that of reductive groups, the usual one nowadays for many applications of the theory.

(a) We assume familiarity with the theory of Lie groups and of Lie algebras. If H is a Lie group, we also denote by H^o the connected component of the identity in H, in the Lie group topology, and by $\mathfrak{h}$ or $L(H)$ its Lie algebra, trusting this will not cause any confusion with the corresponding notation in Zariski topology used up to now.

Let H be a real Lie group. We always assume that H^o has finite index in H, that $L(H)$ is reductive, i.e. direct sum of a semi-simple ideal and of its center, and that $\mathscr{D}H^o$ has finite center. This forces $\mathscr{D}H^o$ to be closed in H^o and H^o to be the almost direct product of $\mathscr{D}H^o$ and of the identity component C^o of its center C. Also $\mathscr{D}H^o \cap C^o$ is finite. Since H has finitely many connected components, any compact subgroup is contained in a maximal one. The maximal ones meet all the connected components of H and are conjugate under Int H^o. [This is in fact valid for any Lie group with finite component group. For connected groups, this is the Cartan–Malcev theorem; the extension to groups with finitely many connected components is due to G. D. Mostow; see [Ho] for a detailed account and references.]

Let K be a maximal compact subgroup of H. A *Cartan involution* Θ_K of H with respect to K is an involutive automorphism whose fixed point set is K. Let Θ_K be one. Then $\mathfrak{h} = \mathfrak{k} \oplus \mathfrak{p}$, where $\mathfrak{p}$ is the (-1)-eigenspace of $d\Theta_K$; we have

$$(1) \qquad [\mathfrak{k}, \mathfrak{k}] \subset \mathfrak{k}, \quad [\mathfrak{k}, \mathfrak{p}] \subset \mathfrak{p}, \quad [\mathfrak{p}, \mathfrak{p}] \subset \mathfrak{k}$$

and $\mathfrak{p}$ is invariant under Int $\mathrm{Ad}_{\mathfrak{h}} K$. Moreover the exponential map exp yields an isomorphism of manifolds of $\mathfrak{p}$ onto a closed submanifold P of H and $(k, X) \mapsto k \cdot \exp X$ is an isomorphism of manifolds of $K \times \mathfrak{p}$ onto H. Note that Θ_K is completely determined by its differential and the requirement that it leaves K pointwise fixed since $d\Theta_K$ determines Θ_K on H^o and K meets every connected component of H.

If H is connected and semi-simple, then the existence of Θ_K is a classical result of E. Cartan. Θ_K is in fact unique, $\mathfrak{p}$ being necessarily the orthogonal complement of $\mathfrak{k}$ with respect to the Killing form. Let now H be connected and commutative. Then $H = K \times A$, where K is a topological torus (product of circles) and A is isomorphic to the additive group of a euclidean space. We may take for A any subgroup $\exp \mathfrak{a}$, where $\mathfrak{a}$ is a supplement to $\mathfrak{k}$ in $\mathfrak{h}$. Then, for any such choice, $\Theta: k \cdot a \mapsto k \cdot a^{-1}$ is a Cartan involution. In that case, therefore there is no uniqueness (however, there will be a canonical choice of $\mathfrak{a}$ in the algebraic group case, see below).

In the general case, we have $\mathfrak{h} = \mathscr{D}\mathfrak{h} \oplus \mathfrak{c}$, where $\mathfrak{c}$ is the Lie algebra of C.

Write $\mathscr{D}\mathfrak{h} = (\mathfrak{k} \cap \mathscr{D}\mathfrak{h}) \oplus \mathfrak{p}_o$, where $\mathfrak{p}_o$ is the canonical complement. We can find a complement $\mathfrak{p}_1$ to $\mathfrak{k} \cap \mathfrak{c}$ in $\mathfrak{c}$ which is invariant under $\mathrm{Ad}_{\mathfrak{c}} K$. Let $\mathfrak{p} = \mathfrak{p}_o + \mathfrak{p}_1$. We claim that it is the (-1)-eigenspace of the differential of a Cartan involution. First it clearly satisfies (1), hence $\mu: k + p \mapsto k - p$ ($k \in \mathfrak{k}$, $p \in \mathfrak{p}$) is an involution of $\mathfrak{h}$. Also, $\mathfrak{p}$ is invariant under $\mathrm{Ad}_{\mathfrak{h}} K$, since $\mathfrak{p}_o$ is so automatically. Then the above remarks show that μ is the differential of an automorphism Θ' of $C^o \times \mathscr{D}H^o$ having $(K \cap C^o) \times (K \cap \mathscr{D}H^o)$ has its fixed point set, which commutes with K, acting by inner automorphisms. Since $\mathscr{D}H^o \cap C^o$ is finite, the kernel of the canonical surjective morphism $C^o \times \mathscr{D}H^o \to H^o$ is finite, too, hence contained in K, and consequently pointwise fixed under Θ'; therefore Θ' goes down to H^o and defines a Cartan involution Θ^o of H^o, again commuting with $\mathrm{Int}_{H^o} K$. It is then routine to check that Θ^o extends to a Cartan involution of H with respect to K.

(b) In view of (1), a subspace of $\mathfrak{p}$ is a subalgebra if and only if it is commutative. It is known that the maximal commutative subalgebras of $\mathfrak{p}$ are conjugate under K^o, and that if $\mathfrak{a}$ is one, then the eigenvalues of $\mathrm{Ad}\, X (X \in \mathfrak{a})$ are all real and $\mathfrak{h}$ is a direct sum of eigenspaces $\mathfrak{h}_\lambda$ ($\lambda \in \mathfrak{a}^*$), where

$$\mathfrak{h}_\lambda = \{ Y \in \mathfrak{h} \mid \mathrm{ad}\, X(Y) = \lambda(X) \cdot Y, (X \in \mathfrak{a}) \}.$$

The λ's for which $\mathfrak{h}_\lambda \neq 0$ are the *roots* of $\mathfrak{h}$ with respect to $\mathfrak{a}$. We have $\mathfrak{a} = \mathfrak{p}_1 + \mathfrak{a}_o$, where $\mathfrak{a}_o = \mathfrak{p}_o \cap \mathfrak{a}$ and we can identify $\mathfrak{a}_o^*$ to the space of linear forms on $\mathfrak{a}$ which are zero on $\mathfrak{p}_1$. Let $W = \mathscr{N}_{K^o}(\mathfrak{a})/\mathscr{Z}_{K^o}(\mathfrak{a})$. Then W operates in a natural way on $\mathfrak{a}$, leaving $\mathfrak{p}_1$ pointwise fixed, hence also on $\mathfrak{a}_o^*$. The set $\Phi = \Phi(\mathfrak{a}, \mathfrak{h})$ of roots of $\mathfrak{h}$ with respect to $\mathfrak{a}$ is contained in $\mathfrak{a}_o^*$, is a root system there and W is the Weyl group of Φ. For $\mathfrak{h}$ semisimple, this is a well-known result of E. Cartan (see e.g. [He]). The extension to the reductive case is obvious.

We can also consider $W' = \mathscr{N}_K(\mathfrak{a})/\mathscr{Z}_K(\mathfrak{a})$. It again defines a finite group of automorphisms of $\mathfrak{a}^*$ or $\mathfrak{a}_o^*$, which preserves Φ, and contains W as a normal subgroup. Note also that $\mathscr{N}_H(\mathfrak{a}) = \mathscr{N}_K(\mathfrak{a}) \cdot \mathscr{Z}_H(\mathfrak{a})$ (see 24.7(i)), so that the consideration of normalizers and centralizers in H or H^o rather than in K or K^o would lead to the same groups W' and W.

The Lie algebra of $\mathscr{Z}_H(\mathfrak{a})$ is stable under $d\Theta_K$, therefore direct sum of its intersections with $\mathfrak{k}$ and $\mathfrak{p}$. The latter is reduced to $\mathfrak{a}$, since $\mathfrak{a}$ is commutative maximal in $\mathfrak{p}$ by definition. This already shows that $\mathscr{Z}_K(\mathfrak{a}) \cdot A$, where $A = \exp \mathfrak{a}$, is an open subgroup in $\mathscr{Z}_H(\mathfrak{a})$. In fact there is equality (see 24.7(1)), so that in particular $\mathscr{Z}_H(\mathfrak{a})/A$ is compact.

(c) In the ensuing discussion, we have to consider the usual topology on the group of real points of an algebraic **R**-group. It is finer than the topology induced by the Zariski topology. We collect here some facts to be used below. G is a connected **R**-group.

(i) *The group* $G(\mathbf{R})$ *has finitely many connected components in the ordinary topology*. This can be deduced for instance from the fact that if X is a complex algebraic variety defined over **R**, then $X(\mathbf{R})$ has finitely many connected components (H. Whitney). A proof for algebraic groups, in the context of Galois cohomology, is given in [BS].

(ii) *Let G be reductive. Then $G(\mathbf{R})$ is compact if and only if G is anisotropic over* $\mathbf{R}$. *If it is, then $G(\mathbf{R})$ is connected.*

Proof. The group of real points of a $\mathbf{R}$-split torus $\neq \{1\}$ contains a closed subgroup isomorphic to $\mathbf{R}$, hence is not compact. Therefore, if G is isotropic over $\mathbf{R}$, the group $G(\mathbf{R})$ is not compact. Assume now G to be anisotropic over $\mathbf{R}$. Then so are all its $\mathbf{R}$-tori. We can find finitely many maximal $\mathbf{R}$-tori $T_i (1 \leq i \leq m)$ such that $L(G)$ is the sum (not necessarily direct) of the $L(T_i)$ and that the product mapping $\mu: T_1 \times \cdots \times T_m \to G$ is surjective (2.2). Then $L(G)(\mathbf{R})$ is the sum of the $L(T_i)(\mathbf{R})$ and therefore the differential of $\mu(\mathbf{R})$ is surjective at the identity. As a consequence, the image of $\mu(\mathbf{R})$, which is obviously compact, contains the identity component of $G(\mathbf{R})$. This establishes the first assertion. To prove the second one we use a result of C. Chevalley stating that any compact linear group is algebraic [12(b)]. View G as embedded in $\mathbf{GL}_n$ by means of a $\mathbf{R}$-morphism. There exists then an ideal J of polynomials with real coefficients whose only zeroes in $\mathbf{R}^n$ are the points of $G(\mathbf{R})^o$. Since the latter is Zariski-dense in G, this shows that J generates the ideal of G, and our assertion follows. We note that Chevalley's argument shows more generally that if $L \to \mathbf{GL}_n(\mathbf{R})$ is a continuous linear representation of a compact Lie group L, then the orbits of L are real algebraic subsets, hence are separated by the L-invariant polynomials on $\mathbf{R}^n$. This assertion follows easily from the Stone–Weierstrass approximation theorem.

(d) Let now G be a connected reductive $\mathbf{R}$-group. Then $H = G(\mathbf{R})$ satisfies all the conditions imposed on H in (b). We assume that $\mathscr{D}G$ is isotropic over $\mathbf{R}$ and fix a maximal compact subgroup K of $G(\mathbf{R})$. We want to describe a natural identification between the relative root system ${}_{\mathbf{R}}\Phi$ and Weyl group ${}_{\mathbf{R}}W$ of §21 and the root system $\Phi(\mathfrak{a}, \mathfrak{h})$ and Weyl group W constructed in (b).

The torus $Z = (\mathscr{C}G)^o$ is uniquely the almost direct product of its greatest anisotropic (over $\mathbf{R}$) subtorus Z_α and its greatest $\mathbf{R}$-split subtorus Z_d (8.15). Then $Z_a(\mathbf{R}) = K \cap Z(\mathbf{R})$ and $\mathfrak{z}_d(\mathbf{R})$ is a supplement to the Lie algebra $\mathfrak{z}_a(\mathbf{R})$ of $Z_a(\mathbf{R})$ in $\mathfrak{z}(\mathbf{R})$. It follows that $Z(\mathbf{R})$ is the direct product of $Z_a(\mathbf{R})$ and of $\exp \mathfrak{z}_d(\mathbf{R})$. The algebra $\mathfrak{z}_d(\mathbf{R})$ is obviously invariant under $\operatorname{Ad} G(\mathbf{R})$, therefore it can be chosen as the $\mathfrak{p}_1$ of (b). We say that $\mathfrak{a}$ or $\mathfrak{p}$ is admissible if it contains $\mathfrak{z}_d(\mathbf{R})$. The discussion in (b) shows that there is a unique admissible $\mathfrak{p}$. We can now associate to K a canonical Cartan involution, namely the one defined by the admissible $\mathfrak{p}$.

(e) We claim now that if S is a maximal $\mathbf{R}$-split torus of G, then $L(S)(\mathbf{R})$ is conjugate to an admissible $\mathfrak{a}$ and conversely any admissible $\mathfrak{a}$ is of the form $L(S)(\mathbf{R})$, for a maximal $\mathbf{R}$-isotropic torus S of G. In view of the conjugacy under $G(\mathbf{R})$ of maximal $\mathbf{R}$-split tori (20.9), it suffices to prove the second assertion. An admissible $\mathfrak{a}$ contains $\mathfrak{z}_d(\mathbf{R})$ by definition. Since G is the almost direct product of Z and $\mathscr{D}G$, a maximal $\mathbf{R}$-split torus of G is the almost direct product of maximal $\mathbf{R}$-split tori of Z and $\mathscr{D}G$ (22.10). This reduces us to the case where $\mathscr{D}G$ is semisimple. Let $\sigma: G \to \operatorname{Ad} G$ be the adjoint representation. It is a central isogeny, whose differential is an isomorphism.

Since the maximal **R**-split tori of Ad G are the images of the maximal **R**-split tori of G (22.7), we may further identify G with $\mathrm{Ad}\, G \subset \mathrm{GL}(\mathfrak{g})$. In suitable coordinates $\mathfrak{a}$ is represented by diagonal matrices. Therefore the smallest algebraic group $\mathscr{A}(\mathfrak{a})$ whose Lie algebra contains $\mathfrak{a}$ (cf. 7.1) is a **R**-split torus. If $L(S)(\mathbf{R}) \neq \mathfrak{a}$, then $S(\mathbf{R})/(\exp \mathfrak{a})$ is not compact, contradicting the fact that $\mathscr{Z}_H(\mathfrak{a})/(\exp \mathfrak{a})$ is compact (see 24.7(1)). Hence $\mathfrak{a} = L(S)(\mathbf{R})$.

It now follows from the definitions that $\alpha \mapsto d\alpha$ defines a bijection of ${}_{\mathbf{R}}\Phi(S, G)$ onto $\Phi(\mathfrak{a}, \mathfrak{h})$. As a consequence, the Weyl groups $W(\mathfrak{a}, \mathfrak{h})$ and ${}_{\mathbf{R}}W$ are the same. By definition ${}_{\mathbf{R}}W$ is the quotient $\mathscr{N}_G(S)/\mathscr{Z}_G(S)$, where S is a maximal **R**-split torus; by the above, we may assume that $\mathfrak{a} = L(S)(\mathbf{R})$. By 21.2, we have $\mathscr{N}_G(S) = \mathscr{N}_G(S)(\mathbf{R}) \cdot \mathscr{Z}_G(S)$. Now an element $g \in G(\mathbf{R})$ normalizes or centralizes S if and only if $\mathrm{Ad}\, g$ normalizes or centralizes $\mathfrak{s}(\mathbf{R}) = \mathfrak{a}$. We can therefore also write ${}_{\mathbf{R}}W = \mathscr{N}_{G(\mathbf{R})}(\mathfrak{a})/\mathscr{Z}_{G(\mathbf{R})}(\mathfrak{a})$. But this is equal to the Weyl group of $\Phi(\mathfrak{a}, \mathfrak{h})$ hence to $\mathscr{N}_{K^o}(\mathfrak{a})/\mathscr{Z}_{K^o}(\mathfrak{a})$. Therefore the inclusion $\mathscr{N}_{K^o}(\mathfrak{a}) \to \mathscr{N}_{G(\mathbf{R})}(\mathfrak{a})$ induces a bijection of $\mathscr{N}_{K^o}(\mathfrak{a})/\mathscr{Z}_{K^o}(\mathfrak{a})$ onto $\mathscr{N}_{G(\mathbf{R})}(\mathfrak{a})/\mathscr{Z}_{G(\mathbf{R})}(\mathfrak{a})$, which completes the justification of the claim made at the beginning of this section. It also follows that these quotients are equal to $\mathscr{N}_K(\mathfrak{a})/\mathscr{Z}_K(\mathfrak{a})$. In this particular case, the groups W' and W defined in 24(b) are therefore equal, although $G(\mathbf{R})$ is not necessarily connected.

24.7. *Two remarks on maximal compact subgroups.* We have used above a fact which has been known for a long time, but for which I do not know a reference. I shall take this opportunity to prove it as well as a sort of counterpart which has been familiar for an equally long time, has been used occasionally, but whose proof has not been published so far to my knowledge.

(i) We again adopt the framework of 24.6(a), fix K and write Θ for Θ_K. We want to prove

Proposition 1. *Let U, V be subsets of $\mathfrak{p}$. Assume there exists $g \in H$ such that* $\mathrm{Ad}\, g(U) = V$. *Then there exists $k \in K$ such that* $\mathrm{Ad}\, k(u) = \mathrm{Ad}\, g(u)$ *for all $u \in U$.*

We claim first it suffices to show that if $X \in \mathfrak{p}$ and $p \in P$ are such that $\mathrm{Ad}\, p(X) \in \mathfrak{p}$, then $\mathrm{Ad}\, p(X) = X$. Indeed, assume this to hold, let g, U, V be as in the statement and $X \in U$. We can write $g = k \cdot p$ $(k \in K, p \in P)$ hence $\mathrm{Ad}\, g(X) \in V$ implies

$$\mathrm{Ad}\, p(X) \in \mathrm{Ad}\, k^{-1}(V) \subset \mathfrak{p},$$

therefore $\mathrm{Ad}\, p(X) = X$ and $\mathrm{Ad}\, g(X) = \mathrm{Ad}\, k(X)$ for all $X \in U$.

Let now $X \in \mathfrak{p}$ and $p \in P$ be such that $Y = \mathrm{Ad}\, p(X) \in \mathfrak{p}$. Since Θ is an automorphism we have $\mathrm{Ad}\, \Theta(p)(d\Theta(X)) = d\Theta(Y)$. But $d\Theta = -Id.$ on $\mathfrak{p}$ and $\Theta(p) = p^{-1}$, therefore $\mathrm{Ad}\, p^{-1}(-X) = -\mathrm{Ad}\, p(X)$ hence $\mathrm{Ad}\, p^2(X) = X$. We have $p = \exp Z$, with $Z \in \mathfrak{p}$ and $\mathrm{ad}\, Z$ diagonalizable (over **R**) in view of 23(e) and 8.15. If (λ_i) are its eigenvalues, then those of $\mathrm{Ad}\, p^2$ are $\exp 2\lambda_i$. In particular, the eigenspace for the eigenvalue 1 of $\mathrm{Ad}\, p^2$ is the zero eigenspace for $\mathrm{ad}\, Z$, and also the 1-eigenspace for $\mathrm{Ad}\, p$. Since X is fixed under $\mathrm{Ad}\, p^2$, it is then also fixed under $\mathrm{Ad}\, p$ and it commutes with Z. From this we see that

(1) $$\mathscr{Z}_H(\mathfrak{a}) = \mathscr{Z}_K(\mathfrak{a}) \cdot \exp \mathfrak{a},$$

as was asserted at the end of 24.6(b).

(ii) Let now H be any Lie group with finite component group. We have already recalled part of the theorem on maximal compact subgroups. It is also known that G/K is homeomorphic to euclidean space. In fact, this can be made much more precise. (cf. [Ho:III, §3, N^0 2]): Given K, there exists a K-invariant subspace $\mathfrak{m}$ of $\mathfrak{h}$, a diffeomorphism φ of $\mathfrak{m}$ onto a closed$K$-invariant submanifold M of H, which is K-equivariant, K acting by the adjoint representation on $\mathfrak{m}$, by inner automorphisms on M, such that

$$\varphi \times Id\colon (X, k) \mapsto \varphi(X) \cdot k \quad (X \in \mathfrak{m}, k \in K)$$

is a diffeomorphism of $\mathfrak{m} \times K$ onto H, which is K-equivariant, K acting by the adjoint representation on $\mathfrak{m}$, by left translations on K and H. As a counterpart to the previous proposition, we have the

Proposition 2. *Let U and V be subsets of K. Assume there exists $g \in G$ such that ${}^gU = V$. Then there exists $k \in K$ such that ${}^kx = {}^gx$ for all $x \in U$.*

Proof. First consider a maximal compact subgroup L and $m \in M$ such that ${}^mK = L$ (which can always be found, by the conjugacy of maximal compact subgroups and the decomposition $H = M \cdot K$). We claim that m centralizes $L \cap K$. Let $x \in L \cap K$. Then $m^{-1} \cdot x \cdot m = y \in K$ hence

$$m \cdot y = x \cdot m = x \cdot m \cdot x^{-1} \cdot x$$

and therefore $x = y$ and $m = x \cdot m \cdot x^{-1}$ by the uniqueness of the decomposition $H = M \cdot K$. Applied to $L = K$, this shows in particular

(2) $$\mathscr{N}_H(K) = (\mathscr{Z}_H K) \cap M \cdot K.$$

Let now g, U, V be as in the proposition. Then $V \subset {}^gK \cap K$. We have just proved the existence of $m \in M$ centralizing ${}^gK \cap K$, in particular V, such that ${}^mK = {}^gK$. The element m^{-1} also centralizes V, whence

(3) $$m^{-1} \cdot g \cdot u \cdot g^{-1} \cdot m = g \cdot u \cdot g^{-1} \quad (u \in U).$$

The element $m^{-1} \cdot g$ normalizes K, hence can be written by (2) as

(4) $$m^{-1} \cdot g = n \cdot k \quad (n \in \mathscr{Z}_H(K) \cap M, k \in K).$$

We have then, by (3) and (4):

(5) $${}^gu = {}^{m^{-1} \cdot g} \cdot u = {}^{n \cdot k}u = {}^ku \quad (u \in U),$$

which proves the proposition.

References for Chapters I to V

1. A. Borel, *Groupes linéaires algébriques.* Annals of Math. (2) **64** (1956), 20–82.
2. A Borel and T.A. Springer, *Rationality properties of linear algebraic groups.* Proc. Symp. Pure Math., vol. IX (1966), 26–32.
3. A. Borel and T.A. Springer, *Rationality properties of linear algebraic groups II,* Tohoku Math. Jour., (2) **20** (1968), 443–487.
4. A. Borel et J. Tits, *Groupes réductifs.* Publ. Math. I.H.E.S. **27** (1965), 55–150.
5. A. Borel et J. Tits, "Compléments à l'article: Groupes réductifs," Publ. Math. I.H.E.S. **41** (1972), 253–276.
6. A. Borel et J. Tits, "Homomorphismes "abstraits" de groupes algébriques simples," Annals of Math. **97** (1973), 499–571.
7. A. Borel et J. Tits, "Théorèmes de structure et de conjugaison pour les groupes algébriques linéaires," C.R. Acad. Sci. Paris **287** (1978), 55–57.
8. N. Bourbaki, Groupes et algèbres de Lie. Chapitre 1: Algèbres de Lie, Hermann Paris (1960).
9. N. Bourbaki, Groupes et algèbres de Lie, Chap. IV, V, VI, 2ème édition, Masson, Paris 1981.
10. N. Bourbaki, Groupes et algèbres de Lie, Chap. VII, VIII, Hermann, Paris 1975.
11. N. Bourbaki, Algèbre Commutative, Chap. 5–7, Hermann, Paris 1964.
12. C. Chevalley, Théorie des groupes de Lie. (a) Tome II: Groupes algébriques, Paris 1951, (b) Tome III: Groupes algébriques, Hermann Paris 1955.
13. C. Chevalley, Séminaire sur la classification des groupes de Lie algébriques (Mimeographed Notes), Paris 1956–58.
14. M. Demazure et P. Gabriel, Groupes algébriques, Tome I, Masson, Paris 1970.
15. M. Demazure et A. Grothendieck, Schémas en groupes I, II, III, Springer L.N.M. **151**, **152**, **153**, Springer Verlag 1970.
16. G. Fano, *Sulle varietà algebriche con un gruppo continuo non integrabile di transformazioni proiettive in sè.* Mem. Reale Accad. d. Sci. di Torino (2) **46** (1896), 187–218.
17. J. Humphreys, Linear algebraic groups, Grad. Text Math. **21**, Springer-Verlag 1975.
18. N. Jacobson, Lie algebras. Interscience Tracts in pure and applied math. 10, New York 1962, Interscience Publ.
19. E. Kolchin, *Algebraic matric groups and the Picard-Vessiot theory of homogeneous linear differential equations.* Annals of Math. (2) **49** (1948), 1–42.
20. K. Kolchin, *On certain concepts in the theory of algebraic matric groups.* Ibid. 774–789.
21. S. Lang, *Algebraic groups over finite fields.* Amer. Jour. Math. **87** (1965), 555–563.
22. D. Mumford and J. Fogarty, Geometric Invariant Theory, 2nd enlarged edition, Ergeb. Math. u. Grenzgeb. **34**, Springer-Verlag 1982.
23. D. Mumford, The red book of varieties and schemes, Springer L.N.M. **1358**, Springer-Verlag 1988.
24. T. Ono, *Arithmetic of algebraic tori.* Annals of Math. (2) **74** (1961), 101–139.
25. M. Rosenlicht, *Some basic theorems on algebraic groups.* Amer. Jour. Math. **78** (1956), 401–443.
26. M. Rosenlicht, *Some rationality questions on algebraic groups.* Annali di Mat. (IV) **43** (1957), 25–50.

27. M. Rosenlicht, *On quotient varieties and the affine embedding of certain homogeneous spaces* Trans. Amer. Math. Soc. **101** (1961), 211–223.
28. M. Rosenlicht, "*Questions of rationality for solvable algebraic groups over nonperfect fields*" Annali di Mat. (IV) **61** (1962), 97–120.
29. J.-P. Serre, Groupes algébriques et corps de classes, Hermann, Paris 1959.
30. J.-P. Serre, Cohomologie galoisienne. Lect. Notes Math. **5**, Springer-Verlag 1965
31. J.-P. Serre, Algèbres de Lie semi-simples complexes. Benjamin New York 1966.
32. T.A. Springer, Linear algebraic groups, Progress in Math. **9**, Birkhäuser 1981.
33. A. Weil, *On algebraic groups and homogeneous spaces.* Amer. Jour. Math. **77** (1955), 493–512.

Additional references for §§23 and 24:

[B1] A. Borel, *Représentations linéaires et espaces homogènes kähleriens des groupes simples compacts*, (1954), Coll. Papers I, 392–396, Springer-Verlag 1983.
[B2] A. Borel, *Properties and linear representations of Chevalley groups*, in Seminar on algebraic groups and related finite groups, Lect. Notes Math. **131**, 1–55, Springer-Verlag 1955.
[B3] A. Borel, *Linear representations of semi-simple algebraic groups*, Proc. Symp. Pure Math. **29**, Amer. Math. Soc. (1975), 421–439.
[BS] A. Borel et J-P. Serre, *Théorèmes de finitude en cohomologie galoisienne*, Comm. Math. Helv. **39** (1964), 111–164.
[C] C. Chevalley, *Certains schémas de groupes simples*, Sém. Bourbaki 1960/61, Exp. **219**.
[D] J. Dieudonné, La géométrie des groupes classiques, Erg. d. Math. u.i. Grenzgeb. (N.F.) **5**, Springer-Verlag 1955.
[He] S. Helgason, Differential geometry and symmetric spaces, Acad. Press 1962.
[Ho] G. Hochschild, The structure of Lie groups, Holden-Day, San Francisco 1965.
[Hu] J. Humphreys, Ordinary and modular representations of Chevalley groups, Lect. Notes Math. **528**, Springer–Verlag 1976.
[IM] N. Iwahori and H. Matsumoto, *On some Bruhat decompositions and the structure of the Hecke ring of p-adic Chevalley groups*, Publ. Math. I.H.E.S. **25** (1965), 5–48.
[J] J.C. Jantzen, Representations of algebraic groups, Acad. Press 1987.
[Ta] M. Takeuchi, *A hyperalgebraic proof of the isomorphism and isogeny theorems for reductive groups*, J. Algebra **85** (1983), 179–196.
[Ti1] J. Tits, *Classification of algebraic semisimple groups*, Proc. Symp. Pur. Math. IX, Amer. Math. Soc. 1966, 32–62.
[Ti2] J. Tits, *Formes quadratiques, groupes orthogonaux et algèbres de Clifford*, Inv. Math. **5** (1968), 19–41.
[Ti3] J. Tits, *Représentations linéaires irréductibles d'un groupe réductif sur un corps quelconque*, J. f. reine u. angew. Math. **247** (1971), 196–220.
[W] A. Weil, *Algebras with involution and the classical groups*, J. Indian Math. Soc. **24** (1960), 589–623.

Index of Definition

Index of Notation

Graduate Texts in Mathematics

48 SACHS/WU. General Relativity for Mathematicians.
49 GRUENBERG/WEIR. Linear Geometry, 2nd ed.
50 EDWARDS. Fermat's Last Theorem.
51 KLINGENBERG. A Course in Differential Geometry.
52 HARTSHORNE. Algebraic Geometry.
53 MANIN. A Course in Mathematical Logic.
54 GRAVER/WATKINS. Combinatories with Emphasis on the Theory of Graphs.
55 BROWN/PEARCY. Introduction to Operator Theory I: Elements of Functional Analysis.
56 MASSEY. Algebraic Topology: An introduction.
57 CROWELL/FOX. Introduction to Knot Theory.
58 KOBLITZ. *p*-adic Numbers, *p*-adic Analysis, and Zeta-Functions, 2nd ed.
59 LANG. Cyclotomic Fields.
60 ARNOLD. Mathematical Methods in Classical Mechanics. 2nd ed.
61 WHITEHEAD. Elements of Homotopy Theory.
62 KARGAPOLOV/MERZIJAKOV. Fundamentals of the Theory of Groups.
63 BOLLABAS. Graph Theory.
64 EDWARDS. Fourier Series. Vol. I. 2nd ed.
65 WELLS. Differential Analysis on Complex Manifolds. 2nd ed.
66 WATERHOUSE. Introduction to Affine Group Schemes.
67 SERRE. Local Fields.
68 WEIDMANN. Linear Operators in Hilbert Spaces.
69 LANG. Cyclotomic Fields II.
70 MASSEY. Singular Homology Theory.
71 FARKAS/KRA. Riemann Surfaces.
72 STILLWELL. Classical Topology and Combinatorial Group Theory.
73 HUNGERFORD. Algebra.
74 DAVENPORT. Multiplicative Number Theory. 2nd ed.
75 HOCHSCHILD. Basic Theory of Algebraic Groups and Lie Algebras.
76 IITAKA. Algebraic Geometry.
77 HECKE. Lectures on the Theory of Algebraic Numbers.
78 BURRIS/SANKAPPANAVAR. A course in Universal Algebra.
79 WALTERS. An Introduction to Ergodic Theory.
80 ROBINSON. A Course in the Theory of Groups.
81 FORSTER. Lectures on Riemann Surfaces.
82 BOTT/TU. Differential Forms in Algebraic Topology.
83 WASHINGTON. Introduction to Cyclotomic Fields.
84 IRELAND/ROSEN. A Classical Introduction to Modern Number Theory.
85 EDWARDS. Fourier Series: Vol. II. 2nd ed.
86 VAN LINT. Introduction to Coding Theory.
87 BROWN. Cohomology of Groups.
88 PIERCE. Associative Algebras.
89 LANG. Introduction to Algebraic and Abelian Functions. 2nd ed.
90 BRØNDSTED. An introduction to Convex Polytopes.
91 BEARDON. On the Geometry of Discrete Groups.
92 DIESTEL. Sequences and Series in Banach Spaces.
93 DUBROVIN/FOMENKO/NOVIKOV. Modern Geometry—Methods and Applications Vol. 1.

94 WARNER. Foundations of Differentiable Manifolds and Lie Groups.
95 SHIRYAYEV. Probability, Statistics, and Random Processes.
96 CANWAY. A Course in Functional Analysis.
97 KOBLITZ. Introduction to Elliptic Curves and Modular Forms.
98 BRÖCKER/TOM DIECK. Representations of Compact Lie Groups.
99 GROVE/BENSON. Finite Reflection Groups. 2nd ed.
100 BERG/CHRISTENSEN/RESSEL. Harmonic Analysis on Semigroups: Theory of Positive Definite and Related Functions.
101 EDWARDS. Galois Theory.
102 VARADARAJAN. Lie Groups, Lie Algebras and Their Representations.
103 LANG. Complex Analysis. 2nd ed.
104 DUBROVIN/FOMENKO/NOVIKOV. Modern Geometry—Methods and Applications Vol. II.
105 LANG. $SL_2(\mathbf{R})$.
106 SILVERMAN. The Arithmetic of Elliptic Curves.
107 OLVER. Applications of Lie Groups to Differential Equations.
108 RANGE. Holomorphic Functions and Integral Representations in Several Complex Variables.
109 LEHTO. Univalent Functions and Teichmüller Spaces.
110 LANG. Algebraic Number Theory.
111 HUSEMÖLLER. Elliptic Curves.
112 LANG. Elliptic Functions.
113 KARATZAS/SHREVE. Brownian Motion and Stochastic Calculus.
114 KOBLITZ. A Course in Number Theory and Cryptography.
115 BERGER/GOSTIAUX. Differential Geometry: Manifolds, Curves, and Surfaces.
116 KELLEY/SRINIVASAN. Measure and Integral, Volume 1.
117 SERRE. Algebraic Groups and Class Fields.
118 PEDERSEN. Analysis Now.
119 ROTMAN. An Introduction to Algebraic Topology.
120 ZIEMER. Weakly Differentiable Functions: Sobolev Spaces and Functions of Bounded Variation.
121 LANG. Cyclotomic Fields I and II.
122 REMMERT. Theory of Complex Functions: Readings in Mathematics.
123 EBBINGHAUS et al. Numbers: Readings in Mathematics.
124 DUBROVIN/FOMENKO/NOVIKOV: Modern Geometry: Methods and Applications, Volume III.
125 BERENSTEIN/GAY. Complex Variables: An Introduction.
126 BOREL. Linear Algebraic Groups: Second Enlarged Edition.

LaVergne, TN USA
05 December 2009
166077LV00003B/21/A